Mapping Cyberspace

Space is central to our lives. Because of this, much attention is directed at understanding and explaining the geographic world. *Mapping Cyberspace* is a ground-breaking book, which extends this analysis to provide a geographic exploration and critical reading of cyberspace and information and communication technologies. *Mapping Cyberspace*:

- provides an understanding of what cyberspace looks like and the social interactions that take place there;
- explores the impacts of cyberspace, and information and communication technologies, on cultural, political and economic relations;
- charts the spatialities, spatial forms and space–time relations of virtual spaces;
- details empirical research and examines a wide variety of maps and spatialisations of cyberspace and the information society;
- has a related website at **http://www.MappingCyberspace.com/**

Mapping Cyberspace draws together the findings and theories of researchers from geography, cartography, sociology, cultural studies, computer-mediated communications, information visualisation, literary theory and cognitive psychology. It is highly illustrated with over fifty black and white illustrations and a colour plate section. This book will be a valuable addition to the growing body of literature on cyberspace and what it means for the future.

Martin Dodge is a researcher and computer technician at the Centre for Advanced Spatial Analysis, University College London. **Rob Kitchin** is a lecturer in Human Geography at the National University of Ireland, Maynooth.

Martin dedicates this book to his Mum and Dad.

Rob dedicates this book to Barry Otterson, a great friend and teacher.

Mapping Cyberspace

Martin Dodge and Rob Kitchin

London and New York

First published 2001
by Routledge
11 New Fetter Lane, London EC4P 4EE

Simultaneously published in the USA and Canada
by Routledge
29 West 35th Street, New York, NY 10001

Reprinted 2001

Routledge is an imprint of the Taylor & Francis Group

Typeset in Goudy by Graphicraft Limited, Hong Kong
Printed and bound in Great Britain by
TJ International Ltd, Padstow, Cornwall

British Library Cataloguing in Publication Data
A catalogue record for this book is available
from the British Library

Library of Congress Cataloging in Publication Data
Dodge, Martin
 Mapping cyberspace / Martin Dodge and Rob Kitchin.
  p. cm.
 Includes bibliographical references and index.
 1. Computers and civilization. 2. Cyberspace–Social aspects.
 3. Communication. I. Kitchin, Rob. II. Title.
QA 76.9.C66 D64 2000
303.48'33–dc21   00-038247

ISBN 0–415–19883–6 (hbk)
ISBN 0–415–19884–4 (pbk)

Contents

Illustrations

Figures

Black and white plates

Colour plates

Tables

Preface

Space is an essential framework of all modes of thought. From physics to aesthetics, from myth and magic to common everyday life, space, in conjunction with time, provides a fundamental ordering system for interlacing every facet of thought. . . . In short, things occur or exist in relation to space and time.

(Sack 1980: 4)

Space is central to our lives. We live and interact in space. Our lives are rooted and given context by the places we live in, the communities we inhabit, our sites of home, work and leisure, and are shaped by complex socio-spatial processes that operate across many scales, from local to global. In turn, spaces are produced and given meaning through social practices creating places. People's daily lives consist of a myriad of spatial behaviours, relationships and movements across and within spaces. From crawling across a playroom, to running around a school yard, to driving to work, to flying great distances for business meetings or a holiday, our daily lives involve hundreds of complex spatial choices and decisions that have to be successfully negotiated – choices and decisions that are socio-spatially situated, and influenced by cultural, economic and political forces. Moreover, the world is geographically demarcated. The surface of the planet is divided into territories at varying scales, from the home to cities to the national and beyond – spaces that are planned, regulated and governed. It is perhaps not surprising, therefore, that considerable attention has been directed at mapping, understanding and explaining the geographic world over the past millennium.

At the beginning of the new millennium, information and communication technologies (ICTs) are reconfiguring space–time relations, radically restructuring the materiality and spatiality of space and the relationship between people and place. Moreover, the conceptual space they support, cyberspace, is extending social interaction through the provision of new media that are increasingly reliant on spatial metaphors to enhance their operation. The combined power of ICTs and cyberspace is changing the way we live our lives, in the same way that the telephone, car and television did in the twentieth century. Moreover, these technological changes affect us regardless of whether we actively use them or ever want to use them simply due to the fact that they are employed by multinational corporations and the institutions which structure daily living. Given the massive projected growth in users of new technologies and online services, and the seemingly constant flow of innovations, it seems certain that the combination of ICTs and cyberspace will become one of the most significant evolutionary developments of the twenty-first century.

In *Mapping Cyberspace*, we provide a geographical analysis and critical reading of ICTs and cyberspace, and their relationship to social, cultural, political and economic life. The book draws on the findings and theories of researchers in a number of disciplines including geography, cartography, sociology, cultural studies, computer-mediated communications, information visualisation, literary theory and cognitive psychology. It is our contention that an *essential* element in understanding ICTs and cyberspace is a comprehension of how they are transforming, and creating new, spatialities, spatial forms and space–time relations. In short, we argue that geography continues to matter, despite recent rhetoric claiming the 'death of distance'. To provide evidence to support our claims, in *Mapping Cyberspace* we detail a literal, conceptual and metaphorical mapping of ICTs and cyberspace. We believe that our analysis will be of interest to social scientists, policy makers, ICTs providers and regulators, software developers, and to information and computer scientists. Hopefully, the book will also be of interest to everyday users of ICTs and cyberspace.

Martin Dodge and Rob Kitchin
February 2000
Maynooth, Ireland

Acknowledgements

This book started life as a coffee-table book. Over time, however, it mutated. First, into a book concerned solely with the spatialities and geometries of cyberspace, and then, for reasons apparent in the Preface, into its present form. During the course of its mutation and writing we have received invaluable advice and help from a number of people. We owe the following a great debt, particularly those who read preliminary chapter drafts, provided maps and administered advice: Paul Adams, Peter Anders, Mike Baty, Mark Blades, Mike Bratt, Tim Bray, Proisnnais Breathnach, Kath Browne, Stan Brunn, Hal Burch, Bill Cheswick, KC Claffy, Jo Cheeseman, Cora Collins, Ken Corey, Jeremy Crampton, Mike Crang, Shannon Crum, Judith Donath, Simon Doyle, Paddy Duffy, Stephen Eick, Sara Fabrikant, Emmanuel Frécon, Peter Fisher, Andy Gillespie, Sean Gorman, Stephen Graham, Muki Haklay, Francis Harvey, Dan Jacobson, Paul Kahn, James Kneale, Marty Lucas, Nico Macdonald, Carl Malamud, Tracie Monk, Tamara Munzner, Shane Murnion, Sam Paltridge, Larry Press, John Quarterman, David Robinson, Peter Salus, Sarah Sheppard, Naru Shiode, Marc Smith, Greg Staple, Nick Tate, Anthony Townsend, Jo Twist, David Unwin, Roland Vilett, Darren W. Williams, Mark Wilson, Matt Zook. We would also like to thank all the other people who generously provided maps and images for the book.

The main cover image is part of a satellite map of AlphaWorld created by Roland Vilett (http://www.activeworlds.com/satellite.html). The smaller image, of arcs on the globe, is from a project visualising the global topology of the MBone by Tamara Munzner, Eric Hoffman, K. Claffy and Bill Fenner (http://graphics.stanford.edu/papers/mbone/).

We promise to visit the temple to Andy Smith sometime in the near future.

Special thanks must be extended to Mike Batty for providing great support and interest in the project from beginning to end.

Beyond the book

Much of the inspiration, and the foundation stone, for this book was Martin's website, 'An Atlas of Cyberspaces' (http://www.cybergeography.org/atlas/), an invaluable resource for anyone who is interested in the geography of cyberspace. Martin also distributes a monthly email bulletin, titled 'cybergeography', which details all the latest mapping and visualisation developments. To have your name added to the distribution mailing list, send an email to: subscribe@MappingCyberspace.com

A website accompanies this book which displays all the illustrations in full colour and provides hypertext links to the projects that we discuss. The address is: http://www.MappingCyberspace.com/

We welcome comments and feedback concerning any aspect of the book. Please send them to: comments@MappingCyberspace.com

1 Introducing cyberspace

In this chapter we provide a brief overview of a number of key aspects central to understanding the scope and importance of the spatialities and geometries of cyberspace.[1] The ideas presented provide a broad context and overarching analytical framework for the analyses and discussion in subsequent chapters. As such, many of the themes are returned to throughout the book, where they are discussed in greater detail. We start by introducing briefly the basic building blocks of cyberspace, followed by a short history of their development. Next, we detail why cyberspace matters – why it is an important focus of study, through a dialectal analysis of how cyberspace disrupts a number of processes and foundational, particularly geographical, assumptions that underpin modernist society and epistemologies. Lastly, we provide a brief outline of the ways in which academics and commentators have theorised cyberspace, before detailing the approach we have used to underpin the analysis in subsequent chapters. In particular, we seek to theorise the role of space and the nature of online spatiality.

What is cyberspace?

The term cyberspace literally means 'navigable space' and is derived from the Greek word *kyber* (to navigate). In William Gibson's 1984 novel *Neuromancer*, the original source of the term, cyberspace refers to a navigable, digital space of networked computers accessible from computer consoles; a visual, colourful, electronic, Cartesian datascape known as 'The Matrix' where companies and individuals interact with, and trade in, information. Since the publication of *Neuromancer*, the term cyberspace has been reappropriated, adapted and used in a variety of ways, by many different constituencies, all of which refer in some way to emerging computer-mediated communication and virtual reality technologies. Here, we refocus the definition back to that envisaged by Gibson, so that cyberspace refers to the *conceptual space* within ICTs (information and communication technologies), rather than the technology itself.

At present, cyberspace does not consist of one homogeneous space; it is a myriad of rapidly expanding cyberspaces, each providing a different form of digital interaction and communication. In general, these spaces can be categorised into those existing within the technologies of the Internet, those within virtual reality, and conventional telecommunications such as the phone and the fax, although because there is a rapid convergence of technologies new hybrid spaces are emerging. Here, we provide a brief outline of the various cyberspaces in existence. More complete descriptions of their form and architecture are provided in subsequent chapters, notably in Chapters 5 to 8.

Internet and intranets

The Internet consists of a global network of computers that are linked together by 'wires' – telecommunications technologies (cables of copper, coaxial, glass, as well as radio and microwaves). Each linked computer resides within a nested hierarchy of networks, from its local area, to its service provider, to regional, national and international telecommunication networks. The various links have different speeds/capacities, and some links are permanent, while others are transient, dial-up connections. Although some networks are relatively autonomous – that is, they are self-contained spaces – almost all allow connections to other networks by employing common communication protocols (ways of exchanging information) to form a global system. Indeed, one key definition of the Internet means computers that are connected by the protocol TCP/IP (Transmission Control Protocol/ Internet Protocol) (Krol and Hoffman 1993). Anyone with a computer, a modem and a telephone can connect to one of the network spaces and through it to the rest of the Internet.[2] The sum of these nodes and their connections is greater than their parts, forming a network that enables people to communicate and share information. This has led some commentators to draw parallels between the human brain and the Internet – the brain comprises of millions of neurons and interconnecting nerve 'wires' that when combined together give rise to human consciousness, thought, memory and the mind; in the same manner cyberspace emerges from the connections of the Internet (Mayer-Kress and Barczys 1995; Heylighen and Bollen 1996). The processing power of the computers and the capacity of the connecting wires varies enormously across the breadth of the Internet but, crucially, information can still be exchanged between the humblest personal computers (PCs) and state-of-the-art servers, through the slowest copper telephone line to the latest gigabit fibre optic cables spanning the oceans.

Within each network space users are normally presented with different modes of interaction, varying in their sophistication and immediacy. Users can browse information stored on other computers, exchange electronic mail (email), participate in discussion groups on a variety of topics, transfer files, search databases, take part in real-time conferences and games, explore virtual worlds (both textual and visual), run software on distant computers, and buy goods and services.

The least sophisticated of these activities involves accessing a remote machine using Telnet and File Transfer Protocols (FTP). Telnet allows a user to log-on (connect) to a host computer outside of their immediate network. In effect, the computer becomes a terminal of the remote host allowing the user to explore the files stored there. As such, it is possible to be geographically located in one place (say London, UK) but be working real-time on files geographically located on the other side of the world (say Sydney, Australia). In general, most host computers require a log-in name and a password before entry is gained but there are some open access sites including many bulletin boards. FTP allows a user to 'download' or copy files from a remote computer host. Unlike Telnet, which makes a user's computer a terminal of the host, FTP only allows external visitors to look at, and download, the files. Again access can be restricted with many sites requiring a username and a password. For open access sites the username is generally *anonymous*. No password is required, but it standard for the user to enter his or her email address.

By far the most common form of Internet activity is sending communications via email. Email allows network users to send messages to each other via their network and in most cases across the Internet. Current estimates suggest that approximately 2.7 trillion emails were posted in 1997, with 6.9 trillion messages predicted for 2000 (*Web Week*, cited in

Treese 1997). To send a message, a person types into the computer the whole memo/letter and then sends the text to the other person's specific, personalised address. In addition to plain text messages, most email applications also allow users to send files of various types across the Internet. Mailing lists are centralised, and in some cases monitored, forums that allow a number of individuals to converse or swap information via email on specific topics. Every email message sent to the list is redistributed to all the other subscribers who then have the opportunity to respond.

An alternative to the mailing list is the bulletin board. Bulletin boards are centralised digital media that allow users to access a number of functions. These can include access to news lists, chat facilities and to connect to, and download information from, other boards. News lists act like 'real-world' bulletin boards and are centralised places to post and read mail. Users can periodically check the board for messages, which are organised under subject headings. Within each subject there are normally several threads of conversation. Users can choose whether to 'un-pin' a message, read it, and reply. As such, the system works in the opposite way to mailing lists. Whereas all mail on a mailing list is posted to all members of the list, all users of a bulletin board must go to the board to check for mail. Usenet is an example of a collection of news lists that is distributed across more than one bulletin board so that people around the world can contribute. Within Usenet there are literally thousands of groups with millions of members discussing the whole spectrum of human activity (see Chapter 7).

The World Wide Web (WWW) consists of multimedia data (mostly text and static graphics but also sound, animation, movie clips and virtual spaces) which are stored as hypermedia documents (documents that contain links to other pages of information) (see Plate 4.1, or this book's website http://www.MappingCyberspace.com). This is a vast and rapidly growing information space, encompassing over 1 billion publicly accessible pages as of January 2000,[3] and is likely to have grown significantly by the time this book is published. Using a browsing program such as Netscape Navigator or Microsoft Explorer, users can connect to a remote computer host and explore and interact with the information stored there. For example, it is now possible to shop and bank online, find out about educational establishments, play interactive games, research places that might be visited in 'real life' and book the trip, to keep abreast of local, national and global news, and much more besides. By clicking the mouse cursor on a link (usually highlighted text or a graphical icon) the user is transported between pages. Thus, the WWW provides a powerful medium in which to explore related subjects, allowing users to easily 'jump' between, and search for, other relevant documents, without concern for their specific location in the network or in geographic space. In addition to displaying hypermedia documents, the introduction of Java and the use of other plug-in applications means that programs can be run and downloaded across the Web. Furthermore, companies are now using the Web as a broadcast medium, channelling radio and television pictures direct to the host machine. The Web has become such a powerful interface and interaction paradigm that it is *the* mode of cyberspace, particularly for the mass of users who only came online since the mid-1990s.

Some Internet spaces allow synchronous interaction with other people rather than stored information. Chat facilities, for example, allow a number of users to converse via the Internet. Interaction takes one of two forms. In the first instance, as a user types their message it simultaneously appears on other users' screens. In the second instance, the text is typed out in full and then sent. Textual, virtual environments such as MUDs (Multiple User Domains) and MOOs (Multiple Object-Oriented environments) provide a themed

Table 1.1 Distribution of Internet users, February 2000

Geographical zone	Total population online (million)
World	275.54
Africa	2.46
Asia/Pacific	54.90
Europe	71.99
Middle East	1.29
Canada and USA	136.06
Latin America	8.79

Source: Nua, http://www.nua.ie/surveys/how_many_online/index.html

context for conversations and a space that users can explore and construct. The difference between these acronyms is mainly to do with how each environment is programmed. Whereas a MUD is hard programmed to contain certain features, which are textually described to a person entering a room, MOOs allow participants to alter and create environments, and to assign meanings and values to objects, which are stored in a large database for future users. These systems do not have the benefits of normal face-to-face conversation, such as voice inflection, eye contact and body mannerisms, but users conversing through chat facilities or virtual environments have developed a number of ways to impart such information through textual and visual means (in visual MOOs such as AlphaWorld[4] animated avatars, representing users, use basic movements to communicate simple greetings and emotions, for example, a nod of the head means hello – see Chapter 8).

The second main form of digital, networked spaces are called Intranets. Intranets have the same functional forms as the Internet, but are private, corporate networks linking the offices, production and distribution sites of a company around the world. These are closed networks, using specific links leased from telecommunication providers, or they employ new virtual private networking technologies, with no, or very limited, public access to files (company employees with knowledge of the correct password might gain entry from a public network). For example, most banks and financial institutions have national, closed Intranets connecting up all its branches, offices and ATMs (automatic teller machines) to a central database facility that monitors transactions. Other systems might monitor orders and bookings, allow email to be sent between different sites, and allow teleconferencing. These systems, as with most networks, are protected by 'fire-wall' systems which ensure that unauthorised users cannot gain access.

Millions of people have discovered the spaces of the Internet and Intranets and regularly inhabit them for personal, organisational and employment related purposes. Present estimates place the number of people online at 275 million (Nua, February 2000) with a growth rate of 30 per cent in 1998 (CyberAtlas 1998). By the end of the year 2000, analysts predict that the number of people online will exceed 400 million. Table 1.1 provides a geographical breakdown of Internet users. In January 2000 there were an estimated 76.4 million computers connected to the Internet (ISC 2000). Not surprisingly, business investment has mirrored user growth. *Boardwatch* magazine reported that the number of Internet Service Providers (ISPs) in the United States and Canada, in August 1997 was 4,133, a big increase from the 1,447 operating in February of the previous year (cited by Treese 1997). By mid-1999 the number of ISPs was 5,078 (Boardwatch 1999). A report by Nortel

Networks and IDC in January 2000 contends that investment in Internet infrastructure will quadruple by 2003 to approximately US$1.5 trillion, while the global Internet economy is forecast to reach $2.8 trillion. The Internet Advertising Bureau's most recent report shows that revenue from online advertising in the third quarter of 1999 was $1.2 billion. It is clear that users and investment will continue to grow exponentially making these figures redundant before this book reaches the shelves. However, it is important to remember that despite the impressive figure for growth and investment, at present only a small percentage of the global population is online. Moreover, there are marked inequalities of access between nations (ITU 1997 and 1998) and also within countries and communities (NTIA 1999).

As Internet and Intranet usage has grown, so too has cyberspace. Everyday, thousands of new webpages are added, hundreds of new users go online for the first time, and new networks, mailing lists and virtual services are created. The Northern Light search engine reported an index of over 128 million webpages in mid-1999 and yet this represented only about 16 per cent of the total indexable Web space (Lawrence and Giles 1999). All indicators of growth such as traffic statistics, the number of posts to mailing lists and bulletin boards, the number of webpages and documents online, the growth and competition between ISPs, and the amount of capital and speculative investment, suggest massive growth continuing for some time.

Virtual reality

Virtual reality (VR) technologies create visual, interactive computer-generated environments in which the user can move around in and explore. It currently takes two forms. First is the totally immersive environment: a user wears head-mounted goggles to view a stereoscopic virtual world that phenomenologically engulfs him/her. When the user moves, the virtual world that surrounds him/her is continuously updated by the computer, providing the illusion that the user is fully immersed in a three-dimensional, interactive space. Access to this space is currently limited as it requires specific hardware that can be expensive and cumbersome. The second form is screen-based and allows the user to interact with a responsive 'game space'. Both forms of virtual reality have three essential attributes: they are inclusive; they are interactive; and the interaction is in real time. The aim of both is to create a sophisticated conceptual space where the experiences are the same as the real world; to make 'cyberspace into a place' (Lajoie 1996). Although, at present, both forms are mainly visual, developers are working on including total sound effects and, in the case of immersive environments, touch. Currently, immersion VR machines are mainly limited to the military, academia, and the arcade entertainment industry, but it is the potential of screen-based VR to provide three-dimensional hypermedia environments, accessible over the Internet, that excites commentators and analysts alike; from their own home, users will be able to enter a virtual world as envisaged by William Gibson. This convergence of technologies is rapidly taking place and in recent years experimental virtual reality sites using virtual reality modelling language (VRML) programming have appeared on the Web that allow people to run virtual reality simulations via the Internet and also to take part in virtual reality MUDs. It is these desktop VR systems that are the predominate focus of our analysis (see Chapter 8). Commercially, it is envisaged that VR will have applications within architecture, planning, the military and surgery as places to explore possible eventualities. Similar to the Internet, virtual reality is characterised by high growth rates and capital investment.

A brief history

To understand fully the growth and geographies of ICTs and cyberspace, one must appreciate their history within the 'information revolution'.[5] The advent of ICTs and cyberspace has not been achieved independently, rather, it is bound within the histories of telecommunication technologies, computing and wider social and political-economic histories. As such, ICTs, as we know them today, can trace their origins as far back as Charles Babbage and the first recognisable basis of a computer, Samuel Morse and the telegraph, and Alexander Graham Bell and the first patented telephone system (Winston 1998). Indeed, the use of wires to connect places together has a long history stretching back to the invention of the telegraph in the first decades of the nineteenth century. There are many striking parallels that can be drawn between the development and growth of the telegraph and the Internet, particularly in terms of how the technologies were perceived by the public, how they were hyped by the media and marketing people, and their impact on society (Standage 1998). The capacity of the wires may have been smaller in the nineteenth century, compared even to today's slowest network links, and there were human operators at the nodes rather than digital computers, but the telegraph's enabling of instantaneous communication over great distances had a profound effect (Stephenson 1996; Thrift 1996). Standage (1998) describes the telegraph as the 'Victorian Internet' and argues persuasively that all the advances in telecommunications made since Morse's famous 1884 message, 'What hath God wrought!', have been incremental improvements rather than revolutionary breakthroughs. Similarly, Neal Stephenson (1996) states that:

> The world has actually been wired together by digital communications systems for a century and a half. Nothing that has happened during that time compares in its impact to the first exchange of messages between Queen Victoria and President Buchanan in 1858.

The important point here is not to take the development of ICTs out of context, accepting them as self-contained and autonomous, treating them in an ahistorical manner, and uncritically accepting the revolutionary hype that surrounds them. As Stein (1999) illustrates in relation to the early development of the telephone in London, technological development is grounded in social, political and institutional geographies and discourses that need to be carefully deconstructed. ICTs, as with all significant technological developments such as the railways, electricity or the motorcar, are portrayed as being revolutionary in nature when in reality the developments are the evolutionary outcome of many other initiatives (Marvin 1998). Perhaps what distinguishes cyberspace, and in particular the Internet, has been the speed of diffusion and growth; it is widely acknowledged that the World Wide Web is the fastest growing communications medium in history. This may well be true, but as detailed below, the unveiling of the Web in 1992 was the outcome of nearly thirty years' of research in digital computing and networking. Moreover, despite its impressive growth statistics, most of the world have yet to experience the Web and many may never gain access, particularly those in the developing world (see Chapter 2).

The recent history of ICTs can be traced to the launch of Sputnik 1 in 1957 and the moon landing of Luna 2 in 1958. To keep pace, and once again overtake the Soviet developments, the US Department of Defense created ARPA (Advanced Research Projects Agency). ARPA's mandate was to rapidly advance technological development. In order to leapfrog the Soviets, ARPA actively sought visionaries who could see clear ways to advance technological development, and where necessary they bypassed conventional proposal refereeing. ARPA sought to fund people who wanted to re-invent computing as it was then practised (Rheingold 1994). The Internet and virtual reality both started life within the ARPA programme, although their subsequent histories differed and they have only recently started to converge (see Figure 1.1). An important point here, and one that is returned to later in this chapter in the section 'Approaching cyberspace', is that the Internet was the outcome of a specific set of political-economic relations (for example, the Cold War): the Internet has a particular historical geography that centres it in the US, and early Internet development was guided by the military-industrial complex (Abbate 1999). Only later was it appropriated into the public and commercial domain.

History of the Internet

The Internet was initially conceived as a method to link several incompatible systems located at various points across the US so that resources could be shared.[6] In late 1967, SRI (Stanford Research Institute) was awarded a four-month contract to study the design and specification of a computer network and in early June 1968 a program plan was submitted to the Director of ARPA. The plan was approved in late June 1968 and offers for tender were sent to 140 potential bidders (O'Neill 1995). The specifications for the system were that all computers in the network were not directly connected to all the others, that interactive response times were good, and that there was added functionality. A think-tank at Bolt Beranek and Newman (BBN) wrote the successful application to develop the first long-distance computer network, ARPANET. BBN designed a system which used packet-switching architecture and an Interface Message Processor (IMP) (O'Neill 1995). Packet-switching involves breaking data or messages into units of equal size for posting through the system. Each packet is labelled with an identifier and the address of its intended recipient. The packet is passed from one packet-switch (node on a network) to another until it arrives at its intended destination. Packets can travel using alternative routes and at their destination are reassembled into their proper sequence using individual identifiers. The IMP was essentially a mini-computer that acted as a translator. The computer would 'talk' to the IMP which would then convey the message to the destination IMP that translated for the destination system. IMPs thus performed the functions of dial-up, error checking, re-transmission, routing and verification on behalf of the participants' computer. In this way, destination systems did not have to be fully compatible and any changes that needed to carried out could be done to the IMPs and not the host machines.

The first ARPANET node was installed at UCLA (University of California Los Angeles) in September 1969, connecting the IMP to a Sigma 7 (Salus 1995). The second IMP was installed at SRI, in Stanford, Northern California, in October, and on 21 November 1969 the computers at UCLA and SRI first communicated. By December 1969 there were four nodes (UCLA; SRI; University of California, Santa Barbara; University of Utah), expanding rapidly to thirteen in January 1971, twenty-three in April 1972, sixty-two in June 1974 and 111 by March 1977 (Hart *et al.* 1992; O'Neill 1995; see also Figure 1.2). In sixteen months ARPANET was a genuine packet-switching network, with at least two available routes between all the nodes (Salus 1995). Access to ARPANET was strictly limited to ARPA contractors. The initial network used dedicated switching protocols called Network Control Protocol (NCP). The Department of Defense, however, wanted a system that would link up machines of different makes and machines that ran at different clock speeds and used different sized packets, and that would link to satellites and packet radio systems which did not have the same format as ARPA packets. The resulting

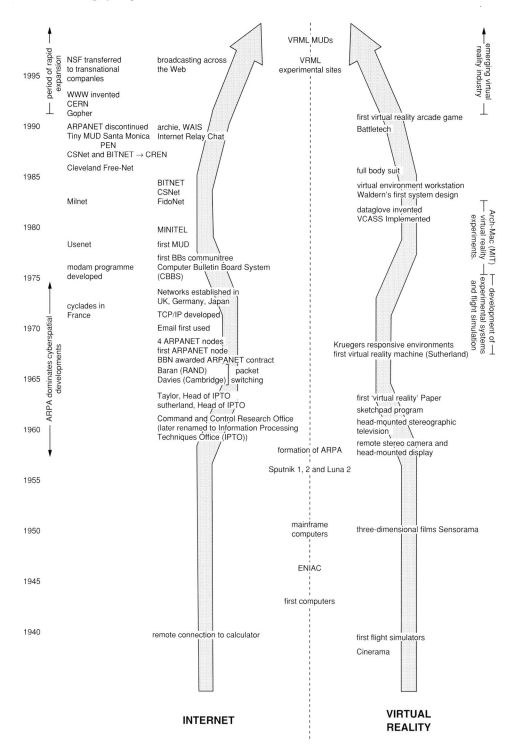

Figure 1.1 Merging timelines of cyberspace

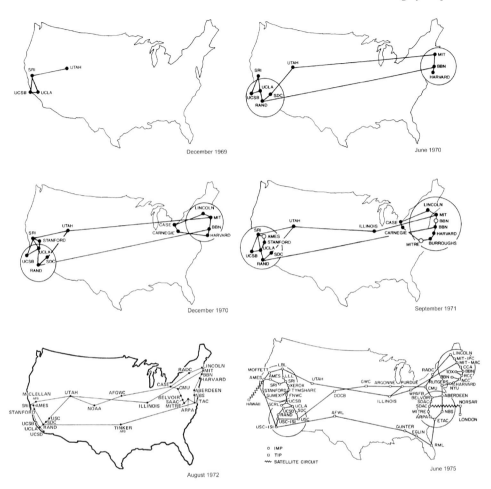

Figure 1.2 The development of ARPANET (1969–1975)

TCP/IP network protocol was developed by Cerf and Kahn in 1973 and has, because of its flexibility, become the standard switching protocols for interconnecting networks, despite recent advances in networking design (O'Neill 1995). These protocols have allowed the Internet to flourish by allowing the transfer of data between platforms that could not be mediated by IMPs.

Email was a key development on ARPANET. Email services had been first used on the large mainframe computers of the mid-1960s. Here, users of a common machine could leave messages for each other and thus exchange information (Cerf 1995). In 1970, Ray Tomlinson at BBN wrote the first program that allowed mail to be sent across a distributed network. The program was quickly circulated between all ARPANET sites and by 1971 the two most widely used applications were electronic mail and remote login services (Hart *et al.* 1992; Salus 1995). Thus, long distance personal computer-mediated communication was born. Other significant technological developments during the 1970s included the development of Ethernet, allowing the first Local Area Networks to come online; the distribution of packets via satellites; the establishing of networks outside the US – in

France, Germany, Japan and the UK; and the arrival of the first non-institutional networks, developed by hobbyists using personal computers, modems and telephone lines to make early bulletin board systems (BBSs).

Bulletin boards were particularly important for two reasons. First, they were the forerunner to general file sharing and public access services. Second, they were the start of non-academics and computer scientists involvement in computer-mediated communication. Many BBSs were unofficial and uncoordinated experiments in what Howard Rheingold (1993) has called 'grassroots groupminds'. These new spaces included all manner of computer conferencing systems (Quarterman 1990), the most famous of which is the WELL (Hafner 1997).[7] Based on affordable home PCs, BBSs developed a vibrant subculture (see Bennahum, 1998a, for personal recollections on this period; and Chapter 3).

The 1980s were characterised by a steady growth in both institutional networks and BBSs, the development of public access Internet architecture, and the construction of virtual gaming worlds. Of the bulletin boards, or discussion groups, Usenet and FidoNet grew to be the most successful, developing national and then international audiences. Usenet's success was dependent on the fact that the information within a discussion group was not limited to one site but forwarded to a many subscribing sites. The system works in such a way that when the next computer in the network receives the message, it checks to see which newsgroups it holds, copies all the relevant messages and disregards the rest. Because each message has a unique code no message is copied twice. By 1986 there were 221 groups and 1,414 sites. By 1988, there were 11,000 sites with four million bytes being posted every day. FidoNet started out as a single BBS, Fido, in San Francisco in 1983. Unlike Usenet, FidoNet links personal computers running DOS rather than university Unix hosts. Fido BBSs started to grow in number because the software for running a Fido system was available for downloading from any other Fido BBS. By 1985, individual FidoNets could communicate with each other using a hierarchical system of gatekeepers to reduce telephone costs. In 1986 there were about 1,000 nodes supporting at least 10,000 users (Hardy 1995). It should also be noted that both AOL and CompuServe started as proprietary conferencing and email services, and only later evolved into Internet companies.

In the early 1980s, two new special purpose networks, modelled on ARPANET, were built in the US. CSNet was designed to give access to electronic mail to non-defence contracting computer science departments and was funded by the National Science Foundation (NSF). BITNET was aimed at the wider academic community and was partially funded by IBM (Hart *et al.* 1992). BITNET was a store-and-forward network which meant that the system was ideal for mailing lists but could not support remote login or general file transfer. In 1989, CSNet and BITNET were merged to form CREN (Corporation for Research and Educational Networking) (Zakon 1996). These networks were accompanied by the Computing for Education and Research program (CER) of the National Science Foundation. In 1980 the NSF received congressional approval to build five supercomputer centres (O'Neill 1995). The selection of sites took place between 1983 and 1984 and the centres were built between 1985 and 1986 at Cornell, Princeton, Pittsburgh, Illinois at Urbana-Champaign, and San Diego. In 1986–87 four more sites were added – at Delaware, Purdue, Washington at Seattle, and Minnesota. To allow scientists from other sites to use the supercomputer sites it was decided to link them together to form a nationwide network. Although originally to be supported by ARPANET, the NSF built their own temporary 'do-it-yourself' network running at 56 Kps, replaced in July 1988 by NSFNET transferring data at 1.5 Mbps using TCP/IP protocols (Hart *et al.* 1992). NSFNET formed the basis for the Internet as we now know it.

In Essex, UK, in 1980, computer communities of a different type first came online. Multi-User Dungeons (MUDs), later to become Multi-User Domains and to mutate into several different acronyms (e.g., MOOs), are gaming systems allowing hundreds of people to take part in an interactive, textual, adventure game (see Chapter 8). The original system was an online version of 'Dungeons and Dragons', allowing individuals to create an identity, explore a textual world and take part in an adventure encountering computer-made people and creatures, and also converse with other adventurers connected to the game from all over the world (Mitchell 1995). In 1989, TinyMUD, a virtual world with the fantasy role-playing element removed, came online, later followed by LambdaMOO, the first to allow users to alter their surroundings and build new spatial domains. Today, hundreds of MUDs of varying degrees of interactivity and content exist, used by thousands of participants.[8]

Usenet continued to grow in the 1990s. It was estimated in 1992 that postings were distributed via 4,129 groups at 23,253 sites, with an readership of about 2.5 million people and over 35 million bytes posted each day. In 1994, there were 58,402 sites and 10,696 groups (Rheingold 1994). In 1999 there were over 32,000 groups (see Chapter 7). In the 1990s, there was a large growth in services and users centred around the World Wide Web and increasing bandwidth. Most important, personal computing expanded rapidly with real advances in both capability and functionality. The Internet has become more user-friendly and accessible to the layperson with a succession of helpful applications. For example, Archie, a device for searching FTP sites, was released in 1990. Prior to this, the contents of anonymous FTP sites could only be determined by searching them personally. Also in 1990, WAIS (Wide Area Information Servers) were released and similarly browsed for information. In 1991, Gopher, an application for searching and retrieving information, was circulated. Also in this period, Internet Relay Chat (IRC) was developed, allowing real-time 'conversations' between users at different machines.

The main breakthrough in terms of generating the exponential growth of Internet use came in 1992 when Tim Berners-Lee at CERN, Geneva, released the protocols and software that created the World Wide Web (Berners-Lee 1999). Here, text, images and sound could combine to provide a range of information. The documents created were also hypertext documents allowing users to directly link to other relevant sites containing attendant information. Mosaic, an application providing users with an easy to use, graph-ical interface for WWW pages quickly followed, as did search engines. In March 1995, WWW became the service with the greatest traffic on the Internet, overtaking FTP. In 1995, the NSF relinquished control of the NSFNET. Main US backbone traffic is now routed through interconnected, commercial network providers. With the growing com-mercialisation of the Internet there has also been an explosion of service providers and a host of spin-off industries such as those that design and maintain webpages, online con-sultancies, and cybercafés where you can connect to the Internet whilst having a coffee. The proliferation of these industries is likely to continue for some time, especially as digitally-based industries collide and merge. Indeed, the late 1990s is characterised by massive commercial investment in cyberspace by investors seeking competitive advantage, particularly with the e-commerce hype and huge stock market valuations surrounding companies like eBay and Amazon.com.

History of virtual reality

Sherman and Judkins (1992) trace virtual reality's inception to Ivan Sutherland's 1965 paper, 'The Ultimate Display', in which he outlined his ideas for an immersive, three-dimensional (3-D) display for computer graphics.[9] As a graduate at MIT, Sutherland had invented Sketchpad, the first interactive computer graphic system, and from 1966 to the early 1970s he created the first three-dimensional head-mounted display, generating crude three-dimensional graphics of outline-style cubes (Rheingold 1991). Sutherland's approach was to try and place humans 'inside' the computer graphics: the head-mounted display used binocular computer screens, each displaying the same image, and a gaze-tracking device which helped mimic 3D. Sutherland's head-mounted display was so heavy that it had to be suspended from the ceiling.

During the 1970s and 1980s, unlike computer networking, virtual reality research progressed slowly (with the notable exception of military flight simulators). The components and computing power were expensive. Many components were one-offs unable to be made commercially with the technology of the time. Virtual reality was an idea waiting for suitable enabling technologies to mature: electronic miniaturisation, computer simulation and computer graphics. Research focused on the technologies that converged to produce virtual reality: interface design, flight and visual simulation, and telepresence technologies (Machover and Tice 1994). Several key ideas emerged from Arch-Mac (later Media Lab) at MIT. In the late 1970s and early 1980s, researchers at MIT combined a wall-sized display, a voice recognition system and a gestural (pointing) input device. The operator sat facing the screen, and using a combination of pointing and voice commands, moved, re-sized and shaped objects displayed on the screen. Other work determined ways to transmit facial expressions and gaze direction via telecommunications, so that facial movements could be altered in synchrony with physical movements. In addition, they created a 'Dataland' using wall-sized screens, eye-tracking, voice input and gestural tracking devices, controlled using a Spatial Data Management System (SDMS) for visually navigating through databases. The aim was to create navigable information spaces.

In the 1980s breakthroughs started to happen in VR research. Researchers at NASA's Ames Research Center in Mountain View created the first affordable, head-mounted displays, three-dimensional audio devices and glove input devices were used to interact with complex three-dimensional computer-generated graphics. By altering the position of the head or hands users of the Virtual Environment Workstation altered their field of vision and could control virtual objects. In 1985, NASA commissioned the Virtual Environment Display System which incorporated a glove-based input device and experimented with three-dimensional audio devices. By the late 1980s and early 1990s, a significant amount of research was underway outside of the United States, and the technologies were starting to attract both commercial investment and media attention.

Virtual reality developed rapidly in the 1990s, with an increasing amount of research and commercial projects. Applications now tend to be simulators, allowing users to experience a particular environment and learn how to react to a series of situations. The applications are diverse, and range from medical training, to learning to drive, to how to handle emergency situations, to overcoming vertigo, plus many others. However, virtual reality technologies are not as advanced in development, or in the numbers of users, as the Internet. At present, they have not crossed the threshold of usability, with display resolutions rendering the user legally blind, head- and hand-tracking devices inaccurate and limited in range, and lag times that can ruin the sensation of immersion (Ribarsky *et al.* 1994). It will probably take ten to twenty years before immersive VR technologies become as advanced as the Internet, and another five to ten years after that before they move out of the labs and arcades and into the home and office. However, there is little doubt that VR will reach the same status as the Internet, and will merge with it. Indeed, we can see the start of such a merger with the increasing prevalence of virtual reality websites that allow Web users to interact with a 'game space'. An example of this convergence of Internet and desktop VR is AlphaWorld, which is discussed in more detail in Chapter 8.

Why ICTs and cyberspace matter

The rapid expansion in investment in ICTs and the exponential growth of users of cyberspace is impressive, but their real importance – why they matter – centres on their transformative agency. ICTs and cyberspace are transformative technologies, changing society in a number of ways. They are presently facilitating a process of restructuring, radically altering social, cultural, political, institutional and economic life. In this section, we explain why they are such effective, transformative agents through a brief examination of how they disrupt a number of processes and foundational assumptions that form the basis of modernist society and epistemologies. Because modernist systems of thought are founded on essentialist, dualistic categories (e.g., real/virtual, natural/technological), we structure our discussion around a series of binary constructs (not all of which are foundational assumptions of modernist thought, but are illustrative of how modernist structures are being undermined).[10] We recognise that by using this illustrative dialectical approach we reproduce the apparent values of such categories, although this is not our intention. Instead we acknowledge that the uncritical use of such categories is highly flawed and essentialist.[11] It should be appreciated that although we divide our discussion into separate sub-sections, the binary constructs we discuss are in fact highly related. It is our contention that an appreciation of the issues discussed here is essential in understanding the spatialities and socio-spatial processes described in Chapters 2, 3, 7 and 8.[12] We return to some of the binary constructs described below in Chapter 10, in our analyses of imaginative geographies of the information society and cyberspace.

Space/spacelessness

In modern society it is recognised that social relations are formed by spatial concerns. People and their sites of production and consumption are organised in relation to a spatial logic dictated by such factors as the friction of distance. Thus, cities developed in order to overcome time with space and were located where they could take advantage of raw materials or routes of trade; sites of production located in relation to materials and markets to minimise costs and maximise sales; and sites of consumption located in places that would maximise sales. Some analysts contend that ICTs render this modern logic of space obsolete (Cairncross 1997). As such, ICTs challenge space–time relations as conceived in modernist thought.

Those who declare the 'death of distance' (e.g., Cairncross 1997) maintain that the instantaneous communications of the Internet and Intranets have led to a collapse in spatial and temporal boundaries, leading to radical space–time compression which frees social and capital relations from modernist spatial logic. Whereas such innovations as the railway substantially reduced communication times, ICTs make them almost instantaneous. Clearly, 'instant' ICTs date back to the same era as the invention of the railway (e.g., the telegraph), but now they are much more sophisticated and support a variety of media used within the service economy. As such, modern ICTs complement the negation of spatial separation with additional services that are more ubiquitous, allowing wider and more ready exploitation. Here, the increasing efficiency of communication is translating into greater and more efficient productivity by permitting the exploitation of a truly globalised economy. For example, increasingly sophisticated interactions and services, which were once predominately place-centred activities, are becoming telemediated (e.g., banking). This has led some commentators, such as Benedikt (1991a), to question the 'significance of geographical location at all scales', and others, such as Gillepsie and Williams (1988), to ponder the significance of classical geographical ideas like the 'friction of distance':

> The idea of telecommunications as 'distance shrinking' makes it analogous to other transport and communications improvements. However, in so doing the idea fails to capture the essential essence of advanced telecommunications, which is not to reduce the 'friction of distance' but to render it entirely meaningless. When the time taken to communicate over 10,000 miles is indistinguishable from the time to communicate over 1 mile, then 'time-space' convergence has taken place at a profound scale. Because all geographical relationships are based, implicitly or explicitly on the existence of the friction imposed by distance, then it follows that the denial of any such friction brings into question the very basis of geography that we take for granted.
>
> (Gillepsie and Williams 1988: 1317)

Correspondingly, the growing use of ICTs as a means to speed-up communication and social/commercial interaction is seen by some to reinforce the significance of time in people's lives. Time, it is argued, is becoming *the* crucial dimension of who is accessible, rather than geographical location. As such, the scarce resource over which commerce competes is not space but human attention and appropriate bandwidth (download times) (Goldhaber 1997; Mitchell 1998). Some critics have called this the 'attention economy', as Mitchell (1998: 21) comments, 'We now ... increasingly live and work within an economy of presence, rather than one of propinquity'. So rather than being constrained by the friction of distance, we are increasingly constrained within a new geography of time which regulates access to people and resources (Harvey and Macnab 2000). Some even go so far as to claim that 'geography *and* time are no longer boundaries' (Hauben 1995, our emphasis), arguing that most aspects of our lives such as shopping and working can now be telemediated.

ICTs' destruction of space by time, it is argued, is revolutionising how business is conducted, transforming patterns of work, and leading to significant levels of urban-regional restructuring. Indeed, preliminary analysis suggests that ICTs, and in particular the use of Intranets, are facilitating globalising processes such as office automation, telework and the adoption of back-offices operations (see Chapter 2) leading to a radical reorganisation of corporations and significant changes in employment patterns within and beyond high-tech companies (see 'Industrial/post-industrial', below, and Chapter 2). Further, there is evidence that urban areas are restructuring to try and gain competitive advantage through cyberspace, and some sections of industry are decentralising to the suburbs and even other areas/countries to take advantage of cheaper rents and skilled workforces, while remaining in constant and instantaneous contact via ICTs. It is thus argued that the transformative agency of ICTs makes geographic space essentially 'spaceless' in that the contingency of space as a determinate of material practices is destroyed; geography no longer matters.

As one of us has argued elsewhere (Kitchin 1998), this is a gross overstatement. There is little doubt that ICTs do significantly disrupt the spatial logic of modernist societies, but they do not render it obsolete. Geography continues to matter – as an organising principle and as a constituent of social relations; it cannot be entirely eliminated. The modern spatial logic can only be done away with if everywhere offers equal opportunities for production and consumption, and everyone has access to them. ICTs, bandwidth and access are unequally distributed both within and between countries (see Chapters 2 and 5). This means that the main spatial processes of modernism, such as centralisation, continue to operate because the use of ICTs as globalising agents are still dependent on real-world spatial fixity – the points of access, the physicality and materiality of the wires. Furthermore, there is a world beyond ICTs and cyberspace in the form of other infrastructures, face-to-face social networks, skilled workforces, access to materials, and local and global markets. One must not overlook the fact that people still live in a material world and require food, shelter and human contact. In cases where services can be decentralised, they still have to locate in areas of suitable skilled labour and conventional transport links. In other words, although ICTs work to destroy space–time relations, to render social relations 'spaceless', other spatial practices, forms and forces resist and work against this attrition. Consequently, we are witnessing simultaneous pressures of spatial fragmentation (decentralisation) and unity (centralisation) (Griswold 1994), working on a number of levels that create a tension between the production of a globalised homogeneity and localised heterotopia. Indeed, a complex interplay between local and global forces exists as some places use and develop their localism, their uniqueness, to try and attract visitors and business. As such, the processes of globalisation should not be seen as totalising since the local is not insignificant. As Morley and Robins state:

> If we have emphasised processes of delocalisation, associated especially with the development of new information and communications networks, this should not be seen as an absolute tendency. The particularly of place and culture can never be done away with, can never be transcended. Globalisation is, in fact, also associated with new dynamics of *re*-localisation. It is about the achievement of a new global-local nexus, about new and intricate relations between global space and local space. Globalisation is like a jigsaw puzzle: it is a matter of inserting a multiplicity of localities into the overall picture of a new global system.
>
> (Morley and Robins 1995: 116)

People, their residences, and their sites of production and consumption are only rendered *partially* footloose by ICTs; the modernist spatial logic is fundamentally disrupted but it does not dissolve into a logic of 'spaceless'. Castells (1996) has suggested that we are witnessing a division of spatial logic into two distinct forms: an emerging 'space of flows' which overlies, and is starting to dominate and control, the old 'space of places'. Geographic space is being supplemented by a virtual space allowing people and organisations to be more flexible in relation to real-space geographies (Kitchin 1998). We believe that this increased, flexible, spatial mobility and modes of accumulation signals that we are now living in an era where the spatial logic is late-modern in nature (Jameson 1991); an era where a new socio-spatial nexus is being constructed (see Harvey 1989).

Place/placelessness

In addition to creating a 'spaceless' world, where space–time relations are meaningless, it is argued that ICTs and cyberspace are creating a 'placeless' world. Here, a combination of cultural globalisation and the spatiality of cyberspace itself is thought to be transforming 'real' world spatiality and the relationship between people and place. We will explore the relationship between cyberspace and placelessness through an application of the work of Relph (1976). Before starting our discussion, it should be noted that the concept of placelessness is not new. Indeed, it was a feature of modern society; for example, Gertrude Stein has referred to the placelessness of cities with the contention that 'there is no there there'. However, its extent has increased and accelerated under the pressures of globalisation.

In *Place and Placelessness*, Relph (1976) explores the relationship between people and places. He posits that there is a powerful relationship between the two, to the extent that 'people are their place and a place is its people' (p. 34). He argues that people develop and need attachments to places. However, the relationship to, and understanding of place, varies. He characterises this through a discussion of experiences of outsiderness and insiderness in places. Peet (1998: 50) summarises these experiences as: 'existential outsiderness, in which all places assume the same meaningless identity; objective outsiderness', in which places are viewed scientifically and passively (as in much quantitative geography); 'incidental outsiderness, in which places are experienced as little more than backgrounds for activities; vicarious insiderness, in which places are experienced in a secondhand way' (e.g., through paintings); 'behavioural insideness, which involves more emotional and empathetic involvement in a place; and finally existential insideness, when a place is experienced without deliberate and unselfconscious reflection, yet is full of significance'. Relph takes these concepts to examine the notion of 'authentic' place-making and inauthentic place-making (placelessness). An authentic sense of place involves a sense of belonging, an inauthentic the converse. For Relph, spatial mobility undermines authentic place-making, leading to the creation of places we have casual and superficial involvement with. Inauthentic places, he contends, are the prevalent mode of industrialised, mass societies and stem from an acceptance of mass values. Placelessness, then, is 'a weakening of the identity of places to the point where they not only look alike, but feel alike and offer the same bland possibilities for experience' (Relph 1976: 90).

In the context of the mapping of cyberspace and the information society, Relph's analysis raises two central questions: to what extent is cyberspace fostering (1) a growth in existential outsiderness and the creation of inauthentic places in geographic space; and (2) the provision of alternative, online authentic places for interaction based around interests rather than geographic location? In other words, does cyberspace help render geographic space placeless? And does cyberspace have places, and if so are they replacing those in geographic space?

For commentators such as Rheingold (1993), cyberspace is fostering the creation of inauthentic geographic places in Western society, leading to a destabilisation in the link between geographic place and identity. He contends, as does Relph, that communities in geographic space are fragmenting and losing cohesion due to cultural and economic globalisation: a coalescing of cultural signs and symbols (also see 'Real/virtual', below, for a discussion of simulacra and hyperreality), increased spatial mobility, a de-significance of the local, and changing social relations (see Chapter 2). Rheingold, however, is much less interested in the causes of destabilisation, but rather in how cyberspace can provide an antidote to placelessness by providing alternative and more attractive authentic places. If we take the definition of place provided by Jess and Massey (1995) – places are characterised by providing a setting for everyday activities, by having linkages to other locations, and providing a 'sense of place' – then there can be little doubt that new places, and new spatialities, are being formed online (see Chapters 3 and 7). Moreover, these places seem to be authentic as they embody a sense of belonging.

However, these places differ substantially from places in geographic space because they can be accessed from anywhere in geographic space (given the right technology), they are based on new modes of interaction, new forms of social relationships, and are centred on common interests and affinity rather than coincidence of location. Whereas social interaction, common ties and location are of importance in traditional notions of community in geographic place, in cyberspace it is suggested that personal intimacy, moral commitment and social cohesion come to the fore. For commentators such as Rheingold, cyberspace thus offers us the opportunity to marry *gemeinschaft* (where community relationships are tied to social status, public arenas and bounded, local territory) and *gesellschaft* (where community relationships are individualistic, impersonal, private and based on 'like-minded' individuals) aspects of community (Fernback and Thompson 1995). This means that individualistic, like-minded people join forces to form public-based communities; cyberspace offers the opportunity to reclaim public space and recreate online the essence and nature of authentic places which are disappearing in geographic space (see Chapters 2 and 3).

Again, we caution against wholesale acceptance of the placelessness thesis. As we discussed earlier, global processes are tempered by local processes with the result that in many instances places retain a 'sense of place'; places are not, and will not be experienced with total existential outsiderness as processes such as territoriality and nationalism continue to connect people to place. As such, whilst cyberspace does undoubtedly fuel a destabilisation in the link between place, identity and community, it does not destroy their interrelation; placelessness is partial. Indeed, many sites and projects seek to provide a reconnection and reinvigoration of place-based communities by fostering interaction among local residents. As documented in Chapter 2, this is a principal aim of many freenets. Moreover, we are sceptical about uncritically treating sites in cyberspace as authentic places. While undoubtedly some users of cyberspace consider themselves to be members of an authentic community, with a shared sense of place, many cyberspace users are transient, moving from space to space. As such, cyberspace for many users consists of inauthentic places.

In addition, we caution against the utopian analysts who promote the creation of places online as an alternative to geographic space, as it is within this space that we bodily reside. Indeed it is paradoxical that providing alternative, authentic places in cyberspace may help to accelerate placelessness in geographic space. Our quest, then, must be for the retention or re-creation of authentic places in geographic space as well as their formation in cyberspace. As we detail in Chapter 2, while online places might seem geographically dislocated, they are recursively connected to real places in a number of ways. One such way is through the individual user who accesses cyberspace from a geographic site. Although an online place may provide a sense of belonging, the user may reside in an inauthentic place offline. The consequence of this could be that a gay man living in an area where there is homophobia or where homosexual practice is illegal may 'belong' online but not offline. The creation of an authentic place online, in this case, is only a partial antidote to offline placelessness – there is still a need to authenticate place for this person offline.

Industrial/post-industrial

For many analysts, the condition of late-modernity[13] is the result of a transference in Western society from an industrially-based economy to a post-industrial one; from a heavy industrial and manufacturing-based economy to a service sector and information economy. Cyberspace, in its incarnation as Intranets and telematic systems, is seen as central to this transformation by providing a media that allow space to be overcome with time (see 'space/spacelessness', p. 13), and allow some forms of labour to be automated. Commentators like Poster (1995), for example, contend that we are moving to a 'mode of information', where knowledge and information replace labour and capital as the central variables of the Western economy. In this economy, ICTs are valuable because they allow businesses to reorganise, reducing costs and increasing productivity, and to merge to form multi-functional, multi-product corporations that operate across all continents, in multiple markets; they foster modes of flexible accumulation (Daniels 1995; Martin 1995). Evidence of the post-industrial society, it is argued, is evident in the unprecedented extent to which the structure and face of Western economies have been restructured during the 1980s and 1990s, with large-scale corporate and employment reorganisation; changing working practices due to the rise in office automation, back-offices and the introduction of teleworking; an increasing division in labour pay and security; the growth in importance of information industries such as banking, credit agencies, insurance, business and legal services, computing services, film and television, and telecommunication providers; the buying-up of smaller companies by larger ones; the merging of multinationals to gain competitive advantage; and increased urban-regional reconfiguration to attract flexible, footloose investment.

Some analysts (e.g., Harvey 1989) believe that all other aspects of late-modern societies, such as cultural transformations, are residue effects as the socio-spatial logic of modern economies is restructured into a new socio-spatial nexus; as capitalist systems of production mutate to take advantage of globalising technologies and flexible modes of accumulation in an attempt to find a new 'spatial fix'. Although we are reluctant to attribute all aspects of the late-modern condition to economic concerns (see 'Approaching cyberspace', below), we acknowledge that they play a significant role. We discuss the geographies of the information economy in detail in Chapter 2.

Public/private

One of the characteristics of the modernisation of society was the division between home (private) and other spatial arenas (public) that accompanied the separation of places of work from places of living (Laws 1994). This was accompanied by the creation of public spaces of social and political interaction, as typified by Habermas's discussion of the importance of cafés as sites of public debate (Habermas 1989). As noted above, many analysts consider that the public sphere, created in the period of Enlightenment, is rapidly disappearing to be replaced by spaces governed by private concerns. This privatisation of public space is probably best illustrated through the well documented transfer of shopping from public streets to privately regulated malls, with the shops remaining on the public street increasingly subject to the gaze of corporate and state surveillance (see Shields 1989; Graham *et al.* 1996). Here, actions in what were largely public spaces are subjected and regulated by private concerns, with people deemed undesirable denied access to these spaces. In addition, some city spaces are closing themselves off from the public sphere, retreating into defensible spaces – gated communities policed by private security forces. Here, modes of surveillance are used to re-assign the public as private.

A swathe of recent writings by academics have noted an increased use in surveillance technologies (e.g., Lyon 1994; Curry 1998; Garfinkel 2000). Whilst noting that surveillance and monitoring are not new phenomena, they report that recent developments in computing and telecommunications qualitatively alter the nature of surveillance by routinising, broadening and deepening it (Marx 1988) through the increased transferability, replicability and availability of records. As a consequence, they contend that it is increasingly difficult to take part in everyday life without leaving a digital trace: individuals and institutional records are digitised and stored in relational databases that are easy to cross-check; CCTV equipment (vision, sound, infrared) monitor public spaces; satellites monitor the Earth from low orbits; and computers monitor credit transactions and movements in cyberspace (Kitchin 1998).

Paradoxically, it is argued that cyberspace is creating new public spaces while being one of the principal means through which public space in the geographic domain is monitored. As such, while commentators such as Rheingold (1994) proclaim that cyberspace replaces rapidly disappearing public spaces with new social spaces (see 'space/spacelessness', p. 13), analysts such as Poster (1995) have begun to question the implications of ICTs in terms of privacy and confidentiality of the individual and collectives, highlighting the emergence of a sophisticated 'surveillance society'. Indeed, there is little doubt that it is increasingly easy to find out about a person's life through the electronic traces contained in digital databases. Government institutions hold our personal records, medical centres our health records, banks our financial records, credit cards record our personal transactions, newspapers record our personal indiscretions (e.g., court appearances) in digital databases that are linked by Intranet connections (Kitchin 1998).

Furthermore, it is now evident that while much of the Internet remains a public sphere, unregulated by private concerns (e.g., Usenet newsgroups), many parts of it, including the initial points of entry, are privatised and regulated by ISPs. Thus, cyberspace raises a number of issues concerning social power including the extent to which it is, or ever has been, a public space, how it is owned and regulated, who has access and who is excluded, how private and confidential the system is, and how it is being used (concerning issues to do with ethics and deviancy).

At the time of writing (2000), there are few legal precedents concerning cyberspace; regulation is dominated by customary law and market-led regulation (where ISPs self-regulate their customers). As a consequence, those who use cyberspace have very few legal rights. Recently, there have been calls from the business community for formalised laws to be introduced by governments and for offences to be legally enforced by a publicly accountable agency such as the police. However, such laws are difficult to enforce in a space which lacks the tangible qualities of geographic space (including a lack of territorial borders that match nation-states). Under market-led regulation, online public spaces largely become, like the shopping mall, privatised.[14]

Moreover, many people fear that online public space is becoming 'polluted' by deviant and undesirable groups. There has been much moral consternation about the uncensored and uncontrolled nature of the Internet. Cyberspace is seen to be a space where some of the more distasteful facets of contemporary society can proliferate and flourish (Squire 1996). For example, it is feared that cyberspace is providing a new space for pornography, messages of abuse, racial and ethnic hatred, anti-social behaviour and crime. There is little doubt that all of these social phenomena have an online presence, although their extent and influence is disputed (see Calcutt 1995; Dery 1996).

Whereas public space both online and offline seems to be coming under pressure as private concerns increasingly regulate spaces, Light (1999) argues that it is important not to overstate the slippage from public to private concerns. She maintains that throughout the twentieth century there have been moral panics concerning the disappearance of public and civic space. As she rightly states, however, there has never been a simple, public space in Western society. She documents that throughout the modern era, public space has been regulated through both legal and cultural forces, excluding and marginalising different sections of society, including most minority groups. For example, what was notionally public space has excluded, and continues to exclude, such groups as women, gays and lesbians, black people, disabled people, and homeless and poor people, through the use of socio-spatial practices and cultural signs employed in that arena. Moreover, although there has been a recent increase in the commercial owning of what was originally considered public space, commercial interests have owned and sought to regulate, through practices and politics, public space throughout the last century. Indeed, cafés and department stores have always been commercial interests regulated by their owners. Moreover, it is possible to argue that the spatial practices of 'ordinary' men and women were far more regulated and confined to particular spatial arenas by political-economic structures in the early part of the twentieth century. Light's discussion highlights the fact that there is not, nor has there ever been a clear distinction between public and private spheres. As such, we should be careful not to fall into the trap of either declaring that cyberspace provides new public spaces or that cyberspace further weakens public spaces in the geographic domain. Instead, we should seek to document the socio-spatial relations of cyberspace, the interplay between public and private concerns, and how these intersect with geographic space.

Broadcasters/listeners

Cyberspace challenges modernist structures of communication in three main ways. First, cyberspace is disrupting the traditional, communication power structures, the mass medium model of one-to-many broadcast, the separation of producer and consumer. In this sense, it is transforming the way we produce and exchange knowledge, and the power systems that underlie such production and exchange (Elkin-Koren 1996). Morris and Ogan (1994), for example, discuss the way in which the Internet disrupts traditional source-message-receiver features of traditional media of communication, sometimes retaining the same configuration, sometimes using different configurations. They group producers and audiences into four categories: (1) one-to-one asynchronous communication, such as email; (2) many-to-many asynchronous communication, such as bulletin boards or mailing lists; (3) synchronous communications on a one-to-one, one-to-few and one-to-many basis as found on Internet Relay Chat, chat rooms and MUDs; and (4) synchronous communication in which the receiver seeks out information from a provider, such as websites, gopher and FTP. Unlike traditional mass media such as television and newspapers, which are based on a one-to-many model where an editor, publisher or producer decides what should be seen and when, the 'consumers' of cyberspace are also the 'producers', so that there is a collapsing of traditional boundaries.

Second, the various forms of cyberspace are providing new forms of communication as written, oral and auditory modes of communication are combined and integrated in different ways. For example, analysts contend that email, chat facilities and MUD-based interactions are creating unique spaces of communication blending together written and oral styles to produce a new linguistic register and create new rules of language (Reid 1994; Cherny 1995). This new lexicon, consisting of a unique brand of shorthand, has emerged to compensate for the time it takes to type out replies given the instantaneous nature of communication.

Third, cyberspace can alter the nature of interaction so that it is possible to sustain multiple, simultaneous conversations. In face-to-face conversation, it is usually only possible to follow one or maybe two discussions at any one time. This is mainly limited by spatial logistics (the number of people within earshot) but also through our need to be able to follow closely the signal we want to listen to from other noises. However, within chat rooms and MUDs it is not uncommon for several threads of conversation to be conducted at the same time. Skilled users can follow these multiple threads and contribute to many of them simultaneously. As noted above, to accommodate the possibilities of changing forms of interaction, modes of interaction have to also change. Multiple conversations are sustained by using new lexicons, keeping conversation to short statements, and utilising the time lag between sending a message and receiving a reply to compose another message. We discuss in further detail the changing modes of interaction engendered by cyberspace in Chapter 3.

Real/virtual

Many analysts believe that cyberspace significantly destabilises a foundational assumption of modernist epistemologies, the separation of real from virtual, genuine from fake (Soja 1997). Benedikt (1991a) argues that cyberspace causes 'warpage, tunnelling and lesioning of the fabric of reality'. Further, cyberspace rapidly increases the blurring of reality and virtuality first started with the printed word, and further developed by radio, television and film. Each of these media, as with cyberspace, provides us with a representation of the real; a copy of the original. Some ICTs, such as VR, seek to extend this representational quality by immersing individuals in mimetic spaces aimed at making the real and virtual indistinguishable. To some commentators, these 'new realms of experience' (McCaffery 1991), television, photographs, cyberspace, have become so integrated into our lives that they have become our 'real space'. Consequently, our memories are now more frequently based on recollections of photos, videos, news footage, and television images, rather than on actual experiences. This acceptance of representational media as 'truth' means that many of us are now willing to accept the copy as original, and put our trust in those that re-represent the world to us (Slouka 1996). This in turn undermines our ability to differentiate between genuine and fake, real and imaginary.

Some commentators maintain that the blurring of real and virtual extends beyond conflation of media images and the real. They argue that 'real' space is itself changing, with aspects of places seeking to copy – in architectural style, in atmosphere, and so on – other locations (e.g., an Irish bar in London attempting to imitate a 'real' Irish bar). In some cases, the bar may be an exact copy of one located in Ireland, but more generally the bar is a simulacrum; a copy without an original. Here, aspects common to Irish bars are blended to produce a place that seeks to 'feel' to the bar's customers as though it would not be 'out of place' in Ireland. In cases where these copies/simulacra seem more real or authentic than the original they copy, then a state of hyperreality exists. For Baudrillard (1983), much of our current world is a hyperreal illusion, full of objects and buildings masquerading as the real – places that replace a territory with a map.[15]

There has always been simulacra; what is different today, however, is their scale and scope given the processes of globalisation (aided by the transformative agent of cyberspace) (Soja 1997). The process of creating simulacra/hyperreal places in present-day society, Baudrillard (1983) contends, is connected to the commodification of society. Here it is recognised that people choose where to visit, shop and so on, in part by the attractiveness or appeal of a location not just the activities contained within; they consume places. As such, places restructure their appearance, using cultural representation for profitable gain by drawing on styles and themes that will attract/seduce custom; real is replaced by

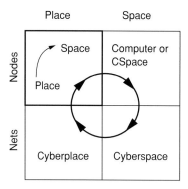

Figure 1.3 Virtual geography
Source: Redrawn from Batty 1997: 341

representations/simulacra which are assigned true value. Zukin (1992: 221) thus suggests that in the latter half of the twentieth century the city has become a 'dreamscape of visual consumption' as capitalism moves from an ethic of production to one of consumption. Baudrillard (1983) fears that the growing number of simulacra, as with media images, undermines our ability to tell genuine from fake, but also that with time hyperreal places extend beyond simulation to replacement. As discussed in an earlier section, 'Place/placelessness', this raises questions of place and identity, as historically a person's identity is rooted in specific locales (Urry 1985).

The blurring of real and virtual also extends beyond the imaginal. Analysts have recently started to argue that our geographic environments are becoming virtualised as computers are used increasingly to manage information concerning places. As such, city structure is becoming composed of and controlled by computers, and a recursive relationship is evolving so that as the city becomes composed of computers, the computer network (the collective power and information of computers across a city) is the city (Batty 1995). Castells (1996) terms this process of merging cityscape and ICTs 'real virtuality'. Here, the virtual spaces of city data and management and the real spaces of buildings and streets become entwined.

Batty (1997) details this process in Figure 1.3. In his conceptual model, cspaces located in individual computers (spaces – applications – within individual machines), and sited in real space, are linked together to form a distributed network: cyberspace. Cyberspace exists within the infrastructure of cyberplace (the infrastructure of the digital world – the actual hardware) and its use mediates the creation of new communications infrastructure and attendant services, which in turn has material effects on the socio-spatial relations of geographic space. Geographic space is thus adapted to take advantage of the possibilities that cyberspace offers, with companies, institutions and individuals computerising their practices to create new cspaces, and so on. Through this process real geographies are gradually being virtualised: composed, controlled and surveyed by ICTs.

Nature/technology

In modernist systems of thought, the natural is considered distinct and separate from the technological. Writers such as Haraway (1991) and Plant (1996) contend that cyberspace is one of a number of technologies, such as cosmetic surgery, biotechnology and genetic engineering, that disrupts this separation by aiding the reconfiguration (blurring) of the boundaries between people, their bodies and the material world. Where once the body was given, god-like, unchanging and fated, now it can be chosen, moulded and contested. Whereas other technologies physically alter the body, ICTs extend and immerse it within technology, particularly when using such devices as datagloves and headsets. Contemporary theorists thus argue that we are becoming hybrid beings, cyborgs, as technology replaces, supplements and recodes flesh; the coalescence of human and machine, nature and technology. As such, theorists are interested in how these technologies affect our identity, suggesting that it transforms it from a concept of stability (or fixity) to one of fluidity (see next section).

The political possibilities of the destabilisation of nature and technology as distinct categories have been particularly explored by feminist theorists. For example, Donna Haraway (1991), in her seminal text, *Simians, Cyborgs and Women*, critically examines the essentialist distinction between nature and culture to articulate a feminist political theory. She argues that if knowledge is constructed, then nature is culturally produced. Thus the distinction between nature and culture, and thus women (nature) and men (culture), is undermined. Haraway contends that we are all cyborgs; chimeras – the 'fabricated hybrids of machine and organism' (Haraway 1991: 150). As cyborgs, she argues, women have an opportunity to reappropriate, contest and enforce new social relations through the recoding of the self and the body. Haraway contends that until recently, 'female embodiment seemed to be given, organic, necessary' (p. 180), geared towards mothering and its extensions. However, as a cyborg, embodiment is fluid, partial and dynamic, not given but waiting to be ascribed meaning. Here, the dominant patriarchal dualisms that underlie and structure our society can be challenged and replaced by a new philosophical basis that re-balances the position of women within society. Cyborg politics, the recognition of the technological hybridity of people, thus represents a metaphorical means for moving beyond the dualisms of gender politics by rendering cultural categories indeterminate and fluid (Armitt 1996; Lupton 1995). Within this cyborg politics, it is hypothesised that cyberspace has the potential to free women of gender relations and sexual hierarchies because the physical differences that underpin social relations in geographic space will be eliminated, or at least radically subverted, within a space where the natural and technological are one (cyberspace). We will return to these contentions in Chapter 3 in our examination of gender geographies online.

Fixed/fluid

The final binary construct we consider relates to identity. A modernist conception of identity is essentialist in its formulation; identity is rational, stable, centred and autonomous. For many commentators, the adoption of cyberspace as a medium of socialisation (and as media that destabilises nature/technology) illustrates that identity is in fact unstable, multiple, diffuse, fluid and manipulable. In this context, cyberspace is seen as an important medium because it allows us to explore who we are and because it is changing who we are. Analysts contend that it achieves this because it offers a disembodied nature of communication accompanied by relative anonymity. In cyberspace, it is argued, people are accepted on the basis of their words, not their appearance or accent. As articulated in a well-known *New Yorker* magazine cartoon, 'On the Internet, nobody knows you're a dog'. Cyberspace thus allows people to carefully construct their self-presentation and to play with their online identity, adopting roles that they would not usually undertake

(Turkle 1995). For example, people can experiment with gender roles, pretending to be male, female or even adopting a gender-neutral position, and also play with personality traits, thus they can be shy or loud and boisterous. An individual's identity becomes fluid, ephemeral and empowering because they can choose how they are represented; the user becomes the author of their life. Further, identity becomes multiple and decentered as different aspects of our lives are disengaged and happen in different worlds, sometimes at the same time.

This conceptualisation of the nature of identity in cyberspace corresponds closely to current psychoanalytical and postmodern theories of identity. These theories maintain that rather than the self being some permanent structure of the mind, or fixed within some genetic code, self is thought of as a discourse in which identity is constructed through multiple experiences. Identity is thus fragmented, decentered and fluid, changing with time and situations, and indeed different for different occasions. With the growth in information industries, and the opportunity to interact in a number of different environments (including cyberspace), analysts now suggest that we rapidly cycle through different identities so that this process of fragmenting and multiplicity multiplies. The self thus becomes a series of roles that can be mixed and matched.

Again, we would caution against uncritically accepting this argument. Although identity is undoubtedly non-essentialist and fluid, and cyberspace provides a media to explore identity, cyberspace should not be viewed as a disembodied media nor should identity be seen as a set of interchangeable, but separate components. As noted in the Preface, one of our central theses is that cyberspace and geographic space are not separate realms, they are interwoven (see also the next section). As a consequence, cyberspace is an embodied media as identities explored and acted out online are always contextualised within experiences offline; our memories, personality, social relations, and so on, are not shed as our fingers hit the keyboard. Going online does not 'flatline' identity constructed in geographic space; we are aware of our lives on and off the screen and our actions are mediated with reference to both: spaces might be distinct and identity might be fluid and fragmented, but they are also situated – our lives in one space are contextualised in relation to life in others. Conversely, our lives offline become embodied through our memories and experiences online, so that a recursive process exists as the virtual is realised and the real virtualised. As such, online identities and experiences can be extremely rich and fulfilling,[16] but they do not take place in a disembodied vacuum, as the following quotations emphasise:

> It is important to remember that virtual community originates in, and must return to, the physical. No reconfigured virtual body, no matter how beautiful, will slow the death of a cyberpunk with AIDS. Even in the age of the technosocial subject, life is lived through bodies.
>
> (Stone 1991: 113)

> There is no loss of body in and through virtual reality technologies. While we may 'lose ourselves' in a good book or in the trance-like state of online interaction, we know that this is a change of consciousness: something in the mind, not the body.
>
> (Argyle and Shields 1996: 58)

We explore the relationship between identity, place and cyberspace in more detail in Chapter 2 and Chapter 3.

Approaching cyberspace

Given that cyberspace is transforming socio-spatial relations it has, in recent years, attracted significant academic and non-academic analyses by researchers and commentators seeking to explain its appeal and its impact. It is clear from the above discussion that this attention has been accompanied by theories to understand the appropriation and implications of cyberspace. In this section, we briefly detail the various approaches analysts have used in formulating their theories of cyberspace and to frame their discussions (for a more detailed discussion, see Kitchin 1998), before discussing the theoretical position that frames our own analyses in the remainder of the book. In particular, given that we are interested in the mapping of cyberspace, in the broadest sense, we discuss how conceptions of space fit within our theoretical frame.

The utopian view

Utopian positions are often atheoretical and empiricist (a simple presentation of 'facts' that are not theoretically grounded) and seek to forecast how technological innovation will affect future societies. Forecasts tend to utilise a form of the 'grand metaphor' whereby 'western society is seen to be moving *en masse* to some new and novel stage in its development as some form of "information society"' (Graham and Marvin 1996). Utopian analysts generally enthuse about new ways of life and construct a future filled with hope and benefits for all. The general ideology is that almost all of our problems – ethical, economic, political – are subject to technical solutions (Aronowitz 1994); that we will use technology to progress and that potentialities will be realised simply because they are possible. The hype surrounding cyberspace is unprecedented and much of it is utopian in character. Much of this writing, however, lacks a theoretical and empirical basis, and in our opinion should be treated with caution.

Utopian positions are sometimes founded on notions of *technological determinism* which posits that social, cultural, political and economic aspects of our lives are to large degree determined by technology. Here, technical innovations are seen as the dominant shapers of society and the way we think and act. Technology is independent, active and determining, and culture and identity is dependent, passive and reactive (Morley and Robins 1995), so that technical change is seen as autonomous, that is, 'outside society' (Mackenzie and Wajcman 1985). Cyberspace thus causes changes in our everyday lives in fairly linear, simple cause and effect relationships; cyberspace *will* lead to the formation of new communities, it *will* lead to changes in business practice, it *will* change how we live our everyday lives. For technological determinists, the questions concerning cyberspace centre on how society can adapt to, and learn to live with, the effects of cyberspace rather than focusing on how we can use, alter and reshape cyberspace to our benefit (Graham and Marvin 1996).

Social constructivism

Social constructivists posit that technology is a social construct and technology and society cannot be separated because they are intimately entwined with each other and with nature (Escobar 1994). As such, technology is mediated by culture, and vice versa. As Penley and Ross (1991) state: 'technologies are not repressively foisted onto passive populations . . . They are developed at any one time and place in accord with a complex set of existing rules or rational procedures, institutional histories, technical possibilities, and . . . popular desires.' Technologies thus do not give rise to themselves but are recognised as the product of our imagination and endeavours, bound in historical systems and dependent on structured relations between people (Haraway 1991). It is argued that not only is the development of technologies socially constructed but so also are their uses. For example, Lemos (1996) contends that 'contemporary technology is embraced, diverted and re-appropriated by everyday life'. Cyberspace is mediated and understood through culture as a social process (Hess 1995), where humans are recognised as being reflexive in nature, with the capacity to choose between alternatives (Lyon 1988). Cyberspace, therefore, is a social artefact as it mediates a series of social interactions and is itself a product of social mediation.

The political economic position

Political economic approaches also contend that technologies are not separated from society. In contrast to social constructivism, however, they suggest that the relationship between technology and society is bound up with capitalist modes of production and the associated political, economic and social relations which underlie capitalism. The political economist maintains that cyberspace's relation to everyday life cannot be understood without considering these broader relations and the capitalist dynamics of advanced industrial society (Graham and Marvin 1996). Here it is posited that technologies are rarely neutral, but are developed in the interests of industrial and corporate profits (Penley and Ross 1991). Analysis thus centres on seeking to identify and explain the relationship between cyberspace and capital, and to chart the social, political and economic manifestations of such a relationship.

Beyond modernism

The theoretical positions of utopianism, technological determinism, social constructivism and political economy remain part of the landscape of modernity. Each is characterised by a search for a unified, grand theory of society and social knowledge. Postmodernists contend that such an unified theory is unobtainable. They seek a new way to understand the world which recognises that the complexity of society is not easily constrained within a single theory that can explain all conditions at all locations. As such, postmodern approaches embody a shift from ways of knowing and issues of truth to ways of being and issues of reality (McHale, cited in Woolley 1992); to offer 'readings' not 'observations', 'interpretations' not 'findings', seeking intertextual relations rather than causality (Rosenau 1992). The postmodernist view is a re-conceptualisation of how we experience and explain the world which includes focusing attention on alternative discourses and meanings.

   Postmodernists are not the only theorists to criticise modernist approaches. Feminists contend that modernist science is dominated by, and reflects the position of men (Rose 1993; McDowell 1999). They argue that analysis tends to be patriarchal, failing to recognise the wider views and needs of society as a whole. They suggest that there needs to be re-negotiation of the role and structure of institutions, and the production of knowledge, so that how we come to know the world is more reflective of the people living in it. Feminist critiques of cyberspace seek to demonstrate the ways in which power relations within this new space are developing and to explore whether a socially just virtual society is emerging (see Cherny and Weise 1996). Whilst one set of critiques have been examining gender roles and whether the imbalances that pervade real world societies are perpetuated in virtual society, another set has been suggesting ways in which cyberspace might develop and the promises offered to groups who are marginalised and oppressed.

It would have been relatively straightforward to have written this book utilising an empiricist stance; simply presenting a description of the mapping of cyberspace rather than seeking to explain such mappings. Whilst our analysis is largely descriptive, we have sought to work within a social theoretical framework that marries aspects of social constructivism, political economy and postmodernism, thereby seeking to avoid the trap, as identified by Bingham (1999), of treating cyberspace as locations of the sublime (as powerful, dislocated, deterministic paraspaces).[17]

Separately, each of the approaches presented seems to us to be limited and limiting. Utopianism blindly predicts how cyberspace will revolutionise the way we live and documents how such changes might occur. Little regard is given to wider social or economic considerations concerning how technologies are re-appropriated and used by society or how technologies fit into the economic landscape. Technological determinism posits that the way we live our lives is largely dictated by technology in simple, cause and effect relationships. Analysis focuses on how society might adapt to cyberspace rather than on how we might use cyberspace to our benefit. From both a utopian and technological determinist viewpoint, technology and society are separate and uncomplicated rather than intimately entwined and inseparable. However, we believe that technologies do not give rise to themselves but are the product of human endeavour and bound within historical structures.

Social constructivism and political economics both recognise the inseparability of technology and society. Again, however, we believe each approach is flawed. For social constructivists, cyberspace is mediated and understood through culture. They are interested in the social processes used in shaping and re-appropriating the interactions between different actors and institutions that construct cyberspatial development and use. As such, they fail to acknowledge the influence of broader social and economic structures of capitalism and the power of political-economic forces. Political economists, in contrast, focus on these larger political-economic structures, but fail to acknowledge the role of social processes in determining how a technology is developed and appropriated. The marriage of these two approaches seems most appropriate, as Stein's (1999) analysis of the development of the phone system in London demonstrates.

In this combined approach, cyberspace usage and development is understood to be socially constructed at the local scale and mediated within the regional/global political economy. It is recognised that there is a recursive relationship between local, social/cultural processes and regional/global, political/economic processes. Here, technological developments, uptake, usage and behaviour (on- *and* offline) is locally constructed through the interplay between individuals and institutions, and tied in with historical systems. These local constructions are, however, linked to larger political and economic matters (e.g., who owns and regulates cyberspace) and affected by such factors as investment, policy, marginalisation, local economic conditions and status (e.g., levels of unemployment, poverty, etc., determine who can be online), and the opportunity to exploit and break into both the local and the global market. This approach acknowledges that the virtual intersects with the 'real'; that cyberspace is not a paraspace, a separate realm to geographic space, but forms part of an experiential continuum in people's lives. As such, Bukatman's (1993: 105) contention that 'a new and decentering spatiality [cyberspace] has arisen that exists parallel to, but outside of, the geographic topography of experiential reality' is firmly rejected. We believe that cyberspace is not a paraspace (other space), but an embodied space, one that allows a situated exploration of othering (see 'fixed/fluid', p. 23).

This integrated approach however, is more than a simple marriage of social constructivism and political economics. Instead, we acknowledge some of the criticisms made by post-modern scholars, particularly their rejection of a search for a grand narrative, universal truths and causality. The framing of our argument within a postmodernist perspective seeks to acknowledge a rejection of modernist values and instead looks to readings, inter-pretations and intertextual relations; it recognises differences between people and places, and highlights that society is progressing through a series of rapid transitions as traditional modernist ideas concerning space, time, reality, nature and so on, are undermined and reconfigured.

This approach means that the geographies of the information society *and* cyberspace are recognised as the product of social, institutional, political and economic processes that shape spatial arrangements and interactions both on- and offline. These arrangements and interactions take a number of forms that differ across space and time. Moreover, as we detail below, it is recognised that space (both geographic and cyber) is itself constructed through social and political-economic forces; space is produced.

Theorising space

Curry (1998: 24) suggests that current conceptions of space lie on a continuum, 'where at one end are those wherein the relationship among objects is strictly contingent, and where at the other are conceptions where the objects in space have very strong, even necessary and intrinsic relations with one another and the space in which they are located.' Space, then, is viewed either as absolute (Aristotelian, Newtonian) – space is understood and treated as a container filled with objects, or as relational – space is understood as the consequence of interrelationships between objects (Leibnizian, Kantian). Curry (1995: 5) thus contends that there has really only been four main notions of space used in geographic thought: Aristotelian, whereby space is 'static, hierarchical, and concrete'; Newtonian, whereby space is 'a kind of absolute grid, within which objects are located and events occur'; Leibnizian, whereby space is 'fundamentally relational and defined entirely in terms of those relationships'; and Kantian, whereby space is conceptualised as 'a form imposed on the world by humans'. Curry suggests that it is the first three notions that have generally been utilised in geographic thought, in particular the Newtonian view, although Kantian views have been adopted in the case of studies of how people perceive and cognise space.

These views are derived from the studies of the physics of space and time. Newtonian and Leibnizian theories posit that space is underlain by physical laws that can be scienti-fically measured. In particular, Newtonian views of space have been adopted in geographic enterprise, underpinning much quantitative work on the search for spatial laws that iden-tify the logic in patterns of human settlement and endeavour. In geographic terms, then, space is defined and understood within Euclidean geometry and for analytical purposes treated as an objective, 'empirical space' (Shields 1997); 'an absolute container of static, though movable, objects and dynamic flows of behaviour' (Gleeson 1996: 390). Here, an essentialist view is often adopted so that dimensions and contents of space are unquestion-ably understood as being natural and given.

In recent years, this absolutist and essentialist view of space has been challenged. This approach, which has minor application to the study of the physics of space and time (other than to observe that this practice is itself a practice of construction, and as such space–time is a 'concrete abstraction' [Lefebvre, 1991]), denotes geographic space as relational. This relational view of space refers to how space is constituted and given meaning through human endeavour. Here, it is argued that space is not a neutral and passive geometry, but rather is continuously produced through socio-spatial relations; the relationship between space, spatial forms and spatial behaviour is not contingent on 'natural' spatial laws, but is the spatial product of cultural, social, political and economic relations; space is not essential but is constructed and produced. As such, space is 'constituted through social relations and material social practices' (Massey 1994: 254). Soja (1985) defines socially-produced space as 'spatiality', suggesting that not all space is socially produced, but all spatiality is. Spatiality, then, is distinguished from space–time physics as it divides space as used and constructed from space as mathematically formulated. The process of (re)producing space, of its spatiality changing through time as a consequence of shifting socio-spatial relations and contestation, Shields (1997) refers to as 'social spatialisation'. It thus follows that in terms of understanding *human* geography, and the boundedness of society and space, it is only spatiality that is worth examining.

The value in treating space as socially produced is clearly advocated in the work of Lefebvre (1991). He details that the production of space, the process of spatialisation, is premised on three, complementary levels. First, he identifies a set of *spatial practices*; processes that influence the 'where' of human endeavours which range from individual routines to institutional creation of sites of investment. These practices are used continuously to (re)produce spatiality. Second, he identifies a set of *representations of space* (e.g., maps) and their power in regulating space and organising development. These representations are thus recognised as social productions with ideological content (see Chapter 4). Third, he argues that there are *spaces of representation*, spaces that are themselves imbued with ideological content, that convey particular meanings, and over which there is often conflict due to contested meanings and values; and spaces in which conflicts occur as individuals and groups seek to claim a territory. These three levels interact to produce a complex spatiality, and thus an understanding of this spatiality must consist of a process of 'unpacking' them. Whilst Lefebvre's analysis is mainly concerned with the production of space under capitalism, Foucault's (1991) analysis of social/power relations in specific sites such as prisons and asylums, and Shields' (1991) analysis of place-myths, demonstrate that space is socially and culturally constructed beyond capital relations (Urry 1985).

Within this conception of space, analysis centres on discovering how and why phenomena vary spatially, and in identifying the socio-spatial processes that operate, at a variety of spatial scales, to produce and reproduce space, and how these are temporally located. The latter point is one that is often ignored in geographic analysis. However, spaces do have a temporal dimension, engendering differing social practices given the time of day, week, month and year (Urry 1985). As yet, however, little consideration has been paid to the spatio-temporality of cyberspace, although we detail a few initial studies of the temporality of Internet usage in Chapter 5.

In this book we are interested in charting both the emerging spatial geometries of cyberspace (identifying its physics of space–time) and the information society, and deconstructing the ways in which these geometries are produced and consumed through socio-spatial practices – examining the spatialities of cyberspace. As such, in our analysis, we conceptualise geographic space from a theoretical framework that is relational (both in geometrical and in social terms). As Curry (1995) notes, whilst much quantitative work claims to be underlain by Newtonian conceptions of space, most is actually relational and

Leibnizian; models ultimately consist of elements that are relative. Therefore, models of the spatial geometry of geographic space are always relational. Indeed, current thought within the physics of space–time also posit that space is relational. As Stephen Hawking (1988: 38) explains:

> space and time are dynamic quantities: when a body moves, or a force acts, it affects the curvature of space–time – and in turn the structure of space–time affects the way in which bodies move and forces act. Space and time not only affect but are also affected by everything that happens in the universe.

To us then, *geographic* space as well as having a spatial geometry (that can be spatially measured), consisting of places of production and consumption, is recognised as a production that is itself consumed. However, the picture in relation to cyberspace, is not as clear. As Clarke and Doel (1999: 297) argue, what we need to do is 'to rethink space–time'.

Cyberspace poses an interesting challenge to philosophers of space. Composed of billions of binary digits, cyberspace exists as a variety of forms including webpages, chat rooms, bulletin boards, MUDs, virtual reality environments, information databases, all with 'their own sense of place and space, their own geography' (Batty 1997: 339). As detailed in Chapter 3, cyberspace can offer worlds that, at first, seem contiguous with geographic and outer space, and yet on further inspection it becomes clear that the space–time 'laws' of physics have little meaning. This is because space in cyberspace is *purely* relational (both geometrically and socially). Cyberspace consists of many spaces that are all constructions – productions of their designers, and in many cases, users; they only adopt the formal qualities of 'geographic' (Euclidean) space if explicitly programmed to do so. Moreover, spaces are often purely visual, objects have no weight or mass, and their spatial fixity is uncertain (spaces can appear and disappear in a moment). Cyberspace has spatial and architectural forms that are dematerialised and dynamic; spaces that are not physically tangible, in that they can only be explored by the mind, yet metaphorically relate to bodily experience. Many spaces have no tangible geographic counterparts – they are spatialisations utilising a geographic metaphor to gain tangibility. Whilst some spaces are productive spaces (tied to sites of work) they are also spaces of consumption (the space itself is consumed), and many spaces are spaces of pure consumption; they only exist to be consumed. It is these spatial qualities that the architect Marcus Novak provocatively explores in his work (see Novak 1991; 1995).

The undertaking of a traditional geographic analysis of cyberspace such as geographic visualisation (cartographic mapping), is a challenge because it breaks the two traditional tenets of Western cartography: (1) space is continuous and ordered, and (2) the map is not the territory but, rather, is a representation of it (Staple 1995).[18] For example, within information spaces the geographic metaphor and territory become synonymous. Here, the use of spatialisations to structure the data becomes the means by which this new territory is negotiated; a VRML webpage is both the territory and the means by which to navigate this territory. Moreover, unlike representations such as maps, spatialisations of cyberspace can be viewed and navigated in forms analogous to the ways in which people habitually navigate geographic space (e.g., wayfinding through streets). This mode of navigation is supplemented by other means such as the use of exterior viewpoints and teleporting. The question this poses is how do we conceptualise and/or theorise space when many facets of the models that have been used to understand geographic space are disrupted in cyberspace? This is not an easy question to answer, but we seek to address it throughout the book.

It should hopefully be clear from this introduction that the mapping of cyberspace consists of both a traditional analysis of the geographies of the information society and cyberspace (e.g., what is where, underlying spatial form and geometry), and an analysis of the social and political-economic forces that operate across and within spaces that shape such arrangements and the people who inhabit them. We are interested in mapping, both literally and metaphorically, on- and offline spatial forms and spatial practices and their interrelation. It is to this mapping that we now turn.

2 Geographies of the information society

Modernity was ending.

(Gibson, 1992: 91)

In this chapter, we discuss geographies of the information society; how the development and use of ICTs and cyberspace effects socio-spatial and material relations.[1] As detailed in Chapter 1, we view ICTs and cyberspace as transformative agents, undermining and re-configuring the processes, spatial structures and institutions of modernist societies (Schroeder 1994). These transformations are contributing to a broader set of changes as we enter the twenty-first century, such as political reorganisation (e.g., the rise of global/continental political structures, the collapse of the former Soviet Union, and the Balkanisation of many areas) and the rise of an economy of signs (see 'Real/virtual', Chapter 1). This is not to say that we believe modernist structures and processes are being replaced. Instead, institutions (e.g., governments and multinationals) are utilising the transformations made possible by cyberspace to further reproduce, reinforce and enhance the mode of capital. As such, the restructuring that is occurring should be viewed as an evolutionary development, and not as a transition to some fundamentally different condition. Here, it is recognised that the development and promotion of ICTs and cyberspace is bound to capitalist modes of production – cyberspace is a commercial product to be economically exploited, used to open new markets of opportunity. As such, the dominant bases of the modernist agenda – enquiry, discovery, innovation, progress, internationalisation, self and economic development – are still principals that underpin Western society (Berman 1992).

Our discussion, then, details the emerging geographies of the information society – how ICTs and cyberspace are transforming cultural, social, political and economic geographies. To date, the exploration of these geographies has largely been confined to investigations of the interrelationships between technology and urban and economic geographies (for excellent overviews, see Castells 1996; Graham and Marvin 1996). It is only recently that geographers have started to consider the social, cultural and political implications of cyberspace. Here, we present an overview of this research, drawing on the findings of those in other disciplines to supplement our arguments.

In reading this chapter, it is important to remember that, as we have argued strongly in Chapter 1, people live in an experiential continuum, running from the materiality of geographic space through to the virtuality of cyberspace. The division of our discussion into two chapters, the first predominately concerned with material space, the second with virtual geographies, is, then, an artificial division of this continuum, in an effort to illustrate how 'old' geographies are being transformed, and how new geographies are appearing.

Cultural and social geographies

> The relation of cyberspace to material human geography is decidedly one of rupture and challenge. Internet communities function as places of difference from and resistance to modern society.
>
> (Poster 1997: 213)

Many commentators argue that the combination of ICTs and cyberspace disrupts a number of factors that underpin traditional forms of cultural and social interaction and thus the relationship between place, community and identity. These critics have identified three distinct modes of disruption. First, ICTs and cyberspace promote the development of global culturalisation, undermining local customs, cultures and traditions. Second, they facilitate global corporate restructuring and foster spatial mobility, which means that people live in several locations characterised by 'incidental outsiderness' (see 'Place/ placelessness', Chapter 1). Third, they provide an alternative space where identity is fluid and disembodied, and community is formulated on the basis of interests rather than on location. As noted in Chapter 1, we do not wholeheartedly subscribe to these viewpoints, and in this section we cast a critical eye over the first two of these contentions, and go on to examine the third more fully in Chapter 3.

Global culturalisation

Cyberspace, it is argued, is a global and globalising media – it can be accessed from anywhere in the world given the right equipment and privileges (e.g., money) (see 'Geographies of exclusion', below), it extends and deepens space–time compression, and it promotes the formation of a global village, as envisaged by the influential media theorist Marshall McLuhan: 'with electricity we extend our central nervous system globally, instantly relating every human experience. . . . This is the new global village' (1964: 358, 93). According to McLuhan's vision, global communication will be used to transcend the differences between cultures and societies, to create a new global village where people will come together and work towards mutual trust and understanding, creating a world that is 'smaller' and more democratic (Waters 1995). Poster (1995) argues that we are on the verge of massive, global cultural reorganisation, as we enter what he terms the 'second media age'. The global village envisioned and being created, it is argued, is one of largely homogenised, Westernised values and cultures of consumption (see Sardar 1995; Peet 1986). Specifically, it is a village largely constructed and dominated by American desires, values and practices. As Sardar notes:

> Cyberspace did not appear . . . from nowhere. . . . It is the conscious reflection of the deepest desires, aspirations, experiential yearning and spiritual angst of Western man, it is resolutely designed as a new market, and is an emphatic product of the culture, worldview and technology of Western civilization. . . . Cyberspace, then, is the 'American dream' writ large; it marks the dawn of a new 'American civilization.' . . . Cyberspace is particularly geared up towards the erasure of all non-Western histories.
>
> (Sardar 1995: 779–81)

Sardar argues that cyberspace fosters an Americanised view of the world. English is the first language of cyberspace and the US dominates its technical developments, innovations, content and usage. Indeed, development and use has largely been contextualised within a Western, white, male, patriarchal and Christian cultural history (Penny 1994). This fostering of an American, globalised homogeneity follows a number of trends evident in other media and in consumerism. For example, television around the world is becoming ever more saturated with American shows, Hollywood dominates cinema, radios play music by American artists to people wearing American-branded clothing eating in American fast-food restaurants. In terms of the Internet, the largest and most popular websites are those owned and operated in the US. The majority of the world's computers also run operating systems and application software created by US corporations, principally Microsoft.

As well as values, the spaces and representations within cyberspace (see Chapters 5 to 8) also portray a highly specific view of the world that is predominantly Western, resting on such conventions as Cartesian space, objective realism and linear perspective (Pryor and Scott 1993). Cyberspace, it is contended, thus accelerates processes of placelessness by helping render obsolete the symbolic and cultural differences between places and by making (American) culture more mobile.

In addition to its content and message, cyberspace also aids the processes of globalised culturalisation in other ways. For example, cyberspace facilitates aids the production and consumption of consumer items by aiding global corporate restructuring and increased market penetration (see below). This, and the redevelopment of places into simulacra – hyperreal sites of consumption using global motifs – is leading to a homogeneity in the urban landscape (the same shops, with an identical look and feel, in identical malls across the globe; see 'Real/virtual', Chapter 1). Cyberspace, it is contended, creates locales that are 'ageographical', lacking in a 'sense of place', and helps to generate cities which are 'placeless', that is, they adopt a global identity (Sorkin 1992). Sorkin (1992) terms these new global cities 'cit[ies] of simulation', and Soja's (1996) analysis of Los Angeles seems to confirm his thesis. He suggests that LA is becoming a pastiche of other locales so that its unique features become obscured in its sameness. This sameness is the result of decentring of production and consumption accompanied by homogeneous, satellite development; gentrification in the form of reworking the old and unique into the new and the same; and new buildings adopting architectural pastiches that do not relate to local, historical styles. Castells and Hall (1995) thus suggest that if dropped by parachute into new 'soft cities' (cities that are marketing themselves as 'wired') such as Cambridge, UK or Massachusetts, or Mountain View, California, or Munich, Germany, that we would not recognise the country, let alone the city, since the universality of the cultural symbols and urban landscape could mean we were located in any place in the Western world. For Jameson (1991), this erasure of difference is the logical outcome of late capitalism – a global, homogeneous marketplace.

As noted in Chapter 1, these globalising tendencies are not, however, totalising. Indeed, there is significant interplay between the local and global spheres, and new forms of localism are being formed, designed to resist the processes of globalisation and to counter placelessness. Some places even seek to balance the global with the local. For example, many areas capitalise on their identity, culture and sense of place. There is a realisation that culture and heritage are desirable commodities that attract consumption, inward investment and tourism (Goodwin 1993). As such, urban cultural resources are being manipulated and 'sold' to provide capital gain (Kearns and Philo 1993) and are becoming part of a growing 'economy of signs' (Lash and Urry 1994). Ironically, cyberspace is one particular medium through which cities are seeking to seduce consumption and refashion themselves, with authorities increasingly creating an online presence aimed at marketing the city. The processes of globalisation, then, do not make the local and historical obsolete, instead, what we are seeing is a re-writing of city spaces developed during the period of Enlightenment and shaped by the processes of modernism, imperialism and colonialism (Jacobs 1996).

Western cities seem to be developing in two directions simultaneously. At one level, they are becoming less distinct, more global and more homogeneous; at another level, they are trying to market themselves as unique locales, set apart from other places, in order to attract consumption. What emerges is a complex interplay between the local and the global; the authentic and the inauthentic, between place and placelessness.

Place, community and identity

In traditional conceptions of community, place is considered of importance alongside factors like common ties and social interaction. *The community*, however, is characterised by such factors as personal intimacy, moral commitment and social cohesion. Luke (1993) argues that territorial communities are now little more than geographically-defined and administered land units which consist of atomised individuals who share little common historical consciousness or beliefs. For commentators such as Rheingold (1993), cyberspace allows the reclamation and development of *the community* without the locale; a space in which people can form strong, cohesive and supportive groupings free of geographic location.

Given the sentiments of Rheingold, and other commentators who believe that the development of virtual communities provides an antidote to social alienation and place-lessness experienced in geographic communities, many analysts are concerned about the implications of ICTs and cyberspace on geographic communities. For example, Robins (1995) contends that in Rheingold's vision, cyberspace becomes an escape hatch. Here, there is a sense of running away from, rather than confronting and finding solutions to, placelessness and social issues. Cyberspace potentially weakens geographic communities by providing a focus centred on interest at the expense of geographic neighbours. The worry is that as people retreat into cyberspace, geographic space will further fragment and society will become increasingly anti-social (as envisaged in much cyberpunk writing – see Chapter 10).

As Wellman and Gulia (1999) point out, however, it is wrong to consider Internet communities as replacements for geographic communities. They note that a person's community (his/her kith and kin) does not necessarily live within walking distance. Instead, geographic communities have been replaced by social networks that are spread over a wide terrain, and which are sustained by letter writing, telephone conversations and now Internet connections. Indeed, they observe that many social networks which do share the same territorial space are often sustained through telephone conversations rather than face-to-face contact. As such, the division between geographic and virtual is not helpful – one is simply an extension of the other. It is the relationship between people that is important, not the medium of communication (Wellman and Gulia 1999). Social networks main-tained exclusively in cyberspace are thus not pale imitations of 'real' networks, or substitu-tions for these networks, they are simply another form of network, a subsection of an individual's total network, similar to pen pals when letter writing was more prevalent.

In addition, cyberspace is often used as a method to 'reconnect' members of a com-munity – a new media through which to maintain or improve a sense of place in a particular locale. Many Western cities now have websites devoted to community relations and development within those cities, many of which allow people to discuss issues among themselves and with local statuary and voluntary agencies. Graham and Aurigi (1997) have examined a number of these websites, referring to them as 'grounded virtual cities'. They found that while some sites undoubtedly did encourage citizen participation and community development,[2] many others were merely commercial and advertising exercises. PENs (Public Electronic Networks) are the most widely discussed form of local government and community development, and we examine these later in the chapter. Moreover, many communities are using cyberspace to develop cross-community and cross-issue alliances to help fight particular concerns. As we detail below, probably the most widely documented political use of the Web was by the Zapatistas of Chiapas (in Mexico), who used it to garner international political support (see O'Tuathail 1994; Froehling 1997, 1999). Other forms of collective action are more localised. For example, Uncapher (1999) charts the development of Big Sky Telegraph, a network designed to link teachers together in the isolated, rural landscape of Montana, and Mele (1999) details the use of the Internet by a local community, Jervay (Wilmington, North Carolina), to gain help in challenging the redevelopment plans of their local area by the city council. Hundreds of other local communities and local protest groups similarly use the Web and other cyberspaces such as newsgroups and mailing lists to garner support, resources and political help. These avenues can be an extremely useful way of creating and sustaining new community-based, political structures that are explicitly tied into geographic locales.

Another way that cyberspace is materially grounded in geographic communities is through the development of identifiable subcultures explicitly focused around ICT technologies. Rushkoff (1994), Schroeder (1994) and Dery (1996) all describe the development of such subcultures, centred on cyberpunk and youth movements which meet in cyberspace, cybercafés, nightclubs and communes, and whose material practices are grounded in computer use, rave, ambient and industrial music, smart or designer drugs, science-fiction writing, and calls for cultural and political change. They suggest that these groups structure their lifestyles into practices which aim to live out and bring about selected aspects of cyberspace's promise. Dery (1996) identifies a number of these subcultures, which include Deadhead hackers, ravers, technopagans, and New Age technophiles, many of which aim to combine 1990s cyberculture with 1960s counter-culture attitudes. In the US, these groups are mainly located on the West Coast, centred on San Francisco, with only a handful of small groups outside of the States, based around major cities such as London and Amsterdam.

Geographies of power

Space is one of the principal mediums through which power is administered and controlled (see Dear and Wolch 1987). Spatial organisation, the demarcation of territories, spatial surveillance and policing, and the expression of cultural ideologies based around such factors as gender, race and disability, all produce complex spatial geometries of power. These geometries seek to maintain hegemonic spatialities, and are constituted and reproduced through a complex interplay of social and cultural forces, situated within wider political-economic forces, so that a variety of intersecting power relations operate across social and spatial scales (from the home to the global). At the local level, power is socially produced and mediated through culturally-defined expressions. Here, different individuals and institutions interplay to create local societal relations that underpin societal interactions. This local scale feeds into and feeds off a much larger set of political and economic relations and dynamics that underlie the global political economy.

As we detail below, cyberspace, for some, disrupts these geometries of power by changing the socio-spatial basis on which they are formed and sustained. In these cases, cyberspace creates either utopian spaces of individual freedom or dystopian futures of 'big brother' with cyberspace operating as a giant panopticon. For others, cyberspace merely reinforces and deepens current geometries, providing a medium through which hegemony is further reproduced. In this section, we explore how cyberspace affects geographies of power, first by examining its implications on political structures, and then by exploring its role in extending modes of surveillance and reinforcing geographies of exclusion.

Political structures

Most commentators agree that the development of ICTs and cyberspace, particularly the Internet, has implications for political systems and modes of governance. ICTs and cyberspace are transforming political structures and organisations, political campaigning, lobbying strategies and voting patterns (Neustadt 1985). Here, cyberspace is viewed as initiating 'qualitatively new political opportunities because it opens new loci of speech' (Poster 1995). This has led analysts to two separate but equally radical conclusions. First, representative government could potentially be replaced by direct government. This means that we will be able to propose, debate and vote on local, national and international issues rather than elected officials acting on our behalf. Second, that the role of place-based political mobilisation will rapidly diminish (Thu Nguyen and Alexander 1996). Instead, political opinion concerning specific topics will be mobilised globally by interested parties as politics fragments and people's perspectives narrow. For example, groups such as the Tibet Information Network, Greenpeace and Amnesty International use webpages to disseminate information and raise political awareness at an international level, by-passing corporate conservatism and censorship by traditional media structures (Warf and Grimes 1997). Readers are given specific information about the best means of effective political action.

As Table 2.1 demonstrates, many groups consider the Internet to be an important emerging political forum through which they can gain support and political strength, with the number of webpages related to progressive causes multiplying at a tremendous rate.[3] As noted above, many of these progressive causes are localised, grounded in specific geographic locales, as concerned residents use the Web to subvert and resist traditional political structures. One notable example is the Zapatistas of Chiapas. Using mailing lists, newsgroups and websites, the rebels' supporters informed the world of the uprising in the Chiapas region of Mexico by the Zapatistas (peasant) army, and mobilised international political pressure on the Mexican government (Froehling 1997, 1999). Although the shooting war lasted only twelve days before a ceasefire was called (largely credited to the campaign waged through the Internet), the 'war of ink and Internet', as it was dubbed by the Mexican Secretary of Foreign Relations, continued for some time (Froehling 1999: 166).

Given the ways in which ICTs open up new political possibilities, Poster (1995) suggests that they represent a challenge to the whole notion of nation-states. New technologies like cyberspace defy the character of power employed by modern governments by undermining the concept of territoriality. Cyberspace, knows no borders. Loader (1997: 9) thus suggests that 'ICT networks . . . facilitate the deconstruction of national, financial and cultural boundaries which are an intrinsic attribute of modernism'. This deconstruction has wide ranging implications given that boundaries are seen as central to current political theories.

Table 2.1 Internet links to progressive causes

Keywords used	Number of links, August 1996	Number of links, October 1997	Percentage increase
Animal rights	258,920	2,650,017	923
Economic justice	94,346	831,406	781
Cultural preservation	60,055	474,452	890
Corporate responsibility	47,620	915,853	1,823
Religious freedom	43,767	536,185	1,124
Anti-discrimination	32,199	386,900	1,101
Labour rights	30,518	2,668,166	8,642
Peace activism	28,902	330,521	1,043
Anti-racism	27,562	340,329	1,134
Youth rights	27,128	2,633,062	9,606
Disabled rights	14,491	2,522,762	17,309
Gay and lesbian rights	12,357	2,598,741	20,857
Minority rights	12,295	2,491,644	20,165
Racial equality	8,900	125,567	1,310
Elderly rights	6,732	2,458,292	36,416
Children's rights	3,452	2,392,095	67,435
Environmental activism	3,397	555,371	16,248
Women's rights	2,744	2,392,095	87,075
Human rights	804	3,330,934	414,195

Source: Warf and Grimes 1997: 265

According to Thu Nguyen and Alexander (1996), cyberspace undermines modernist political discourse which is based on notions such as agency, action, territory, progress and development; they contend that cyberspace helps replace these notions with user-ship, operation, non-linearity, recursivity and chaos. Quoting Emberley (1988: 50), they claim that:

> The old economy of production, of industrial policy, of state initiative, of discrete and singular actors and audiences, of centers and margins, form and contents, in brief, the great order of reverential finalities where the world was compartmentalized, taxo-monically ordered, and prescriptive – all this over.

Thu Nguyen and Alexander (ibid.) argue that although nations still exist, they are progressively losing control over people because cyberspace is undermining polities through the availability of outside information. They describe how nations and organisations keep political control through gatekeepers (people who control and regulate information). With computer networks, they argue, people can by-pass the gatekeepers and get to information directly. This by-pass represents a 'major shift in the nature of embodied power' away from central individuals who 'hold' power (Thu Nguyen and Alexander 1996).

For them, the Internet is a 'conversational, demassified, non-representational democracy that transcends nation-state' (ibid.: 111). It represents the communication system needed to underlie the continued differentiation and complexity of society. As society fractures, the more information social systems need. Cyberspace is providing for that need and, as such, is an active agent of change. An agent that is challenging the traditional notions concerning power, knowledge and information through the altering of conventional power relations. This challenge, they contend, through demassification and atrophy of the polity will eventually lead to chaos through weakened polities unable to contain disruptions. In short, there will be a conflict between the coming Internet society and modernist democratic institutions. They explain that modern politics is grounded in geographical units and communities, and the assumption that individuals have concrete identities and interests. However, cyberspace renders place meaningless, identities fluid and reality multiple. They suggest that democracy based on geographical units is withering and is destined to suffer the same misfortunes as monarchies. However, a number of governments are fighting back by restricting access to the Internet, severely regulating Internet service providers and attempting to filter content (McCabe 1999).

In contrast to Thu Nguyen and Alexander's position, the Internet in many instances, such as free-nets and many commercial websites, rather than challenging convention, actually works to reproduce and reinforce existing hegemonic structures (Interrogate the Internet 1996). For example, local governments in the United States (and in other Western countries) are increasingly experimenting with community-oriented, participatory democracy in cities through the use of free-net systems. Using Public Electronic Networks (PENs), registered users can access city information, complete some transactions, send email to departments, elected officials or other PEN users, and participate in public conferences (Schuler 1995). PENs such as Santa Monica PEN and Cleveland Free-net have grown in use and popularity and hundreds of community organisations maintain and disseminate information. Users can gain access through their own PCs or through those strategically placed in public areas, such as libraries, around the city. In this context, the Internet strengthens the civic, public dimensions of cities/nations by providing a free access, public space for debate and interaction (Graham and Marvin 1996). Indeed, the homeless, a normally poorly organised group, has successfully used Santa Monica's PEN to lobby for shower facilities. In other cases, local governments have adopted ICTs as a means of improving public information provision, using the Web to provide citizens with an improved, more open service (Loader 1997).

Cyberspace, it is hypothesised, is instigating a democratic renaissance based on the notion that an abundance of available data and information is liberating, that it allows greater access to officials, and that it undermines the traditional media bases of democratic institutions such as broadcasters by allowing individuals to be both sender and receiver, thus permitting a more eclectic range of views to be disseminated (Brants *et al.* 1996). Rather than radically overhauling existing structures, the Internet is seen at best as just another tool to help conduct the everyday, democratic functioning of a government, allowing people to express their opinions. As such, the hegemony of nation-states (and other political structures) is reproduced rather than undermined.

Spaces of surveillance

As discussed in Chapter 1 ('Public/private', p. 18), modes of governance are being maintained through the use of surveillance and the creation of defensible spaces. Many commentators argue that ICTs and cyberspace are set to become a key surveillance technologies which will have widespread implications for the individual and the collective, privacy and confidentiality, spatial mobility and access. Indeed, it is increasingly difficult to take part in day-to-day life without leaving a digital trace: 'life in cyberspace generates electronic trails as inevitably as soft ground retains footprints' (Mitchell 1995). What were once private transactions are becoming less so, and information is parted without recourse to violence or trickery but through the filling out of forms or the passing over of a credit card. As a consequence, details of our personal lives are stored in digital databases and can be accessed by anyone with the right privileges (Mitterer and O'Neill 1992; Poster 1995; Garfinkel 2000). Moreover, surveillance technologies are increasingly merging with tele-communication technologies to form modes of surveillance that extend beyond record-keeping. For example, security cameras that are linked together in sophisticated telematic networks, and the use of anklet transponders connected to police computers via telephone modems to monitor people under 'correctional supervision'.[4]

Hegemonic agencies including the police, the military, the government and industry are all actively collecting and exchanging data. For example, most government agencies in Western societies enter details of applications and payments into database systems;[5] transactions such as using a credit card to buy goods are recorded and processed, as are the number and type of goods bought.[6] Education and health records are increasingly being recorded digitally, and there is a growing geo-demographic industry which identifies the characteristics of suitable areas for advertising and marketing based on census and other corporate database variables.[7] Furthermore, Ross (1991) reports that 70 per cent of corporations in the West now use electronic surveillance (and other means) to monitor employee performance, checking on such things as the number of keystrokes performed, the amount of time spent socialising, eating and even going to the toilet. For the most part, the individual has little control over the use of this data, for example who has access to it, and in many cases no knowledge of its existence (Poster 1995). Thus, Critical Art Ensemble (1995) argues that individuals are being complemented with a data body which support different levels of freedom.

In a geographic context, geo-demographic databases are particularly relevant. These databases can be highly sophisticated, dividing up nations into characteristics based on households' postcodes. For example, the UK-based CACI[8] has devised ACORN which divides every postcode in the country into 'life-style' categories. The descriptions of each category typically include information concerning house size, area characteristics, likely earnings profile, car ownership, and so on. Other companies can purchase this information and through the use of relational databases combine it with their own or other purchased data. As a result, in the United States for example, it is possible to purchase digital information from companies such as TRW, Equifax and Trans Union, which reveals facts concerning birth, family, address, telephone number, social security and salary history, credit transactions, mortgage, bankruptcy, tax and legal records, and education and health records (Graham and Marvin 1996).

The fear, as noted above, is that cyberspace 'may well simply turn into an extension of social control through the control of information' (Interrogate the Internet 1996: 129), creating a super-panopticon, a gigantic and sleek surveillance machine that can be used to control and regulate (spatial) freedom:

> The phone cables and electric circuitry that minutely crisscross and envelop our world are the extremities of the super-panopticon, transforming our acts into an extensive discourse of surveillance, our private behaviours into public announcements, our individual deeds into collective language. . . . The individual subject is interpellated by the super-panopticon through the discourses of databases that have very little if anything to do with 'modern' conceptions of rational autonomy.
>
> (Poster 1995: 87)

This is a theme explored extensively in science-fiction novels of the near-future, in particular the use of cyberspace as a surveillance technology to reproduce the dual economy between rich and poor (see Chapter 10).

Although there is nothing new about surveillance and monitoring, having been used for centuries to maintain 'cohesion and coordination of the economic and social order' (Robins and Hepworth 1988: 169), the difference now is that computers, in combination with telecommunications, qualitatively alter the nature of surveillance. Whilst the use of electronic databases has benefits (ensuring that we are paid correctly, that we receive the correct government benefits, that crime protection is improved, that our health is monitored more effectively, and that we have the convenience of paying for goods using plastic debit and credit cards [Lyon 1994]), it also increases levels of surveillance, creating infringements on individual privacy and eroding further the public sphere (including public space).

Geographies of exclusion

> The local geography of cyberspace follows the lines and contours of American racism, sexism and classism. An entire world lives outside those lines, and they've been electronically redlined out of cyberspace. . . . The political geography of cyberspace effectively mirrors the prevailing patterns of global resource distribution.
>
> (Poster 1997: 228)

Cyberspace is often promoted as an egalitarian arena – a space anybody can access and a space of individual autonomy (see Chapter 3). Here, we concern ourselves with the first point: (spatial) access to cyberspace. Cyberspace is patently not accessible to all. For example, in 1996 Moss and Townsend (1996) reported that 50 per cent of all US Internet hosts were located in just five states. This spatial inequality was replicated at the county and district level, with access centred around a few locations. This pattern, although weakening, continues to exist (Zook 2000). These spatial inequalities are clearly visible on a number of the maps of ICTs presented in Chapter 5. Cyberspace usage, and therefore benefits (social, political and economic), are fragmented along traditional spatial and social divisions (see NTIA 1999). In fact, far from creating a more egalitarian society, many commentators have suggested that cyberspace will reproduce and reinforce the rising dual economy within countries (see below) and between the developed and the developing world, and create new inequalities to bring about a world that is more unequal and socially fragmented (Thomas 1995). At present, cyberspace is only accessible to those who have the telecommunication infrastructure (a computer, a telephone line), can afford the equipment, have the skills to operate the equipment, and with the time to interact with it (Fernback and Thompson 1995). Financial parity is of particular importance. Usage is currently dominated by people in the middle- to upper-level income bracket. Access to a computer at home and work is highly correlated with household income and socio-economic grouping (Graham and Marvin 1996; Moss and Mitra 1998). Division has also been noted along racial lines (Hoffman and Novak 1998). As Golding (1990: 90) states:

> Entrance to the new media playground is relatively cheap for the well to do, a small adjustment in existing spending patterns is simply accommodated. For the poor the price is a sharp calculation of opportunity cost, access to communication goods jostling uncomfortably with the mundane arithmetic of food, housing and clothing.

The reality is that cyberspace is dominated by white, middle-class males from Western nations who can converse in English, are computer literate, and are generally in their late teens or early twenties (Warf and Grimes 1997; Clement 1998; see also Chapter 3 for a discussion of gender divisions). This profile is changing, with a broader population coming online. For Miller (1996), deregulation of the market accentuates this social division by allowing companies to target their potential customer base, effectively ignoring groups who are excluded.

Mitchell (1995) reports that the bandwidth-disadvantaged are the new have-nots within Western society. He suggests that if the value of real estate is dependent on location then the value of a network connection is determined by bandwidth. Accessibility becomes redefined so that the 'friction of distance' is replaced by the 'bondage of bandwidth'. A poor, slow network connection provides low levels of access which are more costly to run (it takes more time to download). At present, high bandwidth is confined to information hotspots, mainly focused around key universities, high capacity data sources (e.g., a tele-communications company), and localised centres such as telecottages in rural areas. Unless bandwidth is improved, users with poor links will become increasingly marginalised from the information economy.

Cyberspace also accentuates the division between rich developed nations and poorer developing countries (Holderness 1998; Petrazzini and Kibati 1999). In developing nations, Internet access is limited by infrastructure, cost, and lack of ability to use it. Where there are connections they are likely to be industry or university related. As Hess (1995: 116) notes: 'Cyberspace is an elite space, a playground for the privileged. . . . There *is* a global glass ceiling, and for many in the world a large part of . . . technoculture lies well above it.'

This glass ceiling is maintained not only through poorer connections but through differential costs. For example, Hayward (1998) reports that a modem costs four times as much in India than it does in the US, and Internet access costs twelve times as much in Indonesia compared to the US. Similarly, Holderness (1998) reports that until 1997 India's Education and Research Network (Ernet) was connected to the Internet by a single 64K link (this is only 2.5 times the average *individual* connection in the West), and Bingham (1999) notes that one third to a half of the world's population live over two hours away from the nearest public telephone. Similarly, Petrazzini and Kibati (1999) note that a monthly Internet connection in Armenia costs 485 times as much as in Finland.

According to most analysts, this glass ceiling seems set to remain in place until appropriate levels of industrial, economic and educational development are obtained. As such, whilst some developing nations are seeking to capitalise from the emerging information economy (e.g., India, Jamaica), operations generally remain small by comparison with those in the West. This trend is likely to continue, with the economic gap between developed and developing nations widening even further.

Urban, regional and global restructuring

In the main, geographic contributions to the study of ICTs and cyberspace have concentrated on their role in the emerging information economy, the effect on employment patterns and economic performance, and their involvement in urban-regional restructuring. In particular, attention has focused on the Intranet connections and telematic networks (computer-mediated telecommunications) of transnational companies that are significantly reshaping the post-industrial economic and social landscape (see Castells 1988; Graham and Marvin 1996). ICTs, it is contended, are providing spatial and temporal fluidity that permits and encourages corporate restructuring as capitalism seeks a new 'spatial fix' (see Harvey 1989). It is argued that this fluidity – the transference into a 'space of flows' – is evident in a number of recent trends. In this section, we briefly examine[9] a number of these trends, before providing a case study of Dublin, a city which has received significant inward investment by ICT companies in recent years and undergone large-scale urban restructuring as it seeks to position itself competitively in the global information economy.

Globalisation of trade

For many analysts, the world has been involved in a vast restructuring process since the 1970s, as the capitalist base that underlies and constitutes much of the world's social system mutates into a post-industrial form (see 'Industrial/post-industrial', Chapter 1). The emergence of ICTs are seen as central to this process, forming the basis of a new mode of socio-technical organisation (Castells 1988). Accompanied by deregulatory changes, which introduced new competition, investment and a desire to expand to capture a larger market share, ICTs are seen to be facilitating the internationalisation of production (Langdale 1989), global networking, cross-cultural contacts, the internationalisation of financial markets, and increased international co-operation, joint ventures, strategic alliances and mergers (Robinson 1991). ICTs are being used to increase the scale of production, as driven by the logic of accumulation, and to increase the scale of consumption, as driven by the logic of commodification (Waters 1995). Transnational companies are thus taking on new and greater powers through the effective management of their structures, using information transfer to gain competitive advantage over smaller operations. This means that instead of having a series of largely autonomous sites/plants serving a specific region, companies are centralising services and linking them together using ICTs to form a global-ised system incorporating research and development, marketing, finance, production and the co-ordination of distribution (Graham and Marvin 1996). For example, IBM products have been designed, marketed and sold on a worldwide basis for over a decade, relying on a comprehensive, Intranet system to integrate information from several global sites (see Langdale 1989). Other transnational companies have similarly rationalised their operations on a global basis in order to maximise profits.

Office automation and back-offices

One strategy to help facilitate global reorganisation is the use of office automation and back-offices. Here, manual office work is replaced by automated, computer processing operations, and clerical and administrative work is undertaken in areas with cheaper running costs (labour, rent, etc.) in different regions or countries. This means that docu-ments and files are moved around the world in order to take advantage of a decentralised, global, 24-hour workforce. Particular industries, such as tele-defined businesses – e.g., consumer service centres, telephone operators, telemarketing and market research – and tele-transacted businesses – e.g., airline reservations, banking, insurance and administration – have been quick to explore the possibilities of back-officing (Graham and Marvin 1996). As described in detail below, Ireland hosts a number of back-officing operations for American and European firms, providing highly-skilled and relatively cheap non-unionised, workers. Lower-skilled clerical and data entry work, involving the processing and digitising of vast amounts of documentation, is being farmed out to offshore, back-offices in places such as South-east Asia, the Indian subcontinent and the Caribbean. For example,

British Airways bookings are processed in India, and American Airlines in Barbados. The benefits to transnational companies can be significant. For example, Graham and Marvin (1996) report that wages are typically 20 per cent of those in the West, staff turnover levels are extremely low at 1 to 2 per cent per annum (turnover is 35 per cent in North America) thereby reducing training costs, and military-type discipline ensures high levels of accuracy. Moreover, the world financial corporations are now using offshore banking operations to maximise profits by exploiting the favourable tax laws in certain areas of the world.

Teleworking, telecottages and teleports

In addition to back-offices, the potential decentralisation associated with ICTs has led some companies to explore the possibilities of widespread teleworking. Teleworking refers to the home use of telecommunication services, allowing employees to work from home. Estimates put the number of teleworkers in the European Union in 1996 between 1.25 and 4.6 million (0.8–2.8 per cent of the labour force) (Handy and Mokhtarian 1996). In the main, telework has been heralded as a positive outcome of the new 'information society', allowing greater flexibility both for the worker and for the company. However, several commentators have suggested that any positive effects are counterbalanced by the negative effects of isolation, insecurity, the blurring of the site of leisure with work, the breakdown of worker cohesion and the threat to trade unions. The positive and negative effects associated with telework largely mirrors the social divisions between telework employees. Teleworkers can be divided into two groups. First, there are the well-educated professionals who have satisfying and demanding jobs that frequently require entrepreneurial and specialised skills. Second, there are the less well-educated workers who generally perform less rewarding roles in the form of low-level, menial tasks like typing or data entry (Weijers *et al.* 1992). Adam and Green (1998) report that this division is a gendered one, with women more likely to be in the latter group.

A related idea to telework is the telecottage. These have been promoted as community ventures to link (often isolated) communities to the 'information superhighway', to provide a training base for IT skills, and a place from which to telework. In such ventures, the cost of equipment can be shared along with expertise and advice. Telecottage projects are growing in popularity, especially in rural, unemployment blackspots where they are seen as a potential way to boost the local economy. For example, in the UK telecottage numbers have increased from five in 1989 to 140 in 1995, with nearly all of these situated in rural areas (Selby 1995). A related concept, the electronic village hall (EVH), providing community-based information and communication technologies to remote rural locations and inner-city community groups, is another development (Ducatel and Halfpenny 1993). EVHs were originally developed in Scandinavia as a way to overcome problems of rural isolation. In inner-city areas, EVHs are an attempt to develop and strengthen community ties. In both cases, EVHs are meant to promote cyberdemocracy, and at the same time they help to modernise the local economy by providing a low-cost and low-risk means of learning about and applying telematic technologies.

A more sophisticated, urban equivalent to the telecottage, and one aimed more at small- to medium-sized companies, is the teleport. The teleport is essentially a high-tech office park offering advanced telecommunications links via satellite and fibre optic connections (Warf 1995). The centralisation of facilities means that teleports provide significant scales of economy to smaller users who cannot afford private Intranet connections.

Teleports are seen as the new 'harbour depots' of the information age, performing a similar role as harbours in the age of shipping. They provide their users and the host area with a competitive, economic advantage over other places. Teleport projects are currently appearing in many countries, including some developing nations (e.g., Jamaica and Nigeria).

Organisational and employment restructuring

The combined trends of globalisation, back-officing, teleworking and the increasing commercial value of information and service provision have led to significant effects on organisational and employment structures. Johnston (1993) estimated that in 1993 50 per cent of all jobs in Europe were in information-based services and 80 per cent of all new jobs were being created in this sector. However, many analysts have noted that the relationship between ICTs and employment is not a simple one of job creation. Instead, there are complex relationships relating to both employment quantity (automation/ICTs reducing the number of jobs) and quality (automation/ICTs leading to the deskilling of workers; ICTs/automation providing greater autonomy in decision-making and the gaining of more generic skills). What is clear is that ICTs are leading to corporate streamlining and reorganisation, as jobs are rationalised and centralised. Davis (1993) reports that the shift to an information economy, increased productivity per worker, and a global recession, meant that between 1980 and 1993, the 500 largest transnational corporations in the United States shed some 4.4 million jobs, many of which were professional and technical posts. In only a few cases has labour restructuring within these companies been offset by job creation caused by a higher demand for an improved, cheaper service/product. Contrary to many people's fears, however, ICTs are having a negligible effect on aggregate levels of unemployment (Castells 1988, 1996). Rather, they are fostering productivity, economic growth and expanding employment outside of the high-tech, information industries. As a result, although there is some growth in high-skill professional, engineer and technician employment, the overwhelming growth is in low-skill, low paid, part-time, casual and menial employment. For example, Kumar (1995) reports that 13 million jobs were created in the US between 1973 and 1980, 70 per cent of which were low-level posts in the service and retail sectors, and were predominantly part-time or temporary and occupied by women. Consequently, it is suggested that ICTs are helping to foster an increasingly skewed and polarised occupational structure that is becoming common across Western nations (see 'Dual economy societies', p. 48).

It is predicted that the adoption of ICTs and the accompanying rationalisation of the workforce will result in one of two scenarios. In the first instance, Robins and Webster (1989) foresee a Taylorist (routinising) process of automation, leading to white-collar deskilling. Here, the work of professional and highly skilled technicians, such as architects, bank managers and print typesetters, is becoming 'simplified' and made 'easier' as computer programs undertake all the mental calculations and even decide and implement the best course of action. This in turn will lead to a company with a skewed, hour-glass shaped workforce, with a few executives at the top, no middle management and a wide base of clerks and operatives at the bottom (Kumar 1995).

The second scenario suggests that the traditional hierarchical structures of companies are not well suited to operating in the information economy and as a result will begin to flatten and become more horizontal in nature. Here, it is expected that staff across a company will become more skilled, with powers to synthesise, act on new information and make low-level decisions, rather than all decisions needing to be channelled through the management-hierarchical structure (Cronin 1994). This will lead to a more even distribution of power across the workforce, as employees become more mobile and companies take on 'adhocracy' rather than hierarchical structures in an effort to speed up 'information metabolism' (Malone and Rockhart 1991). However, this situation is also likely to lead to a polarisation of office work with relatively unskilled, low-educated data entry clerks at one end (being replaced by automation), and the remaining workforce diversifying in skills and responsibilities (i.e., a football-shaped workforce). For example, bank clerks are largely being replaced by ATMs and telebanking, with those who remain retrained to sell services, thus upgrading the skills of a smaller number of employees (Castells 1988). This employment restructuring has already led to massive corporate infrastructural changes. In particular, with the shift to automation and telebanking, financial institutions have drastically reduced the number of high street banks, resulting in large job losses. For example, the creation of First Direct, the telebanking subsidiary of the UK's HSBC bank, led to 750 high street bank closures and 15,000 job losses (Graham and Marvin 1996). This situation is likely to increase with the move to online banking over the Internet.

Urban-regional restructuring

There is little doubt among analysts that the processes of globalisation, and employment and organisational restructuring, in part caused by ICTs, are instrumental in the current restructuring of urban-regional fabric. Cities were designed as places to overcome time with space, making communication easier. However, the growth of telecommunications to a certain extent nullifies this function by making communication easier through the overcoming of space with time (Graham and Marvin 1996). As such, there is a juxtaposition between urban places and their electronic counterparts which has led several commentators to speculate that these new, electronic spaces and flows will displace or substitute physical travel and physical urban functions, and lead to the dematerialisation of the city.

At present, however, there is little evidence of a process of dematerialisation and dissolution of city life. To the contrary, evidence suggests that ICTs reinforce city life and urban hierarchies through processes of restructuring. As such, it is apparent that rather than the processes which underlie city development being destroyed, electronic spaces are merely altering patterns of urban development, and changing a city's relationship with its surrounding region and other cities (Graham and Marvin 1996). This restructuring is taking place at different levels as urban fixity and electronic mobility exchange trade-offs. At one level, far from the utopian visions of footloose industries, electronic cottages and the death of the city, ICTs are actually reinforcing, and in some cases increasing, the role of major business centres (Castells 1988, 1996). In fact, the world is only a 'smaller' place for those cities that have attracted a disproportionate share of ICTs (Daniels 1995). For example, Alles *et al.* (1994) report that urban systems are increasingly being dominated by cities that have greater telecommunication infrastructures. Thus, rather than a lessening of the discrepancies between higher order and lower order cities, with a filtering down of economic and social benefits, the differences between 'core' and 'peripheral' cities are becoming more pronounced (Daniels 1995). This centralisation occurs because if corporations are to take advantage of the global reach of ICTs they must locate their command and control centres in areas with a suitable infrastructure, at affordable costs. The Internet itself is spatially concentrated in certain key cities, as evidenced by analysis of infrastructure of wires and domain names (Moss and Townsend 1997a, 1997b, 1998; Zook 1998, forthcoming). As such, a recursive relationship has developed between the ICT industry and those reliant on information: the ICT industry is attracted to the information-rich industries as a source of business, leading to greater density and a range of services; in turn, the information-rich industries are attracted by the availability of cheap, efficient ICTs. Moreover, centralisation is reinforced because many companies are reluctant to give up the close proximity of employees that fosters social and business connections, supplying tacit information considered vital in some industries, notably in the finance sector.

In contrast to this process of centralisation, many office activities, business services and production centres are decentralising. This is because locating in the centre of a major city carries considerable penalties in terms of high rents, high labour costs, recruitment problems, congestion, poor environmental quality and overcrowding (Daniels 1995). Decentralisation takes two forms. First, many companies are decentralising within and across regions to smaller cities and non-metropolitan areas to capitalise on lower workforce and operating costs. Second, decentralisation is occurring from the city to the suburbs, to take advantage of lower worker turnover, worker accessibility and a skilled, cheaper, suburban labour pool, without overly comprising customer accessibility (Castells 1988, 1996; Graham and Marvin 1996). Ironically, the decentralisation of some corporate sectors is actually reinforcing the need for the centralisation of control and co-ordination in the form of global command centres (Moss 1986). That is, the greater the decentralisation of some sectors, the greater the centralisation of others.

Soft cities

> In a world of ubiquitous computation and telecommunication, electronic augmented bodies, postinfobahn architecture, and big-time bit business, the very idea of a city is challenged and must be eventually reconceived. Computer networks become as fundamental to urban life as street systems. Memory and screen space become valuable, sought-after sorts of real estate. Much of the economic, social, political, and cultural action shifts into cyberspace. As a result, familiar urban design issues are up for radical reformulation. . . . we have the opportunity to rethink received ideas of what buildings and cities are, how they can be made, and what they are really for. The challenge is to do this right – to get us the good bits.
>
> (Mitchell 1995: 107, 163)

Many cities have now taken a pro-active role to 'wire' themselves in an attempt to gain a competitive advantage in the global marketplace. This is leading to the formation of 'information' or 'soft' cities centred around a high-tech infrastructure of computer-based networks (Hepworth 1990b). These are cities where the infrastructure is increasingly becoming monitored and controlled by computer networks (Batty 1995). This relationship is deepened by the introduction of smart and responsive technologies in building design, using microchip-based technologies that allow building features to respond to the occupants' movements and desires.

One city that is actively exploring the possibility of becoming a 'soft city' is Singapore. Led by the government's National Computer Board and Telecommunications Authority, its principal aim was to have all places of work, homes and schools 'wired' by 2000. In this example, the city is being re-worked and re-thought as new technological infrastructures are set in place under the auspices of the government's IT2000 masterplan (Choo 1995). The Singapore government hopes to transform the city-state into a networked, intelligent island, using information and communication technology to utilise its intellectual capital and to sustain economic growth. There are five strategic themes which define the intelligent island vision: to become a global business hub; to boost Singapore's economic base; to enhance individual potential; to link its communities both locally and nationally; and to improve the quality and standard of life. Through strong government intervention and strategic partnerships with industry, Singapore hopes to become the world's first, fully networked nation, with every citizen and corporation having access to a range of information services. In turn, ICTs will be used to increase economic competitiveness, attract inward investment and to achieve an 'intelligent society' encouraging learning and innovation (Choo 1995). As a result of this initiative, Singapore has a sophisticated and dense ICT infrastructure, incorporating leased circuits, fibre optic networks, household teleboxes and ubiquitous remote computer access, as it tries to move from unskilled, low wage manufacturing to value-added business services and financial markets (Warf 1995).

Kawasaki in Japan has adopted a similar approach and has been developing a series of eighteen 'intelligent plazas' consisting of smart buildings full of fibre optic and cable services in a bid to create an 'online' city (Batty 1991). The plazas, each one aiming to employ thousands of workers, link new themed areas and are sequentially connected through a main fibre optic grid with subsidiary links to other nodes. The themed areas include a 'techno-venture' park where new telematic applications can be developed, a 'techno-community' where older industries using telematics reside, a 'technopia' representing the existing city centre, and a 'technoport' representing the existing sea port. The hope is that as well as stimulating growth in telematic industries, the plazas will encourage subsidiary industries, such as retail and leisure services.

Similarly, Toyko has become a telecommunication nodal point for the Pacific Rim, with government initiatives encouraging the development and integration of ICTs into the urban fabric to gain telematic supremacy (Alles *et al.* 1994; Rimmer and Morris-Suzuki 1999). In the United States, the main telematic cities are Washington, DC, New York, Boston, Atlanta and Los Angeles. In Europe, the UK with its early deregulation of the marketplace has led the way. After deregulation in the 1980s, London quickly became the European capital for finance, broadcasting, publishing, advertising and other information-rich service industries. As a result, London developed into a major telematic centre representing approximately 80 per cent of the UK data communications market (Graham 1993). To try and keep its dominant edge, London over saw the redevelopment of the Docklands area, with the formation of a teleport that linked a range of ICT technologies. Similar telematic developments are occurring in many European cities, including Manchester, Hamburg, Cologne, Barcelona, Amsterdam and Rotterdam.

Dual economy societies

The outcome of many of these processes is increasing spatial and social divisions and the creation of a dual economy. Such divisions are, according to analysts such as Davis (1990) and Castells (1996), becoming increasingly prevalent in Western society. For example, Mike Davis, in his book *City of Quartz* (1990), details the extent to which Los Angeles is becoming a spatially and socially divided society based on class and racial divisions, and access to different layers of the economy. Widespread organisational and employment restructuring means that jobs are increasingly becoming polarised into well-paid, stable, rewarding and full-time positions in the information economy, or part-time, casual, menial and poorly-paid positions in low-level informational jobs or in traditional industries (Castells 1996). People in well-paid jobs are spatially distancing themselves from the underclass, both by migration to the suburbs and also by regulating their home and work spaces through defensible and surveillance means.

Castells (1996) suggests that increased social and spatial polarisation is an inherent part of the information economy and is becoming more prevalent in all Western societies. He contends that the division between rich and poor is perpetuated and is becoming more pronounced as society divides into two parallel space-economies: the space of flows and the space of places. The space of flows represents the information economy and is characterised by mobility and space–time compression. This space-economy, Castells argues, is starting to dominate the space of places; the traditional modern economy centred on location and the overcoming of time by space (e.g., based on friction of distance). Those who work in the space of flows become wealthy, whilst those in the space of places get left behind. As we detail in the next section, Castells' thesis of a dual economy holds for Dublin, a city that has undergone significant change as it has sought to plug itself into the information economy.

An example of urban-regional restructuring: Dublin

Many of the processes of organisational and urban-regional restructuring are evident in Dublin (capital city of the Republic of Ireland), whose revitalised economy, as with other cities in Ireland, is increasingly centred on ICTs, and industries dependent on ICTs. Proinnsias Breathnach (1998, forthcoming a and b) has provided a detailed account of how Dublin has restructured its economy since the inception of the Irish state in 1922, and in particular the role of ICTs in helping to create and sustain the 'Celtic Tiger' phenomenon of the 1990s. Here, we detail the central findings of his research.

Breathnach (forthcoming a) reports that, historically, Dublin is the economic and administrative centre of Ireland. Post-independence, the city retained its dominant position in the Irish economy due to the policy of import-substituting industrialisation, supported by protectionism. In the 1950s, industrial stagnation caused the Irish government to reject protectionism and adopt an export-led policy, encouraging direct investment by foreign-owned companies through tax incentives and grant packages. The resulting branch-plant economy, aided by entry into the European Union (EU) in 1973, weakened the dominant position of Dublin, as plants which were mainly involved in low-skill assembly and packaging activities located away from the capital. Breathnach notes that these foreign investors dominated the economy by the mid-1980s, accounting for 50 per cent of industrial production and 80 per cent of industrial exports. Hit by the global recession in the early to mid-1980s, Ireland's economy again stagnated, with GDP 63 per cent the EU average in 1987 (Breathnach 1998).

Since the early 1990s, however, Ireland, and in particular, Dublin, has undergone rapid economic change and urban-regional restructuring due to its changing economic base and position in the world economy. Ireland currently has three times the growth rate of any other EU country, a GDP above the EU average, and its unemployment rate has fallen to one of the lowest in the EU (Breathnach 1998). Breathnach (forthcoming a) reports that there are several reasons for this. Most important in the context of ICTs, has been an increasing technological sophistication underlying inward investment. The Industrial Development Agency (IDA) and the Irish government were quick to recognise the advantages of attracting higher-skill, more technology-oriented business to sustain and replace the ailing branch-plant economy. The government invested in increasing the education skills base of the population, and the IDA sought to attract more technologically-oriented manufacturing and to expand the service sector.

Breathnach (forthcoming a) reports that these three new strategies, of offering high-skill but relatively low-wage employees, attracting skills-based manufacturing, and expanding the service sector, coupled with significant investment in the telecommunications infrastructure and existing incentives such as low corporate tax rates (10 per cent for manufacturing and international services), capital and training grants, no local content requirements, unlimited profit repatriation, and ready-made sites, have been extremely effective and have fuelled Dublin's current economic boom. On the manufacturing side, many large multinational companies have located to Ireland, employing large numbers of workers in skilled positions. In relation to ICT companies, these include Intel (3,600), IBM (2,850),[10] Hewlett Packard (2,700) and Motorola (1580), all of which have plants in the Dublin region.[11] These foreign-owned employers are important because for every 100 people they employ they are believed to create approximately 125 indirect and induced jobs elsewhere in the Irish economy (O'Malley 1992, cited in Breathnach 1998). On the service side, there have been three main components: the establishment of an International Financial Services Centre (IFCS) whose licensed firms employed over 6,500 by 1998 and which has started to carve out a niche as second-order node (off-shore banking) in the global financial system; a rapid growth in software operations with employment reaching 9,700 by the end of 1997 (Ireland is the second largest global exporter of software after the US, and many of the world's leading software houses have their European headquarters or offices in Dublin);[12] and rapid growth in call-centres and back-offices.

Breathnach's (forthcoming a and b) analysis concentrates on charting the growth and nature of call-centres in Dublin. This sector of employment has grown steadily throughout the 1990s, so that by 1998 there were fifty centres in Ireland employing over 6,000 people (90 per cent of these centres are in Dublin). This represents some 30 per cent of all international call-centres in Western Europe. He details that this growth has occurred for a number of reasons: relatively low wage costs; an advanced telecommunications infrastructure with competitive rates for high-volume international traffic (Ireland has the cheapest European rates for international freephone numbers); linguistic and cultural commonality with the US; government incentives; centralisation and rationalisation of pan-European operations and the amalgamation of European freephone numbers; and high-calibre bilingual staff. American firms dominate the sector, accounting for 70 per cent of centres and 80 per cent of employment, and include IBM, Gateway 2000, Compaq, Dell, Citibank, Hertz and Oracle. Employees are predominately female (70 per cent), and the centres also employ a large proportion of foreign staff (23 per cent). Due to the repetitive and policed nature of the work, however, the sector is characterised by a young age profile, low wages and high staff turnover (Breathnach forthcoming b). Breathnach (forthcoming a) reports that one currently successful strategy to counter the relatively low-level nature of the work has been to encourage call-centre firms to back-office other aspects of their work to Ireland, such as financial management and software development.

These rapid changes in the economic base have had significant effects on other aspects of the Irish economy. Industries such as construction (50 per cent growth between 1994–1996, Breathnach 1998), housing and retailing, have benefited enormously. Indeed, economic growth has led to significant changes in the urban landscape with the construction of new industrial estates, housing estates, shopping malls, populated by a rapidly expanding population, and serviced by a new and improved transport infrastructure. On the negative side, however, spatial and social divisions have been exacerbated. Dublin's dominant position in the Irish economy has been strengthened due to the need to be able to recruit a suitable workforce (64 per cent of all new jobs in foreign-owned companies have been in the Dublin region, which has 34 per cent of the national population, Breathnach 1998). This is leading to pronounced spatial divisions in the distribution of wealth and wealth-creation across the country. Further, the social division between rich and poor has widened with the creation of a dual economy society. Whilst some sections of the population, namely the skilled and well-educated, have benefited from being plugged into the space of flows, others are trapped in the space of places – low paid jobs, traditional industries, low skill sectors of the information economy, and attendant industries such as catering, retailing and entertainment. A United Nations Human Development Report (1998) indicated that Ireland now has the highest levels of (income based) social polarisation within the European Union, that the number of children living in poverty has increased twofold since 1971, and that the gap between the economic status of men and women is worse than in any other OECD country. Although partly the result in a wage differential, this social division is exacerbated by rising costs of living, best illustrated by Dublin's spiralling house prices. Whilst wages in Dublin on average increased by 17.6 per cent between 1991 and 1997, there was a 231 per cent increase in second-hand house prices between 1994 and 1999 (*Irish Independent*, 24 July 1999, and data supplied by Department of Finance 1998).

Dublin, then, through its investment in skills, corporate incentives, and telecommunication policy and pricing, exhibits many of the trends outlined above, with rapid urban–regional restructuring, organisational and employment restructuring, and increasing spatial and social divisions, as it seeks to be part of the emerging global information economy.

Conclusion

In this chapter we have examined the impact of the conceptual space of cyberspace and the ICT infrastructure that supports it on the spatial and material relations of social, political and economic life. Our analysis suggests that whilst cyberspace and ICTs undoubtedly have a number of material impacts on human activities, the nature and size of these impacts should not be overstated. For example, cyberspace, far from dissolving geographic communities into a state of placelessness, is in many cases being used to foster and support such communities. Similarly, computer-mediated communications are helping to reproduce political structures, not dismantle them. ICTs and cyberspace, then, are aiding a series of evolutionary changes, rather than instigating a set of revolutionary transformations. Whilst we provide an initial overview, it is clear that more detailed studies are needed to further deconstruct the complex continuum of interrelationship geographic space through cyberspace. We now turn our attention to examining emerging virtual geographies.

3   Geographies of cyberspace

In the previous chapter, we considered the emerging geographies of the information society, charting how cyberspace is transforming the spatialities of modernism. In this chapter, we turn our attention to virtual geographies. In the first half, we examine the spatialities of cyberspace. As already noted, for commentators like Rheingold (1993), cyberspace is providing spaces where new authentic places are being created, supporting new forms of community. In the second half of the chapter, we consider the (inherent or produced) spatial geometries, structures and forms of cyberspace. The spatial forms and geometries of cyberspace are entirely socially-produced, and as yet largely undetermined. A key endeavour of many designers, computer and information scientists is to determine or create (using a process of spatialisation) these geometries to aid understanding and navigation.

As stated previously, it is important to remember when reading our account that these spatialities and geometries merely constitute one part of our continuum of experience and as such should not be conceived as separate domains, disembodied and divorced from geographic space (see Chapter 1). Although we can use the medium of cyberspace to play with our identity, our online personae are grounded in our overall experiences and memories. An illustration of this experiential continuum is the extent to which cyberspace explicitly draws on material socio-spatial relations and geographic metaphors to create new spatialities and a 'sense of place'; how material contexts and work practices shape the technological developments which lead to improvements in the 'look and feel' of cyberspace, and often determine its use (e.g., business interactions, telecommunication costs), and affect who gains access; and how existing social networks are augmented by ICTs and cyberspace (e.g., academic social networks) (Wellman *et al.* 1996; Wellman and Gulia 1999). As a consequence, the emerging geographies we detail should be considered alongside those described in the previous chapter.

Spatiality of cyberspace

In this section we discuss the main characteristics of the spatiality of cyberspace, charting the socio-spatial processes and modes of regulation that frame social relations. In Chapter 7 and Chapter 8, we provide a more detailed account of the spatiality of a selection of virtual spaces used primarily for social interaction, illustrating the complex geographies that have started to emerge.

Cyberspace, identity and community

Some critics have argued that the most profound impact of cyberspace is not information processing but how it affects social relations. Thus cyberspace has a number of implications in relation to both identity and community – it allows us to explore who we are, as well as changing who we are; it provides new spaces in which communities can develop.

Analysts suggest that cyberspace achieves a shift in the bases of identity through three means: (i) it aids a process of 'cyborging', extending the body in new ways (Haraway 1991); (ii) it provides a space of disembodiment, as the mind enters a space of interaction free of the body, and its associated codings (e.g., gender, race) (Stone 1991); and (iii) it dislocates the self, as the mind enters a space free of the context of geographic place and community (Mitchell 1995; Adams 1997). Cyberspace is what Foucault has termed a 'technology of the self', a device which effects the social construction of identity by altering the conditions under which it is constructed (Aycock 1995). Indeed, Poster (1995) argues that cyberspace promotes the individual as an unstable identity, an individual bound within a continuous process of multiple identity formation. Robins (1995: 138) notes that 'the self is reconstituted as a fluid and polymorphous entity'. Here, 'the boundaries of the self are defined less by the skin than by the feedback loops' (Hayles 1993: 72). In other words, in cyberspace identity is defined by words and actions not by body and place. As Rheingold explains:

> We reduce and encode our identities as words on a screen, decode and unpack the identities of others. The way we use these words . . . is what determines our identities in cyberspace . . . The physical world . . . is a place where identity and position of the people you communicate with are well known, fixed, and highly visual. In cyberspace, everybody is in the dark. We can only exchange words with each other – no glances or shrugs or ironic smiles. Even the nuances of voice and intonation are stripped away.
> (Rheingold 1993: 61)

In other words, cyberspace is a place where 'the self is constructed and the rules of social interaction are built, not received' (Turkle 1995). In cyberspace, your body is irrelevant and invisible and nobody need know your race, disability, gender, sexuality or material status unless you choose to reveal it (Stone 1991). The experience of dislocation in time and space allows individuals to experiment with aspects of their identity which they conceal in 'geographic' space (Wilbur 1997); the poverty of signals is appropriated as a resource (Kollock and Smith 1999). According to these commentators, identity online is fluid, ephemeral and empowering because people can choose how they want to be represented (Lemos 1996). Identity is thus conceptualised as constructed through multiple experiences, resulting in it becoming fragmented, decentred and fluid, changing with time and with the situation.

There are reports that these online personas can be very powerful, to the extent that being denied the ability to log-on under a designated 'nickname' can leave a person frustrated and bewildered. This is because their online identifier (nickname) has been stripped away, denying the user access to social networks that might have been sustained for some period of time (Bechar-Israeli 1995). In a place where your name is your sole identifier, being denied the ability to use it can lead to a personal crisis.

Mapping identity in 'geographic' space, given its fluid, multiple and fragmented nature, is fraught with difficulties (Pile and Thrift 1995). Given the variety of spatialities online (see below), this task becomes more complex once in cyberspace. Only a few studies have empirically explored online identity, with nearly all concerned with experiments with gender roles, and few extending beyond anecdotal accounts. Most studies fail to consider the role of spatiality in identity construction or the interrelationship between online and offline lives. Further, these studies tend to concentrate on agency at the expense of structure. Perhaps the most comprehensive, academically-grounded study is that by Turkle (1995).

Turkle (1995) studied extensively the relationship between real life and virtual life of a number of people who regularly use MUDs. She found that people use MUDs for a variety of reasons, some for play, others for emotional support that is lacking in their real life. She reports that for those seeking emotional support, MUDs act in two ways. First, MUDs provide a place for self-reflection, a constructive environment to work through or act out real problems. Second, the virtual environment acts as an emotional escape, a substitute for real life but, conversely, one which often deepens real-life anxiety. Both instances illustrate the continuum of geographic space through cyberspace – the use of MUDs is grounded in the real world and MUDs have consequences to life offline (e.g., deepening anxiety). Similarly, Bromberg (1996) suggests that MUDs, far from being just a 'game' or another form of communication, serve four useful social functions. First, MUDs offer an antidote to loneliness, providing solace through communication. Second, they allow users to experiment with identities and personae. Third, they allow users to explore their erotic side by providing a new, 'safe' site for sexual encounters. Lastly, MUDs let users become the masters and controllers of their environment (see Chapter 8).

Clearly, there are many reasons for seeking social interaction online, including, as Donath (1999) and Kollock and Smith (1999) note, a desire to promote one's reputation or gain social acceptance both on- and offline. It must be remembered that the majority of users are not anonymous (e.g., most email addresses clearly identify the person) as most people want to be identified in most of their interactions. Moreover, for those who have invested significant amounts of time developing an identity, cyberspace is not a space of anonymity and inconsequentiality (Ito and Ito 1996). As such, online social interactions exhibit many of the same characteristics as those elsewhere, distinguished by 'expressions given' (how one wishes to be perceived) and 'expressions given off' (often unintentional messages that reveal aspects of one's character) (Donath 1999).[1] In playing with identity in anonymous spaces, many users are seeking to intentionally manipulate 'expressions given' and limit those 'given off'. As discussed below, the messages that are 'given off' almost inevitably translates disembodied spaces into embodied spaces.

Some may question the degree to which cyberspace provides a space of meaningful social interaction (e.g., Robins 1995), but the influence of identity experimentation in cyberspace should not be lightly dismissed. Whilst some users might wish to bracket cyberspace and geographic space into two separate domains, spillover is inevitable as both join to form a single experiential reality (a point we return to in Chapter 10). The social interactions that take place in cyberspace clearly have a significant influence on some people, changing their outlook and values (see McRae 1997). Moreover, the depth and strength of some relationships can lead to radical changes in a person's offline life. For example, leaving one geographic location to go and live in another with a partner met and courted in cyberspace.

This playing with identity through online social interactions means that the users of cyberspace, it is contended, are forging new communities; new social structures that, importantly, are not based on what the participants look like, or where they live, but on what they think, say, believe and are interested in. Indeed, one of the principal effects of cyberspace is the formation of new communities free of the constraints of place and based on new modes of interaction and new forms of social relationships. Rheingold (1993: 5) defines virtual communities as 'social aggregations that emerge from the Net when enough people carry on those public discussions long enough, with sufficient human feeling, to form webs of personal relationships in cyberspace'. Instead of being founded on geographic propinquity, these communities are grounded in communicative practice. Baym (1995) states that: 'community is generated through the interplay between preexisting structures and the participants' strategic appropriation and exploitation of the resources and rules those structures offer in ongoing interaction.'

In part, then, these communities are promoted as an alternative space in which to seek the authenticity (belonging) of traditional locales, which many perceive to be disappearing in the geographic world (see 'Place/placelessness', Chapter 1). As Rheingold (1994: 3) describes:

> People in virtual communities use words on screens to exchange pleasantries and argue, engage in intellectual discourse, conduct commerce, exchange knowledge, share emotional support, make plans, brainstorm, gossip, feud, fall in love, find friends and lose them, play games, flirt, create a little high art and a lot of idle talk.

Rheingold and others believe that cyberspace enables individuals to circumvent the geographical constraints of the material world, allowing people to shape their own communities by providing choices such as who they interact with. As Jones (1995b: 11) writes, 'we will be able to forge our own places from among the many that exist, not by creating new places but by simply choosing from the menu of those available'. Like geographic communities, these online communities have behavioural norms, differing personalities, shared significance and allegiances. Indeed, for McLaughlin *et al.* (1995) the fact that there are commonly agreed protocols, the advent of distinctive referent language (abbreviations, jargon, symbols), and the formation of strong social networks, signifies that online communities, in one form or another, do exist. As Anderson (1983) suggests, at a basic level, all communities are imagined, and so long as members share a common imaginative structure, a community can be said to exist.

The majority of empirical studies indicate that there are many online communities which are rich in their diversity (see Chapter 8). As Rafaeli and Sudweeks (1996) point out, people would not invest so much time and effort in online social interaction if they did not gain some sense of social cohesion or community from their virtual actions. Indeed, they suggest that the form and depth of interaction means that these communities are neither pseudo nor imagined, despite claims made by its critics (e.g., Robins 1995; Sardar 1995). This is because cyberspace fulfils the qualities of what Castells (1996) calls 'real virtuality', a reality that is entirely captured by the medium of communication and where experience is communication.

That said, as we detail below, these communities are not all-inclusive and are subject to geometries of power. Moreover, many people do not take the opportunity to play with their identity online. For example, significant usage consists of supporting social networks outside of cyberspace rather than the creation of new networks. Furthermore, the number of MUD users remains small and the majority of newsgroup and mailing list members are 'lurkers' (readers but not posters). Kawakami (cited in Aoki 1994) reports that in the lists which he surveyed, 83 per cent of members had never contributed and two-thirds of those who had spoken had done so less than three times. Kawakami suggests that there are six main reasons why 'lurkers' outnumber contributors:

1   Reluctance to speak to people whom they do not know.
2   Resistance to participate in a group that has been formed and developed without them.

3   Lack of expertise to participate and a fear of being evaluated by others.
4   Difficulty of deciding to what extent they should disclose of themselves to others.
5   Worry about expressing themselves clearly.
6   A fear of criticism from others.

In other words, people remain silent online for the same reasons they are reserved in 'geographic' spaces. Moreover, most online activity is individual information browsing and searching, not collective social interaction.

Despite these concerns, it is clear that communities of one kind or another do exist in cyberspace. Interestingly, the spatial form, interface design and interaction metaphors, of many online communities bare a remarkable resemblance to 'real' world locales (e.g., AlphaWorld, see Chapter 8). As such, many online interactions are in fact situated within protocols that operate in material space and possess a spatiality that so far has been little explored. This lack of study concerning online spatiality is surprising given that many analysts, such as Poster (1997), discuss cyberspace users' search for a new 'sense of place'; a new spatiality where they can meet and interact.

The extent to which social relations are contextualised by spatiality is illustrated by the work of Adams (1997, 1998). He contends that online interaction is grounded through the use of geographic metaphors. For example, cyberspace is replete with the vocabulary of place – nouns, such as rooms, lobbies, highway, frontier, cafés; and verbs, such as surf, inhabit, build, enter (Adams 1997). Couclelis (1998) details that the use of these geographic metaphors – the spatialisation of cyberspace – is an attempt to translate information and communication spaces into domains familiar and comfortable to users. These spatialisations are particularly powerful when modes of thought and action that work in the familiar domain are also appropriate to the metaphorical domain (as is the case with most cyberspaces, e.g., direction is reversible). The maps in Chapter 6 of this book are attempts to use spatialisation, by mapping non-spatial information into a spatial domain, to provide information spaces that are cognitively accessible, what Cheeseman *et al.* (forthcoming) term 'comprehensible spaces'. Cyberspace, then, is to a great extent built out of the ideas and language of place, and the employment of these metaphors to create sites of interaction engenders an online spatiality. As a consequence, Taylor (1997: 190) states that 'to be within a virtual world is to have an intrinsically geographic experience, as virtual worlds are experienced fundamentally as places'.

Adams contends that one way to understand cyberspace is to examine spatialisations commonly employed within it, drawing parallels between the sense of place imbued in a particular network architecture with those of 'geographic' spaces (Adams 1998). Using combinatorial theory (a method for comparing network forms) he identifies several network typologies that mirror their geographical equivalents in terms of their structure and the social interactions performed. These are:

(a) *Cybercasting* (radial/one-way topology): This arrangement supports communication from one or few to many. This is the topology of radio and television. It is presently used for online magazine and newspaper text and for messages to users from the managers of computer networks.
    Architectural archetype: temples, churches, theatres, lecture halls, auditoriums.
(b) *File search and retrieval* (radial/two-way topology): This is user-driven information search and retrieval in which users extract text, images or sounds from central repositories. Search engines and indexes installed at central or peripheral nodes help locate

material. Online examples include the World Wide Web, wire service reports, online encyclopaedias, education resources.

Architectural archetype: library and archive stores.

(c) *Email* (one-to-one or one-to-many/one-way topology): This connection resembles regular postal mail except the messages travel much faster, and sending mass mailings to all members of a group is easier and cheaper.

Architectural archetype: mailbox and the seclusion of a private room or office.

(d) *Computer bulletin board* (radial/two-way topology): This arrangement is topologically identified with file searching and retrieval, except that the database that users search acts also as a repository for users' contributions. Users interactively search the messages of others and leave their own inquiries, comments and replies.

Architectural archetype: bathroom stall with graffiti or a message board.

(e) *Computer forum* (many-to-many, two-way): This arrangement is often referred to as the 'chat room'. It involves real-time discussion among spatially separated participants, all of whom are logged on at the same time and view each others' contributions instantaneously.

Architectural archetype: auditorium and central square, places that allow one to listen to a public speaker and also exchange comments with other listeners.

(f) *Multi-user environments* (many-to-many, two-way): These contexts are similar to computer forums but have textual descriptions automatically inserted by a computer program which narrates the experience of being in an invented place.

Architectural archetype: role-playing theme park/public space.

(Abridged from Adams 1998: 92–93)

Adams argues that relationships between physical/social structures and human agency are replicated online. Places along the continuum of experience are multiple, diverse, and linked by complex paths that need to be traversed. He contends that an analysis of how spatial/place metaphors, in combination with a comprehension of network topologies, affect communications within cyberspace will lead to an understanding of the social interactions that take place there. Whilst we would question the deterministic link between socio-spatial interactions and spatial forms, Adams' analysis does illustrate that spatial structures and forms in 'geographic' space are often used to provide context for interactions. The work of Correll (1995) illustrates this well.

Correll's study of an online lesbian café describes how patrons constructed an elaborate café setting in which they could contextualise their interactions (for example, patrons 'bought' drinks and hung out round the jukebox). She suggests that the construction (spatialisation) of this shared setting created a communal sense of reality which grounded communication. In essence, the locale needed for community in 'geographic' space was replicated online, so that place and setting remained important. Indeed, for her, the spatialisation of the online meeting space was the secret to the community's success, suggesting that without the shared reality of the bar, the community might have dissolved. This bar, however, differed in significant ways from gay bars in 'geographic' space, 'where the games are for real' (Correll 1995: 281). In cyberspace, patrons could explore their ideas and thoughts without fear of physical retribution. Thus, the bar served to augment offline lives by providing a surrogate community for a group who tend to be marginalised in geographic communities. In this case, the café provided a relatively safe space, something which is often denied women in the geographic world, in which they could express and explore their sexuality. Similarly, Foster (1997) documents an attempt to create a virtual community which he regards as unsuccessful because it failed to achieve a 'sense of place'. In this instance, the community was a PEN (Public Electronic Network) seeking to revitalise a geographic locale. Instead of fostering social interaction, however, the PEN disintegrated in monologues and separate spaces.

Online geographies of power and exclusion

In the same way that societies in 'geographic' space are organised through a series of power relations (e.g., political and legal structures; cultural ideologies such as gender and race), so too are social relations online. Although much hyperbole maintains that everyone in cyberspace is equal, a study of any online community reveals that this is not the case. Power relations continue to exist, manifested both through the regulation of various cyberspaces and also through the expression of cultural ideologies. This in turn affects the spatial freedom of users, policing and limiting their use of different spaces.

The traditional form of regulation in cyberspace has been through an informal set of customary laws. Online interactions have been policed through the consensual actions of users accessing and interacting in cyberspace (Bilstad 1996). For example, it is expected that people using MUDs respect each others' characters and recognise the unwritten rules of engagement, and it is assumed that people taking part in newsgroups will stay within communally agreed protocols ('netiquette'). If participants of MUDs or newsgroups transgress the bounds of customary laws, they must accept community administered 'punishment'. Many MUD and newsgroups have protocols which members must abide. These protocols are usually available in a public FAQ (Frequently Asked Questions) file and are generally common across MUDs and newsgroups. For example, customary laws relating to newsgroups include: sticking to the topic of the group; no advertising or commercial postings; no/limited cross-posting; no sending of private mail, or mail of little interest, to the rest of the group; no excessive flaming; no spamming.[2] Community administered 'punishments' consist of mail bombing or mass flaming the offender's mail account,[3] or a vigilante programmer deleting offending messages from a list on behalf of everyone else, or other users ignoring the offender.[4] Kollock and Smith (1994) suggest that these customary laws are an attempt to manage the 'virtual commons' with users seeking to balance individual and collective rationality.

Kitchin (1998) lists five reasons why customary law is failing online users, leading to unjust spatialities. First, while customary law is useful in enforcing a basic netiquette, users are 'tried and convicted' without appeal and by 'mob rule'. A limited number of communities, such as LambdaMOO (see Mnookin 1996), have tried to address this issue by introducing a system of petition and appeal, but most communities simply rely on netiquette. Second, customary law concerns violation of group concerns not individual concerns, and there are a number of offences, such as discrimination, harassment, silencing and interception, that it does not address (see below). Third, offenders can simply log back on to the system using a different identity and continue to break customary laws. Fourth, customary law does not address 'ordinary decent criminal' activity and in many ways implicitly condones it by opposing more formal laws. Fifth, customary laws were established when the Internet was largely publicly-owned and regulated by 'friendly' system operators. The Internet is now increasingly owned and run by private companies and large institutions driven by financial concerns.

Although many online spaces still rely on customary law, power rests ultimately with the system operator or owner. Operators and owners, using a system of market-led regulation, can remove and permanently exclude, without appeal, users who they deem inappropriate from services and online communities. Indeed, cases have been reported where users have had their ISP accounts withdrawn by the owners because of 'improper' behaviour, sometimes with seemingly little provocation and no right of appeal (Lessig 1999). In other instances, users have had their accounts terminated for criticising an ISP's service or policies.[5] For victims of self-regulation by ISPs the consequences can be great. For example, many users build up complex social networks based on an identifiable ISP email or website address. Under market-led regulation, ISP providers run a fine line between trying to balance the needs of many customers – appearing welcoming, accessible and inclusive – and wanting to stay in control (Mitchell 1995). As such, users are given the impression that they are the ones in control, that they are free to express views and opinions, but in reality it is the ISPs that are in control, dictating access and censorship. Without formal laws, the legal landscape of cyberspace favours owners not users, producing places with clearly demarcated geographies of justice: inclusion or exclusion, without the right of appeal.

The alternative to customary law and market-led regulation is a more formal legal framework to protect cyberspace users. Although various governments have sought ways to legislate cyberspace, they have encountered a number of difficulties. First, there has been resistance by many cyberspace users, who see formal legislation as running against the ethos of the Internet. Second, cyberspace knows no borders and their extents do not marry with those of nation-states and territorially-based legal systems. Third, systems such as Usenet do not have identifiable agents to direct actions against as many discussions are unregulated and many individuals use anonymous usernames. Fourth, attempts to legislate cyberspace often run counter to other pieces of legislation. For example, the Communications Decency Act (CDA) which was, passed in February 1996 in the United States, was overturned in June 1997 because it infringed on the First and Fifth Amendment rights. As a consequence, whilst there will be further attempts to formally legislate cyberspace, it will probably be a number of years before users' interests are protected in law. Currently, most effort in this area is expended on copyright or trademark protection, at the behest of large corporations.

In contrast to formalised systems of regulation, such as customary laws or market-led rules, access to online spaces are also regulated through informal and formal social relations. As such, communities often operate through a system of insider/outsider relations, with new members having to 'earn' the right to join the inside clique. For example, Correll (1995) in her study of an online lesbian café found that patrons were divided into four broad groups: regulars, who chatted in a familiar, smooth conversational style; newbies, who were trying to learn the protocols, gain trust and develop friendships (thereby becoming regulars); lurkers, who tended to observe interactions and occasionally post notes; and bashers, mainly men who sent abusive posts. In this example, newbies must play by the rules of the regulars in order to gain entry into the central clique. Similarly, Tepper (1997) describes how some Usenet newsgroups create a set of subcultural games that define the boundaries of group insider/outsider. One such example is a game known as 'trolling', whereby a newsgroup member posts a message containing a deliberate (and often humorous) flaw. Those 'inside' the group recognise the troll and ignore it, those outside often correct the flaw. The person who corrects the flaw unwittingly positions him/herself outside the central clique and the correction is itself often subject to further trolls, subjecting the person to the ridicule of those 'in the know'. Only with time and experience will the poster learn the art of trolling and be able to move from the outside to the inside. Tepper details that some of these groups can be decidedly unfriendly to those who are not willing to put in the time to learn the rules. In other cases, the subcultural boundaries are maintained by in-jokes, particularised lexicon sets (e.g., 'WAFU YN', which means, 'We're alt.folklore.urban you're not'), and other games. For example, there is one hacker newsgroup that can only be joined by hacking into it. Tepper, drawing on the writings of Homi Bhabba, suggests that it is through these performances of inclusion and exclusion that community gains its collective identity.

In many cases, online social relations are maintained through the reproduction of cultural ideologies that operate in 'geographic' space. Here, instead of providing a liberatory space, where cultural identifiers such as race, gender and disability are stripped away, cyberspace helps to reproduce those same ideologies, using these codifiers as the basis of exclusion. There are numerous examples demonstrating that patriarchal relations are reproduced online. Women in cyberspace still attract the unwanted attention of men, are still sexually harassed and receive abusive messages, and are still expected to adopt the same gender roles as in 'geographic' space. Clerc (1996) suggests that patriarchal relations exist online because men often communicate for status whereas women communicate to develop and maintain relationships. Further, the dominant group in cyberspace, and those who decide on appropriate online behaviour, are men (Sutton 1996). Sutton suggests that some men have even decided that flaming and individualism are acceptable. The fear for some feminists, and there is some evidence that this is already happening, is that women will try out cyberspace but will leave after unsolicited attention or abuse, or else they will only post self-censored messages for fear of reprisal (Brail 1996). In fact, Brail cites one woman who stated that, 'I think (the Internet) is the last bastion of real ugly sexism because it's unmoderated and faceless'. Rather than cyberspace erasing issues of gender, Hall (1996) suggests that it is being intensified discursively. In Foucaultian terms, online women (in general) still remain 'docile bodies' – bodies which are regulated by, and subjected to, men.[6] As such, many analysts have pointed to the fact that gendered identities, rather than undergoing a process of re-negotiation in cyberspace, are becoming more clearly demarcated through heightened sexism and the creation of women-only spaces. These issues have led a number of researchers to claim that far from being a *disembodied* space, cyberspace is an *embodied* space, codified by spillover from the material world (Ito 1997; Kitchin 1998; O'Brien 1999; Warf, forthcoming). As O'Brien (1999: 95) states:

> when persons enter cyberspace they bring with them preformulated cultural scripts which they use to map the new territory. In other words we use existing cultural representations to give meaningful order to uncharted netscapes. As social creatures, our maps or scripts consist primarily of categories for defining and distinguishing self and other and the context for interaction.

Whilst the geometries of power are largely determined by forms of regulation and the expression of cultural ideologies, access and acceptance is also determined by skills and knowledge. Those who are well versed in online community relations and possess technical skills such as typing or programming are likely to be more readily accepted. Most users, however, are not computer experts and many online interactions are not intuitive – you have to learn how to log on and have the courage to experiment. As Badgett and Sandler (1993) note, 'learning your way round IRC [Internet Relay Chat] is a lot like learning another language, finding your way around a new town, or playing blind man's bluff'. Therefore, people who are not computer literate are at a disadvantage.

The combined consequences of these forces, plus the ability to self-select which communities to join, means that cyberspace promotes 'uniformity more than diversity, homogeneity more than heterogeneity' (Healy 1997: 63).

Spatial geometries, forms and structures

In this section, we turn our attention away from the spatialities of cyberspace to consider its spatial geometries, forms and structures. This discussion provides a context for Chapter 4 and Chapters 6 to 8, and our analysis of how people have tried to map cyberspace using the process of spatialisation (see above).

As detailed in Chapter 1, determining the spatial geometries of cyberspace is a difficult task for two principal reasons. First, cyberspace consists of many different domains, each one within its own form and structure, and second, the spatial geometries and forms of cyberspace are entirely produced. Even those cyberspaces with an inherent spatial form, such as virtual reality simulations and visual MUDs, are qualitatively different to 'geographic' space in a number of fundamental ways. As Memarzia (1997) states:

> In cyberspace there are no physical constraints to dictate the dynamics or spatio-temporal qualities of the portrayed virtual space. Gravity or friction does not exist in cyberspace unless it has been designed and implemented. . . . cyberspace is not limited to three dimensions, since any two-dimensional plane or point may unfold to reveal another multi-dimensional spatial environment. . . . There are no ground rules concerning scale consistency in a virtual environment. Furthermore, the scale of the environment, relative to the user or viewer, may be altered at will. . . . Cyberspace can be non-continuous, multidimensional and self-reflexive. . . . In general, all principles of real space may be violated in cyberspace and the characteristics and constraints are only determined by the specifications that define the particular digital space.

Novak (1991: 251–2) argues that cyberspace has a 'liquid architecture':

> Liquid architecture is an architecture that breathes, pulses, leaps as one form and lands as another. Liquid architecture is an architecture whose form is contingent on the interests of the beholder; it is an architecture that opens to welcome me and closes to defend me; it is an architecture without doors and hallways, where the next room is always where I need it to be and what I need it to be. Liquid architecture makes liquid cities, cities that change at the shift of a value, where visitors with different backgrounds see different landmarks, where neighbourhoods vary with ideas held in common, and evolve as the ideas mature or dissolve.

Here it is recognised that cyberspace has a spatial and architectural form that is dematerialised, dynamic and devoid of the laws of physics; spaces in which the mind can explore free of the body; spaces that are in every way socially constructed, produced and abstract. Indeed, Holtzman (1994: 210) refers to the designers of virtual worlds as 'space makers'. While some cyberspaces do have an explicit spatial form (e.g., three-dimentional virtual worlds), they exist only in code; a combination of zeros and ones – objects are merely surfaces, they have no weight or mass (Holtzman 1994). This leads Morse (1997) to describe cyberspace as an infinite, immaterial non-space, suggesting that it takes the form of a liminal space:

> Virtual landscapes are liminal spaces, like the cave or sweat lodge, . . . if only through their virtuality – neither here nor there, neither imaginary nor real, animate but not living and not dead, a subjunctive realm wherein events happen in effect, but not actually.
>
> (Morse 1997: 208)

Indeed, as Mitchell explains, cyberspace is:

> profoundly *antispatial* . . . You cannot say where it is or describe its memorable shape and proportions or tell a stranger how to get there. But you can find things in it without knowing where they are. The Net is ambient – nowhere in particular but everywhere at once. You do not go *to* it; you log *in* from wherever you physically happen to be . . . the Net's despatialization of interaction destroys the geocode's key.
>
> (Mitchell 1995: 8–9; emphasis in original)

As Memarzia (1997) explains, the digital landscapes of cyberspace only possess geographic qualities because they have been explicitly designed and implemented. This means that 'there's no there there' (Holtzman 1994: 197), and yet, as we have discussed above, the space has a spatiality, a tangible geographic quality that fosters social relations. A space without space, 'a nonplace' (Gibson 1987), and yet it possesses a spatiality and virtual places. Moreover, a space where geographic 'rules' such as the friction of distance can be broken through the creation of what Dieberger (1996) terms 'magic' clauses, for example, teleporting. Benedickt (1991b: 128) thus argues that virtual realities need not, and will not, be subject to the principles of ordinary space and time, which will be:

> violated with impunity. After all, the ancient worlds of magic, myth and legend to which cyberspace is heir, as well as the modern worlds of fantasy fiction, movies, and cartoons, are replete with violations of the logic of everyday space and time: disappearance, underworlds, phantoms, warp speed travel, mirrors and doors to alternate worlds, zero gravity, flattenings and wormholes, scale inversions, and so on. And after all, why have cyberspace if we cannot (apparently) bend nature's rules there?

Benedickt argues that, cyberspace is a 'common mental geography' (in Gibson's (1984) famous phrase – a 'consensual hallucination'), a medium in which 'ancient spaces' (mythical or imaginal spaces) become visible; the abstract spaces of the imagination freed from Euclidean geometry and Cartesian mapping; spaces where the 'axioms of topology and geometry so compellingly observed to be an integral part of nature can . . . be violated or re-invented, as can many of the laws of physics' (Benedickt 1991b: 119). Indeed, many of the descriptions of cyberspace by novelists, describe in detail its spatial qualities (see Chapter 10). In nearly all cases, Cartesian rules do not apply to the virtual spaces being envisioned, except in relation to the body, where the mind/body distinction seemingly benefits significantly from cyberspace: the mind literally becomes free from the 'meat' (as discussed above, this quality has important implications for online identity).

Thus, Castells (1996) refers to cyberspace as a space of flows characterised by timeless time and placeless space; a space where the formal qualities of time and space are qualitatively different. Castells argues that temporality is erased, suspended and transcended within cyberspace. Stalder (1998) extends this idea to its logical conclusion by arguing that the defining characteristic of timeless time is its binary form. Timeless time has no sequence and knows only two states: presence or absence; now or never. Anything that exists does so for the moment and new presences must be introduced from the outside, having immediacy and no history. As such, 'the space of flows has no inherent sequence, therefore it can disorder events which in the physical context are ordered by an inherent, chronological sequence' (Stalder 1998).

In a similar way, Castells suggests geographical distance dissolves in the space of flows so that cyberspace becomes placeless. Movements within cyberspace are immediate, at no cost, presences can be multiple, and distance, as we currently understand it, is meaningless. There are no physical places in cyberspace, only individual digital traces that are all equally distant and accessible (traces, however, might be considered metaphorically a place, e.g., AlphaWorld – see Chapter 8). Every location is each others' next-door neighbour; everything is on top of everything else; everywhere is local (Staple 1995). Stalder (1998) extends the placeless space argument, suggesting that cyberspace is a binary space where distance can be measured in only two ways: zero distance (inside the network) or infinite distance (outside the network); here or nowhere.

Despite being able to have any geometry desired, rather than break out of traditional conceptions of space, many cyberspaces which do not have a geographic referent or explicit spatial quality do in fact adopt standard geographic metaphors such as proximity to improve navigability and usage. Here, geographic and topological concepts are used in a process of spatialisation to aid the presentation of complex relationships or even construct new relationships that have never existed to open new analytical and didactic vistas (Cheesman *et al.* forthcoming). In some cases, as detailed above, cyberspace is designed to look like 'geographic' space and common 'real world' modes of interaction are deployed to make cyberspace more intuitive to use. For example, the use of the geographic metaphor structures most MUDs, both textual and visual. Indeed, most MUDs consist of a labyrinth employing a solvable maze metaphor which consists of a linear route through connected spaces with a series of limited choices that lead either to dead ends or to the exit (Murray 1997). Such spaces are relatively easy to navigate through and map because they typically adopt a Euclidean geometry (although in the case of cyberspace the metaphor tends to be broken – see Chapter 8). In contrast, the tangled rhizome[7] metaphor creates a non-linear navigable space. Hypertext structures, such as those on the World Wide Web, take a rhizome form. Rhizome spaces can be disorientating to navigate one's way through and are more difficult to map (see Chapter 9). Murray (1997) explains that solvable mazes are given to purposeful navigation, whereas tangled rhizomes to wandering, as there is no beginning or end. Often, however, we wish to navigate purposefully through the Web, and this is where spatial conflict can arise.

From this discussion it should be clear that the spatial geometries of cyberspace are made up of a complex collection of domains, some explicitly spatial with direct geographic referents (e.g., VR models of a geographic location), some explicitly spatial without a geographic referent (e.g., MUDs), some with real world referents but no explicit spatial form/attributes (e.g., a list of names, a webpage), and some with no geographic referent and no spatial form/attributes (e.g., computer file allocation tables). In the latter cases, these can be visualised using spatialisations. In all cases, any existing geometries are productions, existing only as spatialised constructions. In light of this complexity, in Chapters 5 through to 8 we consider the ways in which researchers have sought to map ICTs or use the application of geographic metaphors (spatialisation) to make data comprehensible.

Summary

In this chapter we have examined the spatiality and spatial forms of cyberspace itself. Our three central arguments are that cyberspace possesses a spatiality that needs to be examined; that the socio-spatial relations of cyberspace are produced; and that cyberspace is an embodied space. Given these three points, it is important that cyberspace is not treated as some kind of paraspace. Instead, analysis should concentrate on mapping the complex ways in which new geographies are emerging and are extending and affecting traditional socio-spatial relations. We endeavour to provide such a charting in the following chapters.

4 Introducing the cartographies of cyberspace

Maps of cyberspace are almost as rare as 16th century portalans. [Cyberspace] explorers practice their trade without maps. . . . Few among this frontier fraternity have both the navigational and drafting skills of a Ferdinand Magellan or a James Cook. Even for those that do, the challenge of mapping cyberspace is in some ways more formidable than that faced by the sea captains of the past.

(Staple 1995)

One aspect of cyberspace that has received relatively little attention is the literal and metaphorical mapping and geographic visualisation of the new spaces that are being created. For centuries, cartographic maps have been used to store and represent geographic knowledge about the world and beyond. They form an integral part of how we understand and explain the world. Maps are powerful graphic tools that classify, represent and communicate spatial relations; a concentrated database of information on the location, shape and size of key features of the landscape and the connections between them (Hodgkiss 1980); a method to visualise a world that is too large and too complex to be seen directly (MacEachren 1995). Well-designed maps are effective sources of communication because they exploit the mind's ability to see relationships in physical structures, providing a clear understanding of a complex environment, reducing search time and revealing spatial relations that may otherwise not be noticed (Kitchin and Tate 1999).

This chapter provides a brief history of how geographic space has been cartographically mapped. It examines the reasons for, and challenges in, geographically visualising ICTs (that is, mapping infrastructure and demographics onto a geographic framework) and metaphorically mapping cyberspace (using techniques of spatialisation to map information into spatial and temporal domains; Couclelis 1998). As noted in Chapter 3, the latter technique of using geographic metaphors to provide a tangible structure to cyberspace is a particularly challenging exercise. Next, we outline a topology of maps of ICTs and cyberspace, which is used to structure our discussion in Chapters 5 through to 8. We follow this with a discussion of two traditional concerns, data quality and levels of understanding, and three critiques of geographic visualisation – namely, issues of representation, the power of mapping, and ethics – that are instructive when considering the maps and spatialisations detailed in the following chapters. In the final section, we briefly examine how the representation of geographic space is affected by cyberspace.

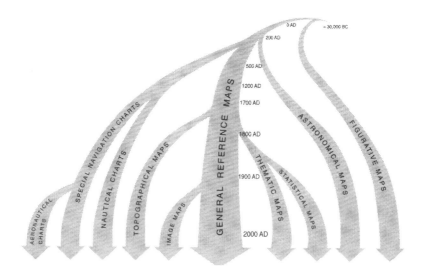

Figure 4.1 The evolution of maps
Source: Robinson *et al.* 1995: 22

A brief history of cartography

> the domains that explorers chart, and the maps they produce, open up territories to inter-
> ests that view them differently ... [B]e they goldfields, stands of timber or ... human
> cultures ... maps serve as the groundplan, the blueprint, the graphic agenda for subsequent
> exploitation.
>
> (Hall 1992, cited in Staple 1995)

Western map-making has become progressively more sophisticated over the centuries (see
Figure 4.1). The basis of the traditional Cartesian maps that we use today (for example,
Rand McNally or Ordnance Survey maps) have their origins in the work of Ptolemy, a
librarian in Alexandria in the second century AD. Combining geography with mathematics,
Ptolemy documented instructions for making map projections, devised a system of latitude
and longitude, and constructed a series of maps of the known world. His ideas, while
influential in Islamic science, failed to be adopted in Europe until the thirteenth century,
when refugees fleeing from the Byzantium capture of Constantinople reached Italy in 1410
(Thrower 1996). Ptolemy's ideas changed the face of European cartography, underlying
the mapping revolution of the Renaissance period and igniting the shift from symbolic
representations towards scientific representations. Prior to the Renaissance period, land
maps were primarily cartographic representations of culture rather than accurate geographical representations of spatial relations, and sea charts, which although relatively
spatially accurate, were charted from experience rather than formal surveying techniques
(Livingstone 1992).

The mapping revolution of the Renaissance was fuelled by the exploration of new
worlds. Explorers from Europe set sail to verify Ptolemy's map and discover, claim and map
new lands. Maps became increasingly important because they both aided navigation and
delineated new territories and borders (and the riches therein). At first, maps of the world

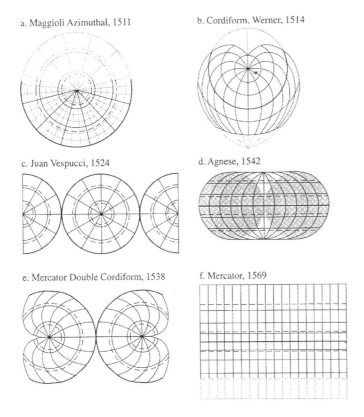

a. Maggioli Azimuthal, 1511

b. Cordiform, Werner, 1514

c. Juan Vespucci, 1524

d. Agnese, 1542

e. Mercator Double Cordiform, 1538

f. Mercator, 1569

Figure 4.2 Map projections pre-1569
Source: Thrower 1996: 74

were patchy and incomplete due to poor knowledge and weak surveying and mapping techniques but with time map-making became more sophisticated as cartographers became better skilled and versed in geometry and overcame the problems of surveying and representing a spheroid on a single sheet of paper. For example, by 1507 Martin Waldseemüller had produced a map of the world that included North and South America, and from 1511, a number of map projections had been developed culminating in the now commonly used Mercator projection in 1569 (see Figure 4.2) (Thrower 1996). Although maps became more geodetically, planimetrically and topographically accurate,[1] the role of maps as symbolic agents continued (as it does today) (Blakemore and Harley 1980).

The period following the Renaissance was an era of formalisation with increased absolute and relative spatial accuracy due to improved and greater surveying, advances in the mathematical proficiency of cartographers, and the development of much more precise instruments. Throughout the next two centuries, map-makers achieved an intellectual respectability as 'men of science' (Livingstone 1992). Accompanying this formalisation was experimentation with mapping techniques and visualising different types of data. For example, Halley used thematic mapping to chart the trade winds and different types of hatching was used to show topographic relief. Map-making in the eighteenth and nineteenth centuries was characterised by the formalisation of cartographic conventions (e.g., scale, orientation, placement of text, layout) and the symbols that were used.

Modern cartography uses a number of standardised methods for mapping the geographic world. These extend beyond standard or thematic Cartesian maps to include cartograms, and conceptual and topological maps that seek to provide relative rather than absolute spatial accuracy. As Wood (1993) notes, there is no one accepted typology of maps. Indeed, he provides a number of examples: *The Map Catalogue* (Makower 1986) provides a threefold taxonomy of land, sky and water; The *US National Report to ICA* (Loy 1987) categorises maps into three classes – government mapping, business mapping and university cartography; Southworth and Southworth (1982) provide an eightfold taxonomy that includes land form, built form, networks and routes, quantity, density and distribution, relation and comparison, time, change and movement, behaviour and personal imagery, and simulation and interaction.

At the end of the twentieth century, cartography has undergone three major evolutions. First, there has been the digitalisation of cartography and the widespread use of computer systems such as Geographic Information Systems (GISs) and Computer-aided Design (CAD) that are able to store, process, manipulate and transform spatial and attribute data. GISs provide maps that can be manipulated, transformed and selectively interrogated by their users (Burrough 1986). Whereas the paper map itself is the database, in GISs any one map is simply one particular spatial representation of the database (Burrough and McDonnell 1998). GISs allow the construction of maps on different scales from the local to the regional and beyond, which combine different data variables, and use different classification schemes, all from the same database and achieved in a matter of seconds, whereas paper maps need to be redrawn at great cost, in terms of money and time. Furthermore, maps once static, qualitative representations that needed to be drawn by cartographic 'experts' become dynamic, quantitative models (although not necessarily interactive) in a GIS, and are easily constructed by people with little cartographic training. Moreover, geographic data in a GIS can be queried in ways that are difficult using a paper map (for example, the question 'where are all locations of feature A' would involve a large manual search of one or several paper maps, but in a GIS this information is easily recovered and mapped).

The second evolution in cartography is the creation of new methods of geographic visualisation (see Hearnshaw and Unwin 1995). This has seen a move away from static maps to interactive, dynamic and animated geographic visualisations that can be designed by anyone with access to software and data. For example, it is now possible to create three-dimensional terrain models that can be viewed from different angles, flown through, and draped with various, relevant, variable overlays. Increased computing power and sophisticated algorithms create displays that are difficult and time consuming to perform by hand. This allows the data to be interrogated in ways that were nearly impossible before the advent of the computer.

These two evolutions, widespread access to mapping technology and geographic visualisations, extend the power of mapping in qualitative and quantitative ways: (1) opening up new ways to comprehend the real world; (2) providing an effective means of structuring immaterial phenomena and material that has no geographical referent, thereby increasing comprehensibility; (3) replacing static representations with multiple representations that are interactive and dynamic; (4) allowing non-cartographers access to data and to produce their own maps, thus breaking one of the major principles of traditional map-making theory, that there is a clear separation between the cartographer and the user (Crampton 1999).

The third evolution is a change in attitude over how maps are conceptualised. As we have argued in Chapter 1, an approach that combines social constructivism and political economy provides a suitable framework in which to analyse cyberspace. Such an approach is also useful in the context of cartographic mapping. As Harley (1989) has persuasively argued, maps are not merely scientific artefacts which faithfully represent that mapped, they are also social constructions. Maps are produced for a purpose (whether economic, political or social), and the features included on them and the message communicated by them is imbued with the cartographer's or the sponsor's (corporate, state) intent and values (see 'Power of mapping', p. 75). Many cartographers/institutions now accept that while cartography presents itself as an objective, scientific pursuit, like all sciences, it is a process of construction, shaped by many forces. Throughout this, and the next four chapters, our analysis seeks to move beyond traditional cartographic concerns such as accuracy, precision, verisimilitude (having the appearance of truth, realistic depiction), and mimesis (imitation, mimicry), to acknowledge that the process of mapping is situated, embodied (even if computer-generated, as the process was originally programmed) and partial (Gregory 1994). As such, we recognise that maps themselves are not neutral, unproblematic artefacts but constitute systems of power-knowledge (Harley 1989).

Why 'map' ICTs and cyberspace?

> We knowledge workers have always been thirsty people. Tired of shuffling about in dusty clouds of guesswork and operating on approximations and hunches because we couldn't get the data we needed, we begged for a cool glass of information. Instead, the World Wide Web gave us a firehose of digitized papers, reports, images, statistics, and formulas. Now we're drowning in the torrent of words and pictures, desperately in need of some high ground from which to survey the flood.
>
> (Technological Partners 1997)

As detailed, it has long been recognised that geographical visualisations, in all manifestations, form an integral part of how we understand the world. In the case of information that has a geographic referent and spatial attributes (e.g., ICTs), constructing a map or spatialisation provides a means by which to visualise and describe that form. It also reveals important insights into who controls the infrastructure, who has access to cyberspace, how the system can be surveyed, and how and from where cyberspace is being used.

Cyberspace is a complex and multifaceted medium. As discussed in the previous chapter, some domains are explicitly spatial with direct geographic referents (e.g., a VR model of a real-world place), other domains have an inherent spatial form without a geographic referent (e.g., textual and visual MUDs), or a real-world referent but no spatial form/attributes (e.g., a list of names, a webpage), or no geographic referent and no spatial form/attributes (e.g., computer file allocation). In addition, domains with a real-world referent may have a materiality (e.g., has a mass) or be immaterial in nature (e.g., gravity, heat).

The spatial form/attributes of cyberspace data that have no geographic referent can be mapped by a process called spatialisation.[2] Here, a spatial structure is applied where no inherent or obvious one exists in order to provide a means of visualising and comprehending space; to utilise the power of spatial representation in order to describe complex informational spaces in a new, more easily interpretable form. Information attributes are transformed into a spatial structure through the application of such concepts as hierarchy and proximity (nearness/likeness). Mapping in both a literal and a metaphorical sense can thus provide a means of facilitating the comprehension of, navigation within, and documenting the extent of (marking out territories) these varying forms of cyberspace. Indeed, without spatialisations, topological structure data are almost impossible for humans to interpret because they are held in large textual tables that provide no tangible referents other than attribute codes. As we document in Chapter 6, a number of researchers are now experimenting with the process of applying spatialisations to cyberspace (see Gershon and Eick 1995b; Card *et al.* 1999).

The challenge of spatialising cyberspace

At a technical level, ICTs are relatively easy to map. The physical architecture and topology of the networks can be mapped onto geographic space and the traffic through this network can be represented using an appropriate form of visualisation. Similarly, the physical location and characteristics of hardware, software and wetware (human users) can be mapped using traditional cartographic methods (see Chapter 5). Cyberspace, however, provides a much greater challenge: the effective mapping of visual spatial forms (but with no materiality, such as virtual worlds, see Chapter 8), and the use of spatialisations to provide comprehensibility for non-spatial or immaterial information that is difficult to navigate through and understand due to its complexity and mutability. As such, visualisers of cyberspace must find ways to map spaces with differing spatial forms and geometries, including some with no recognisable geometrical properties; and find ways to map spaces that break two of the fundamental conventions of geographic visualisations. These conventions are that (1) space is continuous and ordered, and (2) the map is not the territory, but is a representation of it (Staple 1995). As previously noted (see Chapters 1 and 3), cyberspace can be discontinuous and organised in a non-linear way, and in many cases the spaces are their own maps. In a deeper sense, a session in cyberspace is a rhizomatic map of itself (Staple 1995); rather than being external to a representation of data, we are navigating links within data. However, as Novak (1991) notes, this is not to deny that cyberspace has an architecture (geography), contains architecture, or even is architecture, just that this architecture is its own.

The challenge for cartographers, however, is only partly a matter of spatial form. As Staple (1995) notes, mapping cyberspace is just one part of a wider project that aims to map places that cannot be seen, such as distant galaxies, DNA and brain synapses (see Hall [1992] for fascinating examples of these). Cyberspaces, though, are 'infinitely mutable' (Staple 1995), changing daily as new computers are added, the infrastructure is updated, and the content refined, expanded and deleted; cyberspaces are transient landscapes – spaces that are changing constantly but where the changes are often 'hidden' until encountered. As time unfolds and more and more data is uploaded, the mutability of cyberspaces will increase accordingly to create spaces that are constantly evolving, disappearing and restructuring. Geographic visualisations of geographic spaces are out of date as soon as they are published, as the landscape portrayed is modified. The vast majority of information portrayed, however, remains stable and the shelf-life of a map can be many years. The shelf-life of a map of cyberspace, given the current and projected dynamic nature of cyberspaces, is likely to be very short. Furthermore, as yet, unlike geographic space, there are no agreed conventions in relation to how a space is designed or how it is traversed, providing a diverse set of spaces which differ in form, geometry and rules of interaction.

To complicate matters, if spatialisations of cyberspace are programmed to be interactive, the spatial representation (map) becomes the territory – map and territory become synonymous; rather than being external to a representation of data, we are navigating links within data. Here, the use of a geographic metaphor to structure the data becomes the means by which this new territory is navigated. For example, a VRML webpage is both the territory and the means in which to navigate this territory.

The wider challenge to cyber-cartographers, then, is to construct dynamic maps and spatialisations of a variety of cyberspaces, some with no explicit spatial relations, some with an in-built relational (topological) geography (e.g., textual and visual virtual worlds), and to map out the intersections between virtual and geographic spaces; to produce maps and spatialisations that will aid navigation within, and comprehension of different cyberspaces at both the theoretical and the practical level.

At present, it is probably fair to state that in relation to this wider challenge, carto-graphers of cyberspace are at the same stage as the cartographers at the start of the Renaissance period. Although armed with a knowledge of traditional mapping and soph-isticated computing, they do not, however, have the blueprints that Ptolemy provided European cartographers plotting the new world. As a consequence, cyber-cartographers tend to extend the methods used to visualise geographic space in order to try and map and visualise the spaces of cyberspace, albeit within different media. Although not satisfactory, there is little alternative until a radically new conception of mapping is put forward. As Bukatman notes, it is perhaps inevitable that both science-fiction writers (e.g., William Gibson, see Chapter 9) and cyber-cartographers use Cartesian mapping forms since 'it reduces the infinite abstract void of electronic space to the definitions of bodily experience and physical cognition, grounding it in finite and assimilable terms' (Bukatman 1993: 152). Indeed, whether such a new conception will emerge is still not known, but given the need to constrain visualisation to two-dimensional and three-dimensional displays, the limits of our perceptual systems and spatial understanding, and that we are conditioned think in Cartesian terms, it seems a little way off. As we will see, most cyber-cartographers are experimenting with new methods of visualisation but, as yet, they are related to methods of visualising geographic space. However, this is not to deny the inventiveness with which traditional cartographic techniques are being used.

A typology of mapping cyberspace

To our knowledge, there have been only two attempts to create a typology of the visual-isation and spatialisation of cyberspace. The first is by Dodge (1997), who divides what he terms 'cybermaps' into a number of classes: geographical metaphors, conceptual maps, topology maps, land use maps and landscape views, virtual cities and navigation tools. The second, by Jiang and Omerling (1997), classifies maps of VR spaces using a threefold, system centred on function: navigation, cyberspatial analysis, and persuasion. Both, how-ever, adopt a classification scheme which fails to recognise the differences between data source/type, and the complexities and differences between what the maps/spatialisations are seeking to represent in spatial terms. As discussed, the mappings of cyberspace vary as a function of geographic reference, spatial form/attributes, and materiality of the informa-tion that is mapped. We therefore propose a classification that varies along three axes: (1) geographic referent (ICTs/cyberspace); (2) immateriality/materiality (material, spatial form/immaterial spatial form); and (3) map/spatialisation form (static, animated, inter-active, dynamic). Each axes is discussed in turn.

It is clear that in the practice and process of mapping there is a strong difference between maps that concern ICTs (infrastructure, hardware, demographics, etc., and other immaterial aspects such as gravity and heat), and those that concern cyberspace. In the first case, there is a geographic referent that the map is seeking to represent. In the second case, no geographic referent exists and a process of spatialisation is applied to make comprehensible data that would otherwise be too complex to understand. In the first case, then, issues such as the degree of spatial equivalence are key – the extent to which the visualisation corresponds with reality. In the second case, such comparisons are impossible.

Mappings can also be defined along axes of materiality and their spatial attributes. For example, maps of ICTs can have a material geographical referent (e.g., infrastructure) or an immaterial referent (e.g., heat). In the cases of a material referent, cartographic qualities most often match those of geographic space in terms of conventions and design. Essentially, data is mapped onto a geographic base data (e.g., mapping network architecture onto a world globe). Another form of visualisation is a virtual model of a geographic location. In the case of the immaterial referent the data are mapped metaphorically, placing the values into a two-dimensional or higher display. In relation to cyberspace, mappings can portray digital data with no geographic referent but with spatial attributes (e.g., visual MUDs), data with a geographical referent but no spatial attributes (e.g., webpages), and data with no geographical referent and no spatial attributes (e.g., computer file allocation). At a basic level then, mapping can be divided into one of four categories:

1   'geographic' space/material (conventional mapping, e.g., infrastructure)
2   'geographic' space/immaterial (spatialisation, e.g., heat)
3   cyberspace/spatial (conventional mapping, e.g., AlphaWorld)
4   cyberspace/non-spatial (spatialisation, e.g., file structure)

Each of these four categories has the potential to be further sub-divided by various forms of mapping. Mapping can take one of four forms: static, animated, interactive or dynamic. Static mappings are the equivalent of traditional cartographic maps in that they are snapshots in time. The difference is that they vary in visualisation technique, for example extending to three-dimensions. Animated mappings portray a sequence of static maps to provide a time-series. Interactive mappings move beyond static mappings so that the user can move through and interrogate the map from different viewpoints (essentially 2.5-dimensional+ virtual models). Whilst the 'map' itself is static, the user becomes dynamic. Dynamic maps are where the mapping automatically updates as the information used in its construction is updated. These forms of mapping can be combined so that a map can be static and interactive or dynamic and interactive.

This process of categorisation could continue. For example, it is possible to categorise the maps on the basis of function as Jiang and Omerling (1997) have done, identifying those used for navigation, comprehension and persuasion. We feel, however, that to continue broadening our typology would be confusing and unhelpful given that we think that the relevant differences are captured by the typology already outlined here. We use our typology (particularly axes 1 and 3) to guide the discussion in Chapter 5 through to Chapter 8, so that Chapter 5 concerns mapping ICTs and Chapters 6 to 8 considers the spatialising of cyberspace, with the discussion of mapping in each ordered from static to dynamic.

Issues to consider

This section details some of the key issues that need to be considered in relation to the maps and spatialisations presented in subsequent chapters. To date, these maps and spatialisations have largely been produced without any critical analysis. Whilst they are often informative, it is important to consider five key questions when viewing them:

- How 'accurate' is the map?
- Is the map interpretable?
- What does the map *not* tell us?
- Why was the map drawn?
- Is the map ethical?

Data quality and availability

Spatial visualisations are only as accurate as the data used to underpin the representation. Therefore, a key question for those seeking to construct maps of ICTs and spatialisations of cyberspace is access to accurate and plentiful information. This has been a specific concern of cartographers, particularly since the Renaissance, but has become a major issue since the widespread adoption of GISs. In particular, spatial data users are concerned about such issues as data coverage, completeness, standardisation, accuracy and precision. Here, accuracy refers to the relationship between a measurement and 'reality', and precision refers to the degree of detail in the reporting of a measurement. It is generally recognised that all spatial data are of limited accuracy due to inherent error in data generation (e.g., surveying) or source materials (digitising, geocoding) (Goodchild 1989). Concern about quality has led to the development of national transfer standards in spatial data with regard to geographic space, for example the Spatial Data Transfer Standard (SDTS) in the United States and the National Transfer Format (NTF) in the UK.

Similar standards do not exist for cyberspace and sources are scarce and fragmented, with no definitive, comprehensive databases. Consequently, most efforts to spatially visualise ICTs and cyberspace have been undertaken by academic researchers, dedicated enthusiasts (often from within the technical Internet community) and commercial consultants. The results, while fascinating, are often limited in scope, geographic coverage and currency when compared to the wealth of statistics gathered and mapped for geographic space by government agencies like the United States Geological Service, Ordnance Survey, and bureau of censuses. This is compounded by the fact that both ICTs and cyberspace lack central planning and a controlling authority that monitors and gathers statistics on their operation and use. In addition, the provision of both infrastructure and content services has become an intensely competitive and profitable business. As such, corporations are wary of giving away details that may aid competitors or threaten security.

The fast growing and dynamic nature of ICTs and cyberspace means that the issue of data quality and coverage is of critical importance. Online spaces such as the Web are in constant flux. We have encountered many dead-ends and re-routings because data and the links between them no longer exist or have changed address. Without pointers, trails become cold and it is difficult to locate sites because of the inability of current search engines to search effectively for specific pieces of information. (In our experience, searches often produce lists containing over a million pages in response to a request.) Mappings will become increasingly important in terms of understanding the connections between cyberspace spaces and geographic spaces, and for the comprehension of, and navigating through, cyberspace, but without the suitable high-quality data to underpin their construction, they will only be of limited use. Foote and Hubner (1995) provide a useful list of questions that need to be considered when assessing the quality of the spatial data displayed, and these questions can be applied to data concerning cyberspace.

- What is the age of the data?
- Where did the data come from?
- In what medium was it originally produced?
- What is the areal coverage of the data?
- At what map scale was the data digitised?
- What projection, coordinate system, and datum were used in the source maps?
- What was the density of observations used for its compilation?
- How accurate are positional and attribute features?
- Do the data seem logical and consistent?
- Do cartographic representations look 'clean'?
- Is the data relevant to the project at hand?
- In what format is the data kept?
- How was the data checked?
- Why was the data compiled?
- What is the reliability of the data provider?

Level of user knowledge

As the work of cognitive cartographers (e.g., Lloyd 1997; MacEachren 1995) has illustrated, maps, while effective at condensing and revealing complex relations, are themselves sophisticated models. For example, Liben (1991) has noted that most maps are not 'transparent' but are complex models of spatial information that require individual's to possess specific skills to process them. Using a map means being able to read a map, which requires a distinct set of skills that must be learnt. This implies that a novice will learn little from a professionally-produced map unless they know how the map represents an area. This also applies to the mappings of cyberspace, particularly in the case of three-dimensional interactive mappings and spatialisations. Thus, care needs to be exercised in relation to the design of mappings so that the target audience can understand and use the information that is given. As far as we are aware, whilst there has been some work on the legibility and design of visual virtual worlds (e.g., Darken and Sibert 1996a) and hypertext (e.g., Bachiochi *et al.* 1997; Kim and Hirtle 1995; Nielsen 1990), there has been little or no work on the legibility of maps of ICTs or spatialisations of cyberspace (see Chapter 8). Indeed, many of the maps presented in Chapters 5 to 8 in this volume are difficult to interpret without reference to the explanation in the text. The need for such reference means that the maps hold poor communicative properties, which need to be improved.

Representation

Spatial visualisations are representations. They aim to represent, in a manner that is consistent, some particular phenomenon. An age-old question, therefore, relates to the extent to which spatial visualisations adequately represent data portraying key features. Mappings necessarily depict a selective distortion of that which they seek to portray – they generalise and classify. It is now well known that alternative methods of generalising and classifying can reveal differing spatial relations. For example, it is known that the same data collected from differing sets of spatial units (e.g., wards, districts, counties, states) can produce significantly different spatial patterns. This is known as the Modifiable Areal Unit Problem (MAUP) (see Openshaw 1984). Problems also exist in the use of crisp boundaries to represent what are in reality continuous phenomena. The use of zones, or representation of features in a choropleth map, assumes not only a model of spatial uniformity within each zone, but that we can delineate the boundary between each zone in a precise and meaningful manner. For many geographical phenomena this is simply not the case: natural spatial variation leads to gradual change, and the difference between reality and the model can lead to erroneous interpretation (Burrough and McDonnell 1998).

These are issues, well known to cartographers, which affect all maps, including those of ICTs and cyberspace. As Harpold (1999) notes, strategies of aggregation used in demographic maps of the Internet hide variation within units. Moreover, he contends that mappings of ICTs reproduce particular hegemonic messages because of the way in which they adopt traditional cartographic map units (e.g., political boundaries) in order to display data. As yet, however, these issues have been little considered in the preparation of spatial visualisations of ICTs and cyberspace.

Furthermore, as noted earlier, debates concerning representation are often centred around modernist concerns such as accuracy, precision, verisimilitude and mimesis. However, for data with no geographic referent or materiality, this poses the question, 'By what standard do we judge these factors?' When data and mapping become synonymous, how do issues of representation apply? Here, cyberspaces may become meaningless outside of their representation. The need for standards to be set and for issues of representation to be addressed, then, is of paramount importance. As noted below, however, the context within which standards are set needs to be recognised and constantly critiqued.

Power of mapping

As Harley (1989) and others (e.g., Wood 1993) have argued persuasively in relation to traditional cartographic maps, maps are not objective, neutral artefacts. Mapping is a process of creating, rather than revealing, knowledge, as a result, decisions are made about what to include and what to exclude, how the map will look, and what the map wants to communicate (MacEachren 1995). Maps are never merely descriptive – they are heuristic devices that seek to communicate particular messages (Harpold 1999). Maps are imbued with the values and judgements of the individuals who construct them and they are undeniably a reflection of the culture in which those individuals live. As such, maps are situated within broader historical contexts, and according to Harpold (1999) reflect hegemonic purposes through the use of historically and politically inflected metageographies (sign systems which organise knowledge into visual systems that we take for granted, for example, a world territory map). Geographic visualisations are thus situated, embodied and selective representations. Whilst they might pertain to be objective, mimetic devices, they are ultimately constructed for a particular purpose. Commonly, the messages are those of the powerful who pay for the maps to be drawn, and the ideological message is one of reproducing hegemonic power.

Similarly, mappings of cyberspace are the products of those responsible for coding their construction algorithms. They are mappings designed for particular purposes. As such, they too are systems of power-knowledge, and we should be careful to look beyond the data generated, and to question, in a broad sense, who the map was made for, by whom, why it was produced, and what are the implications of its message; in other words, following the advice of Harley (1989), deconstruct its creation.[3]

The power of maps can be illustrated by the extent to which they are being used to market various aspects of cyberspace enterprise. The provision of Internet services and infrastructure is a highly competitive business, dominated by large corporations, many of which operate globally. These corporations make significant use of maps in their marketing strategies. Indeed, the Internet marketing map is an important tool to demonstrate the power of a company's network to potential customers. There are many examples available on most Internet network provider's websites. They employ all manner of cartographic styles to represent the topology of the network, but the most common is some form of arc-node representation on a geographic base map. The companies invest considerable effort in producing high-quality maps that present their networks in the best possible light. The two main ways they do this are, first, to demonstrate the geographic reach of the network, emphasising all the distant places that are linked together, and, second, to illustrate the tremendous capacity of the 'pipes' of the network to cope with huge user demands. In this way, Internet marketing maps fit into a long tradition of maps used by companies to promote their networks, be they shipping, airlines, railroads, or postal carriers (Fleming 1984; Monmonier 1991).

In the light of this discussion, when analysing the maps in Chapters 5 to 8 it is important to understand the rhetorical power – that is, the 'second text' – of each map.

Ethics

One final issue to consider relates to the ethics and responsibility of researchers producing spatialisations of cyberspace. As Smith (1999) notes, these forms of spatial visualisation (discussed in Chapters 6 to 8, this volume) open up cyberspace to a kind of panoptic surveillance, revealing interactions within it that were previously largely hidden. If the appeal of some of these spaces is their anonymity, then members may object to the group being placed under wider scrutiny, even if individuals are unidentifiable. Here, public analysis may well represent an infringement of personal rights, especially if the individuals were not consulted beforehand. In some senses these maps may work to shift the spaces they map from what their users consider semi-private spaces to public spaces (Smith 1999), and thus the maps may actually change the nature of cyberspace itself. Here, it is important to consider the ways, and the extent to which, maps of cyberspace are 'responsible artefacts'.

Geographic space online

This chapter has so far considered the spatial visualisation of ICTs and cyberspace. Before we move on to catalogue the ways in which researchers are attempting such visualisations, in this final section we detail in brief, through a selection of examples, the ways in which cyberspace is changing how we map/represent geographic space and how it can augment interaction with geographic space.

The digital-based, cartographic revolution of the late 1980s and early 1990s has seen geographic space come online. Maps of all geographic scales are accessible over the Internet and new types of maps have started to appear. Companies such as MapQuest and MapBlast![4] have made maps available online, with users able to define the areal extent of the data they require. Between 1996 and September 1999, MapQuest supplied over 1.6 billion maps to people accessing its site, plus it provided travel directions to countless others (Novak 1999). Peterson (1999) estimates that in 1999 approximately 40 million maps were created per day on the Internet. Other companies offer access to aerial photography

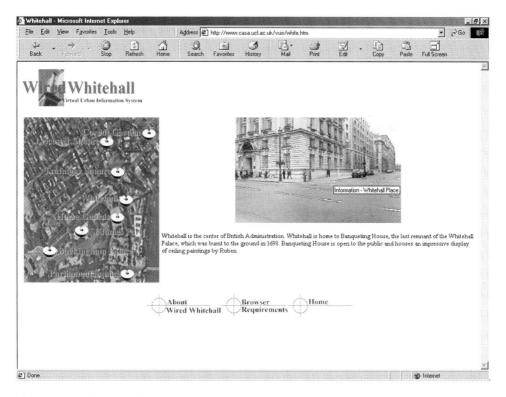

Plate 4.1 Wired Whitehall
Source: Andy Smith, http://www.casa.ucl.uk/vuis/

and satellite imagery on the Web (e.g., Microsoft's Terraserver[5]). Moreover, a number of companies now offer landscape visualisation software and data across the Internet. For example, Truflite[6] market software that allows you to view and fly through landscapes composed of texture (Tiger census files) and elevation (DEM) data. Similarly, experiments at CASA, University College London (UCL), have linked digital photography to street maps which means that a user exploring the map can also witness the vistas from positions within the map to familiarise themself with an area (Batty *et al.* 1998). (A prototype called 'Wired Whitehall' is available on the Web,[7] see Plate 4.1.) Other UCL-based projects provide an interactive virtual reality model of the UCL campus accessible over the Web.[8] Users can explore the virtual environment from many angles and vistas and can also view accompanying video images and slide shows. All the buildings in the environment provide hotlinks to websites which give information about the departments and the people who work there. A further development is the use of a video interface so that users can move and pan around exteriors and interiors of the College with icons such as globes and pictures providing links to other sites both internal and external to the College (Batty 1997). Plate 4.2 provides an example of this system where the user starts in the main quadrangle, enters the North Cloisters through a video window, finds a globe which is then spun highlighting Internet sites around the world that can be accessed by clicking on the globe (in this case, the Geography Department at SUNY Buffalo). Similar projects exist for other places and cities around the world, creating what Taylor (1997: 185) terms 'museums of the real'.

Plate 4.2 Virtual UCL
Source: http://www.ge.ucl.uk/vucl/

Another development has been the utilisation of VR technology to create basic and sophisticated geographic modelling and simulation systems. These range from projects where the VR system is used to view possible scenarios (e.g., what a cityscape might look like when a new tower block is added, or how an area will look if deforested), to projects employing sophisticated modelling algorithms on real world data (e.g., the visual modelling of coastline erosion, or the throughput of visitors given a certain spatial design, or the calculation of viewsheds from particular locations) (see contributions to Fisher and Unwin [forthcoming] for examples). Here, VR builds on the functionality of GISs to provide advanced degrees of spatialisation that have mimetic qualities.

Interactive, hyper-media maps are now quite common vehicles for structuring navigation through websites. As well as displaying cartographic information, these maps provide hyperlinks to related information concerning places on the map. Websites such as The Virtual Tourist[9] provide a set of hierarchically structured hypermedia maps, from the local to the global scale, providing users with geographically referenced links to information concerning any location in the world. Furthermore, users can zoom in to enlarge portions of the map. Figure 4.3 displays the clickable map for UK universities, providing a geographically formulated directory of accessing academic websites.[10]

One novel approach is that adopted by Dan Jacobson, an Assistant Professor at Florida State University in the US. He has created a series of hierarchically linked, hypermedia sound maps for visually-impaired people that are accessible via the Web (see Jacobson and

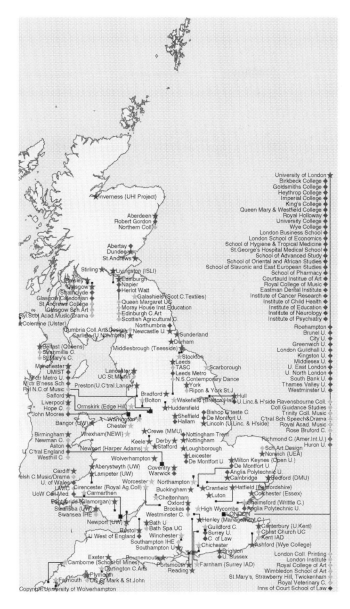

Figure 4.3 UK clickable website directory
Source: University of Wolverhampton and Peter Burden, http://www.scit.wlv.ac.uk/ukinfo/uk.map.html

Kitchin 1997). As the mouse cursor is moved over the screen different audio files are played which correspond to the areas it is crossing. Sound files consist of spoken word and representative sounds (e.g., the sound of traffic for a road, or waves crashing for the sea). Each map is accompanied by large font text, an enhanced cursor and screen magnifying software. The maps are simplified and displayed with strong borders and defining colours allowing people with low vision a degree of visual access. The hierarchical structure of the maps allows users to explore a whole area, from its global form to individual details, before visiting the location. The maps can also be used to link up with other information. For example, maps of a university campus would have hotlinks that could transfer the map-user to the webpages of a department in the building over which the cursor resides. These could then be read aloud using a speech synthesiser. Trial tests have proved successful at conveying spatial information, and have been popular among users. Further work is seeking to implement a combined haptic, aural and visual interface using a haptic mouse (see Jacobson *et al.* forthcoming).

It is not only representations of geographic space that are appearing online. Images of the real urban environment can now also be viewed in real-time via the Internet. For example, one site[11] allows Internet users to watch real-time CCTV images from around the world. In an ironic statement about the surveillance society we live in, the CCTV link displayed via the Metropolis art project[12] places underneath the image a fax crime report file with a fax number for the relevant police force. Other sites allow real-time viewing of weather satellites.[13] Geographic views and images that were once impossible to display are now realised, captured and transmitted instantly. To paraphrase Marshall McLuhan, the medium becomes a map.

5 Mapping information and communication technologies

Despite the massive growth of cyberspace in the last twenty years, the materiality that supports it, ICTs, is largely invisible (Batty 1990; Moss 1986).[1] Much of this infrastructure is hidden underground, located in anonymous server rooms, placed in conduits and roof voids, and housed in grey boxes that quietly hum under people's desks. Given this invisibility, it is easy to assume that the infrastructure of cyberspace is as ethereal and virtual as the information and communication that it supports. However, the infrastructure has a materiality that can be mapped onto geographic space and displayed using cartographic techniques. The type of data, and its origins, destinations and the paths it travels through the various networks that compose the Internet, can also be mapped. Furthermore, this infrastructure is used by people located within geographic space, sat at a desk, accessing cyberspace through the screen, keyboard and mouse. Thus, we need to consider ICTs in relation to demographic matters – the characteristics and location of the people who use the various spaces and services. This, too, can be mapped to geographic space to reveal the demographic profile of ICTs and cyberspace.

In this chapter, we detail and examine maps that have been created to market and aid our understanding of the structure, organisation, operation, use and demographics of ICTs. As outlined in Chapter 4, this mapping is important because it reveals insights into the power structures of the material (and, in turn, immaterial) aspects of cyberspace in terms of who controls the systems, who has access to them, how the systems can be surveyed, and how and from where cyberspace is being used. The maps we examine are all in the public domain and the majority are freely available on the Web. They have mostly been produced by academic researchers (in an attempt to understand the scope and development of ICTs) and consultants (marketing maps for corporations). As such, these maps represent a fraction of all the maps of ICTs that have been produced, but they do provide a valuable overview of the cartographic approaches and visual metaphors currently being employed. Many of the maps that we do not see are confidential to the companies that own and operate the infrastructure, and either reveal sensitive information or are of a practical nature, primarily used by system engineers (e.g., a schematic map of a cabling arrangement).

The chapter is divided into three sections: maps of infrastructure; maps of traffic; and attempts to map the temporal aspects of ICTs. Each category also reveals aspects of the demographics of ICTs. In each section we order the discussion in relation to our map typography detailed in Chapter 4: static, animated, interactive and dynamic. When viewing the maps in this chapter, it is important to consider them critically – to question them in relation to the issues detailed in the previous chapter (i.e., how 'accurate' is the map? Is the map interpretable? What does the map *not* tell us? Why was the map drawn? Is the map ethical?). Whilst we subject the maps we present to this critical gaze, our discussion is mainly descriptive, detailing the cartographic principles behind the map.

Maps of infrastructure

Static and animated maps of infrastructure

The original maps of ICTs were static maps of network architecture, a classic example of which is the network linkages of the original ARPANET network (see Figure 1.2, p. 9).[2] These maps are simple line drawings, but they have historical importance because they chart the growth of the network which formed the first part of the Internet. Their cartographic style is purely functional, using an arc-node representation, where nodes represent sites and paths the links. Although the nodes are located relatively accurately, the paths are merely relational, designating a link between nodes, and not the exact geographic path of this link – a style common to many network maps. Another technique, that of magnification, is used on later maps to reveal greater detail in densely networked regions. The most revealing feature of these maps is that they demonstrate how ARPANET's growth delineates the geography of military installations and research universities involved in defence contracts in the United States.

A large number of maps of government-sponsored and commercially-owned networks have been produced, most for the purposes of marketing.[3] In this instance, maps are deployed in order to vie for potential customers. The provision of Internet services and infrastructure is highly competitive and dominated by large, usually global, corporations. The Internet marketing map demonstrates the power of a company's network to potential customers in two main ways: first, it demonstrates the geographic reach of the network, emphasising all the distant places that are linked together (as shown by the arc-node map); second, it illustrates the tremendous capacity of the 'pipes' of the network to cope with large user demands. An example of this second strategy is employed by John Neystadt, a computer consultant, in his maps of Internet backbones in Israel.[4] One of his maps shows the Israel Academic Network and its connection to IBM's network (see Figure 5.1). Here, an arc-node representation is again deployed except this time the width of the paths is used to encode information about its capacity (e.g., 10 Mbps connections are drawn as thicker lines than lower capacity 256k links) and lines are colour-coded to represent network owners. In other cases, rather than just line width, different colours are used to represent bandwidth. For example, Plate 1A presents the backbone connections of the UUNET network in the United Kingdom and its external links.[5]

The alternatives to static arc-node maps are choropleth maps and dot maps. Choropleth maps attempt to move beyond mapping the extent and capacity of single or multiple Internet networks to provide a wider picture of geographical location of infrastructure. They do this by presenting aggregate-level statistics on the amount of Internet infrastructure (such as the capacity of international links or computers per capita) within a defined territory. One of the most widely distributed set of choropleth maps is produced by Larry Landweber, and endorsed by the Internet Society. Since 1991, he has published sixteen 'International Connectivity' maps, charting the global diffusion of the Internet. Plate 1B displays his earliest available map from September 1991.[6] These are simple choropleth world maps with countries classified into four categories ranging from no public network connectivity to full Internet link.

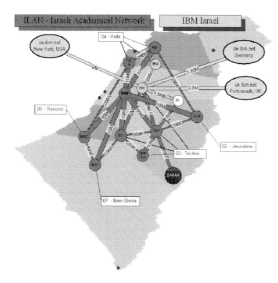

Figure 5.1 Israeli Internet infrastructure
Source: John Neystadt, http://www.iGuide.co.il/

Although visually striking, Landweber's maps provide a conceptually simple, one might even say simplistic, picture of the geography of the Internet. They also have an inherent flaw, that of ecological fallacy. Ecological fallacy is a well-known phenomenon in geography whereby the aggregate characteristics of an entire population are inappropriately ascribed to individuals within the population. For example, Landweber's maps ascribe a single value to a whole territory, suggesting that everywhere within this area has equivalent levels of Internet connectivity. This is clearly not the case, with the maps promoting an artificial sense of homogeneity and masking variation and inequality within territorial units. In many respects it seems illogical to create maps that rigidly demarcate cyberspace into national borders or other territorial units. The network technologies of cyberspace are forging connections and virtual groups that are beginning to subvert the primacy of national boundaries, which are represented on maps by crisp lines. In reality, however, these borders are meaningless to data connections and flows. One of the major reasons why these ecological fallacies arise is that researchers rely on 'off the shelf' data that is readily available at the country level, and which can be turned into choropleth maps with ease and little thought. In many studies of Internet diffusion the same data sources – for example, the World Bank, OECD, International Telecommunications Union, CIA world database, and Network Wizards Internet data – appear time and again. The usefulness of choropleth maps are usually further compounded by poor classification selection to categorise the underlying attribute data, resulting in distorted visual patterns (Jenks and Caspall 1971).

These factors, ecological fallacy and a poor choice of categories, mean that Landweber's International Connectivity maps provide a distorted view of the global spread of the Internet. Holderness (1998: 39), commenting on Landweber's last map, notes that: 'Almost the whole world, it seems from a casual inspection of this map, has turned Internet-coloured. The sun never sets on the Internet; it appears to reach everywhere except some war-torn corners of the world.' Holderness has attempted to reconfigure the Landweber map, to

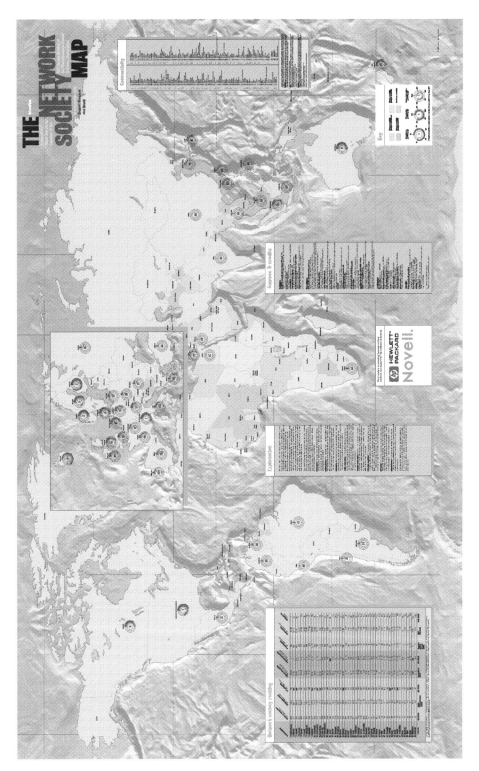

Plate 5.1 Network Society Map
Source: World Link, http://www.worldlink.co.uk/

remove some of the more obvious distortion by fading non-metropolitan regions outside of the OECD countries and greying out the uninhabited deserts, tundra and ice fields.

Another example of a map that falls into the same trap, but one which presents the information for commercial purposes – that is, cyberboosterism – is the Network Society Map (see Plate 5.1).[7] This map contains considerable, conscious and unconscious, bias and distortion. First, it uses a simple classification system based on territorial units to construct a choropleth map of 'how well prepared 49 of the largest and most dynamic economies are to compete in the network society'. A typical array of measures of national technological 'progress', such as phone lines, PCs and Internet hosts per capita, are used to rank the forty-nine countries that are highlighted. Second, an Americentric view is presented, which leaves much of the world unmeasured and unmapped. In reality, many of these territories are key to the sustenance of the Network Society, providing sites of low-paid, low-skilled back-officing and the manufacture of computer and telecommunication components which are almost exclusively exported.

Dot maps have also been used to show the availability of different Internet infrastructure and services in specific locales. Figure 5.2[8] displays the number of hosts connected to computer networks across the world in 1997, as determined by MIDS.[9] MIDS refer to this style of presentation as an 'icon map'. Figure 5.2 presents data on four distinct networks – the Internet, BITEARN, FidoNet and UUCP.[10] Symbol type and position are used to represent the count of particular network hosts at each location. Different symbol shapes (circles, squares and ellipses) and colours (green for Internet, black for BITEARN, etc.) distinguish the four networks, while the size of the symbol is proportionally scaled to show the number of hosts present at that location (usually a city). The major problem with the MIDS icon map is the density of symbols plotted on the map in the heavily networked regions of North America and Western Europe. Inappropriate symbol scaling and data aggregation have led to severe problems of over-plotting, with certain regions disappearing under a confusing mess as symbols are piled on top of each other. This makes it almost impossible to read any useful information from the map in these regions.

Another method that is used to map Internet infrastructure, beyond specific network topology, is to map the geographic location of virtual addresses. Just as postal addresses identify a unique place to which letters are delivered, Internet addresses perform the same function in cyberspace. In order for information to be transferred between computers, each computer must be identifiable. At present, each computer on the Internet has a single virtual address that is recorded in two ways, one for use by other computers and the other for use by people.

Computers identify other computers on the Internet using a unique number called an IP address (short for Internet Protocol address). IP addresses are 32-bit numbers consisting of four numbers ranging from 0 to 255, separated by a period, for example, 144.82.100.130. This particular IP number identifies a single Unix workstation at University College London, providing it with a globally unique location in the Internet. Numeric IP addresses were originally the only form of virtual address. However, they proved difficult for people to remember. Consequently, an alternative system of addressing called 'domain names' was developed in 1987 by Paul Mockapetris, which was subsequently deployed across the Internet (Postel 1994; Barkow 1996). Domain names are short textual names (e.g., www.ucl.ac.uk) that are structured in a hierarchical system known as the DNS (Domain Name System), which can be thought of as a tree that grows from the specific to the general. The domain name can be easily decoded to reveal details about the site. For example, 'www.ucl.ac.uk' reveals that the computer is a Web server, 'ucl' means that the

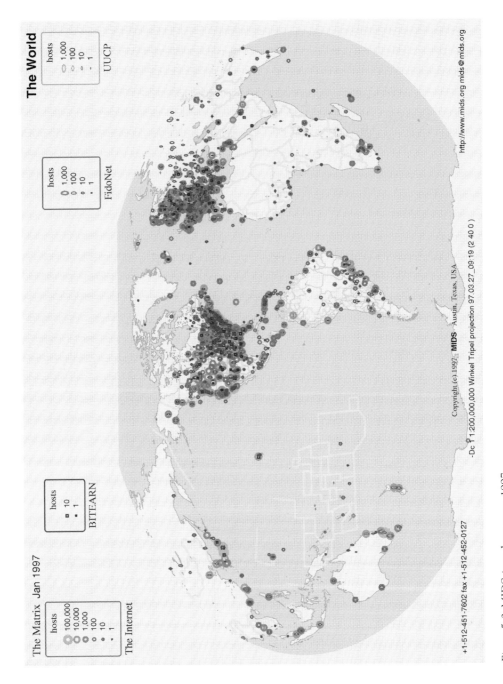

Figure 5.2 MIDS icon host map, 1997
Source: MIDS, http://www.mids.org/

Figure 5.3 Three-dimensional surface of IP address space density in the UK for government agencies

server is located at University College London, 'ac' means that the host organisation is a university,[11] and 'uk' refers to the fact that the organisation is located in the United Kingdom.[12] It is important to realise that both the domain name, www.ucl.ac.uk, and the IP address, 144.82.100.130, identify the same location in the Internet.

This system of addressing means that cyberspace does have limitations (a finite number of addresses). Even though there are a very large number of possible locations these address schemes define a finite chunk of cyberspace, making it possible to use them to compute the extent to which the Internet is growing (simply by counting IP addresses and domain names). The numeric IP addressing system, in its current version, defines a massive expanse of virtual land, with over four billion unique locations.[13] In practice, however, the Internet address space is considerably smaller than this because of how it has been partitioned and allocated to companies and organisations (Hubbard *et al.* 1996; Semeria 1997). Allocations policies, combined with the rapid growth of the Internet, have led to the realisation that address space is a valuable resource that needs to be conserved to avoid dangers of exhaustion (Huston 1994, 1997). When one considers the increasing scarcity of IP addresses and their vital importance in the underlying infrastructure of the Internet, it is surprising that they have received such scant attention from researchers.[14]

It is, however, possible to visualise the real-world geography of IP address space. For example, Shiode and Dodge (1999) have undertaken such an exercise by placing onto a base map the geographic location of the registered owners of each block of IP address space.[15] Using data from Réseaux IP Européens (RIPE)[16] from March 1997, they mapped the geographic location of 44.7 million IP addresses allocated to companies and organisations in the UK. The map is a continuous density terrain map of government bodies, created using a GIS (see Figure 5.3).

Domain names are an increasingly valuable commodity as corporations seek memorable addresses for their websites, and there have been a number of attempts to map and analyse their distribution (e.g., Moss and Townsend 1997a, 1997b, 1998; Zook 1998, forthcoming). This task, however, has two main problems. First, names are allocated by

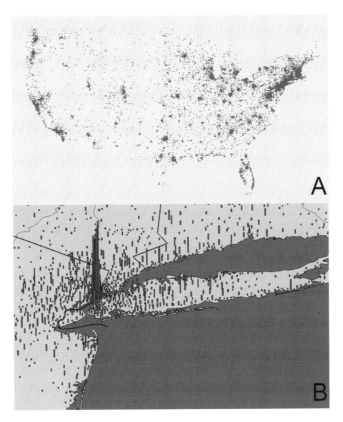

Figure 5.4 Imperative! domain name maps: (a) dot map of the USA, (b) 2.5-dimensional map
of New York City
Source: Imperative!, http://www.internet.org/

many different agencies and, second, it is difficult to pin down the exact size of the global
domain name-space because of its delegated structure and rapid growth (Landers 1997).[17]
Given the restrictions on name length and the rapidity of growth, there has been concern
that name-space may become exhausted as all possible names are used. This is unlikely,
however, as there are many billions of possible names (Holtzman 1997).[18] To date, most
maps have concentrated on plotting the ownership of generic top-level domains in the
US. The geographic location of the owner of these domains can be determined from the
registration database, which has a billing postal address, containing zip codes that can
easily be mapped to street-level locations using GIS software and map data (Longley and
Clarke 1995).

One company based in Pittsburgh, Imperative!, has mapped domain names in the US
as a marketing device to promote its value-added 'Market!IT' domain name database
(Imperative! 1999).[19] This database is built using data harvested on a regular basis from
the InterNIC registrations of generic domain names. Figure 5.4 shows two of its maps. The
first, (a), is a simple dot map showing all US domains, with each one represented by a
single dot, clearly showing the uneven distribution of domain names. The second, (b),
displays a sub-sample of this data for the New York City region, with the height of
columns representing the total number of domain names in each zip code. The tallest

Figure 5.5 Domain name maps for San Francisco: (a) SF and Bay region, (b) SF downtown, (c) SF multimedia district
Source: Matthew Zook, University of California, Berkeley

vertical bars are in Manhattan, which has some of the highest densities of domain names in the world (Moss and Townsend 1997a). In many ways this map of a virtual infrastructure matches the real-world infrastructure of immense skyscrapers in Manhattan.

Similarly, Matthew Zook, in the Department of City and Regional Planning at the University of California, Berkeley, has produced a series of domain name maps, which range in scale from the regional through to street level, of a number of selected cities.[20] Rather than rely on Imperative! data, however, he gathered his own data on the ownership location of .com, .net, .org and .edu domains from InterNIC using a computerised survey that ran for five weeks in summer 1998. Figure 5.5 displays three of Zook's maps of San Francisco. He uses simple dot and proportional symbol maps, with background road and town data to add context. Each symbol represents the number of domain names at a particular zip code. The first map, (a), is at the regional level, showing the Bay area with dense clusters evident in San Francisco, down Silicon Valley, and in the Oakland/Berkeley area. The next map, (b), displays the domain names in San Francisco itself, with the densest concentrations found in the financial district and the 'South of Market' area (famed as the Multimedia Gulch). The final map, (c), displays only a few city blocks around South Park, in the heart of the multimedia district. This mapping led Zook to conclude that the 'Internet industry exhibits a remarkable degree of clustering despite its

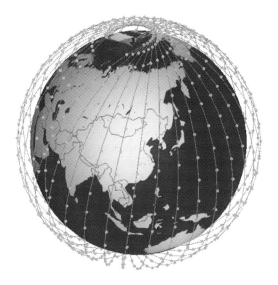

Figure 5.6 Frame from a SaVi animation of the Teledesic satellite constellation
Source: Patrick Worfolk and Robert Thurman

reported spacelessness' (Zook 1998: 18). A conclusion similarly reached by Moss and Townsend (1997a, 1997b, 1998) in their extensive analysis of Imperative!'s data.[21]

While these maps reveal the underlying geography of Internet domains, the data and mapping technique are problematic. With regard to the data, the billing address may or may not be the place where the domain is actually used, and as such does not necessarily indicate the location of computers and other hardware. For example, someone located in London may register a domain name for a website that is physically hosted on a machine in Washington, DC. Also, it is important to realise that not all domains are created equal. For example, the Microsoft.com domain is much bigger than MappingCyberspace.com, with many sub-domains as compared to a single website. Yet, when mapped and analysed, they are generally considered of equal importance. Moreover, like the rest of the Internet, domain name space is dynamic and growing rapidly, as a result, maps are out of date as soon as they are published, although they clearly reveal general trends. In relation to the cartographic technique, a major problem is that the data is not normalised in relation to the density of population or firms residing in a locale. Therefore, it is not possible to determine if clusters merely follow population density or whether a specific cluster exists for a specific reason.

Static maps provide a snapshot of a particular set of infrastructure at a given point in time. One method to gauge the extent and nature of change through time is to compose these snapshots into an animated sequence. One such set of animations have been produced by Robert Thurman and Patrick Worfolk at the Geometry Center at the University of Minnesota. Created using SaVi (satellite visualisation), their animations show the changing positions of constellations of broadband communication satellites across the globe.[22] Figure 5.6 shows a single frame[23] from an animation of the orbital paths of the Teledesic satellite constellations.[24] The animation shows the 'birds' crossing the sky in an orderly procession. Teledesic is a multibillion-dollar low Earth orbit constellation of several hundred satellites, circling at an altitude of 435 miles, designed to provide broadband data transmission for networks including the Internet, and financed in part by Bill Gates (Wood 1999). Satellites and mobile communications are becoming significant elements in the ICT infrastructure, and it is vital that we understand the geography of this infrastructure as they are vulnerable, and difficult and expensive to repair or replace. The complexity of the patterns of orbital positions and surface coverage produced by these satellite constellations means that they cannot be represented effectively on static maps alone.

Interactive maps of infrastructure

The static and animated maps we have discussed so far convey only a limited amount of information in a form that cannot be altered or questioned by the map reader. In contrast, some researchers are exploring more sophisticated representations that allow the map reader to interact with the map and to explore the visualisation further. Interactive maps can take a number of forms. Some allow the map reader to change his/her viewing position in both geographic and temporal space, others to modify the visual presentation of the data by changing the classification or symbology, or through the subsetting of an area of interest by zooming or database selection, and still others allow the user to interactively enquire about individual data objects. In some cases, these features are combined. All of these approaches are pushing the conventional boundaries of map representation into the realm of cartographic visualisation (Kraak and Ormeling 1996). Moreover, they often utilise the power of the World Wide Web to deliver data and interactive representations to a user's desktop where he/she is then free to explore (MacEachren 1998). Crucially, these maps shift some of the power of representation from the map author to the map viewer, allowing the user to explore the data (although this is within the parameters set by the map creator).

Mapnet[25] is an interactive ICT mapping tool that allows Web users to interactively map and examine the Internet backbone infrastructure of over thirty different commercial and education and research networks (Claffy and Huffaker, n.d.). Mapnet was developed by Brad Huffaker at CAIDA[26] and is programmed in Java so that it can be run using a standard browser. Figure 5.7 displays a screen shot of Mapnet in action, mapping the network infrastructure of two important commercial Internet backbones, UUNET/Worldcom and Telstra (based in Australia). Mapnet visualises networks on a flat, terrain style, geographical base map, with the networks represented using the arc-node form. The interface is made up of four major elements. In the centre is a large map window, and above this is a series of control buttons and menus that provide a number of useful interactive functions. At the bottom are two text boxes, the one on the left provides a selection list of networks to map and the one on the right displays the results of interactive enquiries.

The Mapnet application offers a significant degree of control over data representation and enquiry. One can set the line thickness and colour code the arcs by company or by bandwidth (capacity of the link). It is also possible to zoom from the global view to regional levels in order to see detail more clearly. Moreover, using the enquiry mode, it is possible to click on a link of interest and receive details about that link (in terms of the end nodes, capacity and the company that owns it). Here, the user has the flexibility to explore the network infrastructure of his/her choice, mapped onto a common, familiar framework of real-world geography. This is important because it allows a comparison of networks of different companies. Usually it is difficult to compare networks because information is generally only available in marketing maps, which are presented in all manner

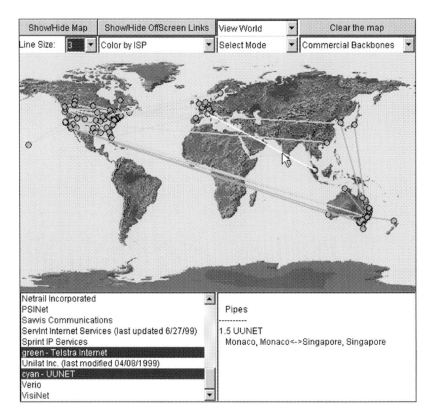

Figure 5.7 Mapnet interactive map of Internet infrastructure
Source: http://www.caida.org/Tools/Mapnet/

of different formats and cartographic styles. However, Mapnet is reliant on commercial networks to provide up-to-date information, something many are reluctant to do.

In contrast to the two-dimensional nature of Mapnet, Figure 5.8[27] displays a three-dimensional, interactive map of the backbone network of CESNET, the education and research network of the Czech Republic.[28] The network links between nodal towns and cities are represented by the vertical and horizontal tubes. The width of the tube indicates the capacity of that link. The central focus of the network is the capital Praha (Prague), which is identified as the centre of the star shape. Unlike the static maps presented earlier, this map is a VRML model,[29] constructed from a data file that describes the content and shape of the map. When displayed using a VRML client program, the data is rendered locally into the configuration which can be viewed from any angle. In this case, the user can rotate, zoom, pan and flip the three-dimensional map by using the control buttons down the side of the browser window. In a sense, this allows the map reader to get 'inside' the map. Constructing the map data as a three-dimensional model means that it can be examined from an almost unlimited number of positions and angles, giving more power to the viewer compared to static or even animated maps. This model is also interactive in another way – by clicking on a link of interest, a line graph of monthly traffic statistics is shown.

Another set of interactive VRML maps, but ones with a greater level of geographical sophistication and method of visualisation, are those produced by Tamara Munzner, K.

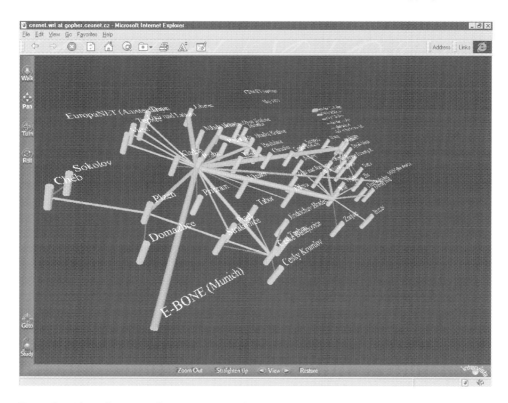

Figure 5.8 Three-dimensional VRML map of the CESNET backbone

Claffy, Eric Hoffman and Bill Fenner, detailing the global topology of a special subset of the Internet known as the MBone (Munzner *et al.* 1996).[30] MBone consists of a special set of routes, known as 'tunnels' in Internet-speak, which run on top of the ordinary Internet network and are used to deliver multicast data. Multicasting is an Internet protocol designed for delivering efficiently a single copy of a portion of data to many different people. It is especially useful for distributing real-time audio material and video, such as live events like concerts or Space Shuttle launches, to a large audience without the need to send individual copies to everyone. The tunnels themselves are created between special routing computers, forming a dedicated Multicast back*bone* network.

The relative ease of creating MBone tunnels through the Internet and the increasing popularity of multicasting live events via the Web, has led to the rapid growth of the MBone network. However, its growth was largely unplanned and no one authority has the responsibility or power to co-ordinate its development. Consequently, the infrastructure of the MBone has grown into a very complex and inefficient topological structure, with many duplicate, redundant tunnels. This is a problem because it causes the unnecessary waste of scarce Internet capacity and undermines the principle of multicasting. Existing tools for trying to understand the structure of MBone infrastructure were of limited use because they produced long text listings of tunnel routes (some seventy-five pages in length in June 1996) from which it was very difficult to determine the topology, and hence the need for MBone maps (Munzner *et al.* 1996).

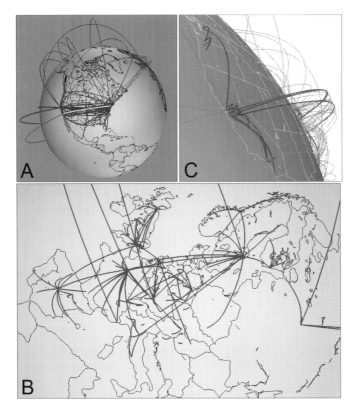

Figure 5.9 Three-dimensional MBone maps: (a) Global View, (b) Europe, (c) USA
Source: Tamara Munzner and IEEE

Figure 5.9 (a) shows a striking view of Munzner and colleagues' MBone visualisations at the global scale, while Figure 5.9 (b and c) presents two regional views of Europe and America.[31] These maps use a method of plotting links between node routers as curving arcs onto a globe displaying country and state boundaries.[32] Line colour and thickness were used to show characteristics of the tunnels, while the height of the arcs above the surface of globe was simply a function of distance – so the longest tunnels were drawn as the highest arcs. This makes sense in the context of understanding the MBone topology because long links that span large parts of the Internet are the most important. The maps displayed in Figure 5.9 are static shots of VRML models which can be viewed interactively, allowing users to explore fully the topology. Clicking on an arc reveals details of that tunnel. Whilst the maps are effective, one of the problems in constructing them was determining the geographic location of the MBone routers. This proved to be a time-consuming process, involving considerable amounts of manual effort. This lack of data, and a reliable, automatic means of determining and locating Internet infrastructure, is the key impediment to mapping ICTs. The visualisation approach of constructing three-dimensional VRML models composed of arcs across a globe has also been used to map other elements of Internet infrastructure including the topology of global cache networks (Huffaker *et al.* 1998).

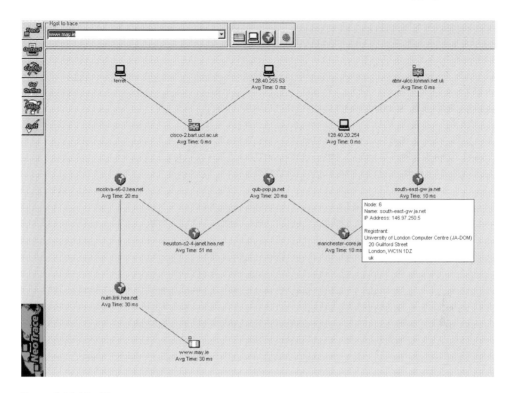

Figure 5.10 NeoTrace traceroute

Dynamic maps of infrastructure

Given the rapidly evolving nature of the Internet, maps employing static data (all those discussed above) quickly become dated. To counteract this limitation, research is being undertaken to create dynamic maps that are automatically updated in real-time. One way this can be achieved is through the mapping of traceroutes. Traceroute[33] is an Internet utility which reports the route that data packets travel through the Internet to reach a given destination, and the length of time taken to travel between all the nodes along the route. Although a trace between two locations will potentially vary each time the utility is run (due to new network connections, traffic conditions and failures), this data allows researchers to probe the real-time structure, complexity and performance of the Internet (Rickard 1996; Thoen 1997). Figure 5.10 displays the route taken between University College London (UCL) and Maynooth, Ireland, showing the intervening nodes and the average round-trip time of data packets using NeoTrace.[34] Each node can be questioned for more information.

One of the significant limits of conventional traceroute utilities like NeoTrace for dynamically exploring the Internet's infrastructure is that the origin of the trace is fixed to one location, namely the computer on which the trace is run. This limitation of fixed origin point has been partially overcome by the development of Web-based traceroutes. These enable people to use any suitably configured Web server they are connected to as the origin point of their trace. There are several hundred publicly available Web traceroute

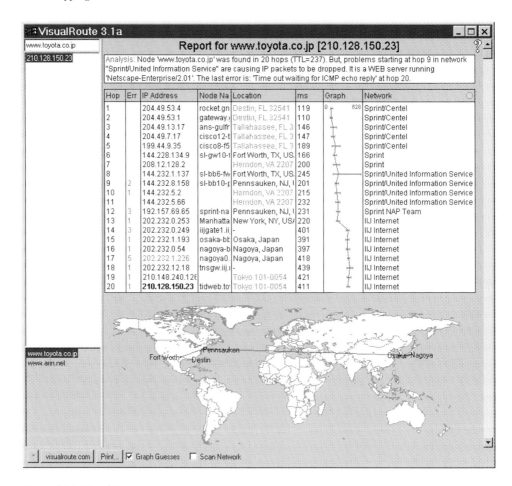

Figure 5.11 VisualRoute traceroute

sites.[35] These too can be mapped both logically (as with NeoTrace) or by mapping the locations of the nodes in the trace onto geographic space. There are several commercial utilities being developed that map ICT usage using geographical traceroutes.

VisualRoute[36] developed by Jerry Jongerius at Datametrics Systems Corporation, presents traceroute results both in table form and plotted onto a world map (see Figure 5.11). In this case, the route from Florida to Toyota in Japan is plotted, showing that the data packets took 20 hops to reach their destination. Each row in the table represents one hop. The columns in the table provide much useful information such as the domain name of the machine at each hop, its approximate geographic location, and the company name of the networks that are being traversed (in this instance, Sprint and IIJ Internet). The round-trip time for each hop is shown as a numeric value and also on a graph. A limited amount of user interaction with the map is possible through panning and zooming. VisualRoute is limited by its geographic database of node locations, although it allows users to add to this database and contribute their own, more detailed, base maps. A new version of the utility, VisualRoute Server, can be used over the Web.[37]

GeoBoy,[38] another geographic traceroute application from NDG Software,[39] maps a trace using a node-arc display as either a flat two-dimensional world map or, more interest-

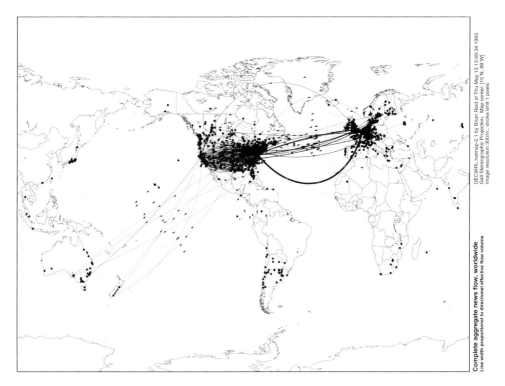

Figure 5.12 Usenet traffic flows
Source: Brian Reid, http://reid.org/brian/

ingly, a three-dimensional global view. The trace results are also presented in the familiar tabular form below the map. Plate 1C displays a screen shot of the three-dimensional global map interface, with a trace being plotted from the city of Perth in Australia across the Pacific Ocean to California and then to the destination, www.microsoft.com, near Seattle. The user is able to interact with the map using panning and zooming functions, and in the three-dimensional mode it is possible to rotate the globe along its longitudinal axis. As with VisualRoute, the application determines the geographical location of each hop using a static database, with its inherent problems of scale and currency in the fast changing world of the global Internet.

Maps of Internet traffic

Like the maps of infrastructure, Internet traffic and activity has been mapped in a number of different ways. Rather than detail the logic of applying different modes of mapping (e.g., static or dynamic), in the following sections we merely provide overviews of different map techniques and detail their utility.

Static and animated maps of Internet traffic and activity

Figure 5.12 is one of the earliest examples of a static map of Internet traffic flows still publicly available. Produced by Brian Reid, then a researcher at Digital Equipment Corporation (DEC) Western Research Lab in Palo Alto in Silicon Valley, it displays news traffic flow

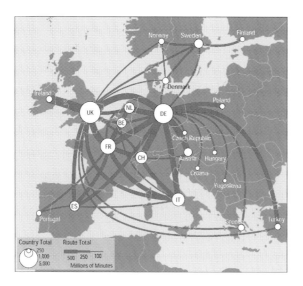

Figure 5.13 European traffic flows, 1997
Source: TeleGeography, Inc., http://www.telegeography.com/

on the Usenet network in 1993.[40] Reid has measured the number of hosts in the Usenet network and the amount of traffic (number and volume of posted articles) flowing between them, along with estimates of the number of readers, from 1986 to the early 1990s. As part of this project, a mapping tool, Netmap, was developed to visualise the geography of the Usenet network infrastructure and traffic flows (Reid 1988).[41] The maps produced by Netmap are somewhat rudimentary in style, consisting of a simple arc-node display, where the width of the arcs are proportional to the volume of traffic flowing over the links between backbone nodes. This mode of representation has led to severe over-plotting in the US and European regions. In an attempt to overcome this problem, some researchers have eliminated the geographical framework in favour of a wholly topological one (see Chapter 6).

Figure 5.13 further develops the cartographic style employed in Reid's Usenet maps. This map displays European telecommunications traffic flows using public telephone networks, measured in millions of minutes, in 1997, and was produced by TeleGeography Inc. (Staple 1997).[42] The nodes and the links are used to represent two aspects of the data. The nodes, positioned on the approximate geographic location of the capital city of each country, are drawn as circles with their diameter proportional to the total volume of outgoing traffic for that country (as labelled). The traffic flows between country nodes are represented by lines sized in proportion to the total annual traffic between a pair of nodes. Only those links over a minimum threshold are shown to limit the problems of data volume and over-plotting.

Plate 1D, one of the most widely reproduced visualisations, is a single frame from a visually powerful animation of Internet traffic flow on NSFNET for September 1991.[43] It was produced in 1992 by Donna Cox and Robert Patterson, visualisation researchers at the National Center for Supercomputing Applications (NCSA), University of Illinois at Urbana-Champaign. In it, a visual metaphor of a virtual network floating above the US is used to represent the NSFNET backbone, the core of the Internet in the early 1990s, with the source of the inbound data and its connecting node shown by the coloured vertical lines. The lines are colour-coded to indicate the volume of traffic being carried, ranging from low (purple) to high (white).

Interactive maps of internet activity

Plates 2A, 2B and 2C are examples of the work of the visualization research group, led by Stephen G. Eick, at Bell Laboratories-Lucent Technologies,[44] in Naperville, Illinois. Since the early 1990s, this team have been pioneers in the mapping of communication network traffic onto geographic space using a number of two-dimensional and three-dimensional spatial metaphors (Becker *et al.* 1991). They have developed customised visualisation applications called SeeNet and SeeNet3D to produce their interactive maps (Becker *et al.* 1995; Cox and Eick 1995; Cox *et al.* 1996; Eick 1996). The SeeNet visualisation toolbox produces link-nodes of network traffic mapped onto a two-dimensional geographic framework (see Plate 2A). The nodes are represented by proportional symbols (such as rectangles) drawn on the map at their geographic location. The size, shape and colour of the node symbols can be used to represent aggregate statistics for that node. The links are mapped as straight lines between nodes. The lines can be colour-coded and scaled to convey data values, such as capacity, flow or utilisation for the link. Importantly, SeeNet provides a range of interactive tools that allow the end-user to manipulate directly the visual presentation of the data (Becker *et al.* 1995). This allows for intuitive and powerful exploration of the data to reveal potentially interesting and unforeseen patterns and structures. In addition, it is possible to use animation to visualise the changing patterns of network data over time.

There are severe limits to the size and complexity of the networks that can be visualised with the two-dimensional link-node maps of SeeNet due to visual clutter. To overcome the inherent information clutter caused by two-dimensional map displays, Eick and his colleagues have explored three-dimensional spatial metaphors to map networks in their SeeNet3D application. SeeNet3D implements a number of novel three-dimensional network visualisation metaphors using a geographic framework (Cox and Eick 1995; Cox *et al.* 1996). For example, it is possible to produce a three-dimensional arc-node map as displayed in Plate 2C which portrays Internet traffic flows between fifty countries, as measured by the NSFNET backbone, in the first week in February 1993 (Becker *et al.* 1995). The colour, thickness and height of the arcs are used to convey the traffic statistics for that link. As with SeeNet, the user has considerable interactive control of the arc map in SeeNet3D, for example the opacity of the arcs can be varied, and the arcs in Plate 2C are partially translucent so as not to completely obscure lines at the back of the map. The map can also be rotated and scaled, so that the user can view it from any angle. Plate 2B shows the same data as displayed in Plate 2C, but it is mapped onto a globe view, in a manner similar to the MBone maps. The user is able to vary the degree of opacity of the globe to enable arcs traversing behind the sphere to become visible and to rotate the globe to view the data from any position. The SeeNet3D system can also produce animated sequences showing the changes in traffic over time.[45]

Dynamic maps of Internet activity

Plates 2D and 2E display a moment in the dynamic geographic visualisation of Internet traffic, and were created by Stephen E. Lamm, Daniel A. Reed and Will H. Scullin, in the Department of Computer Science at the University of Illinois (Lamm *et al.* 1996). These

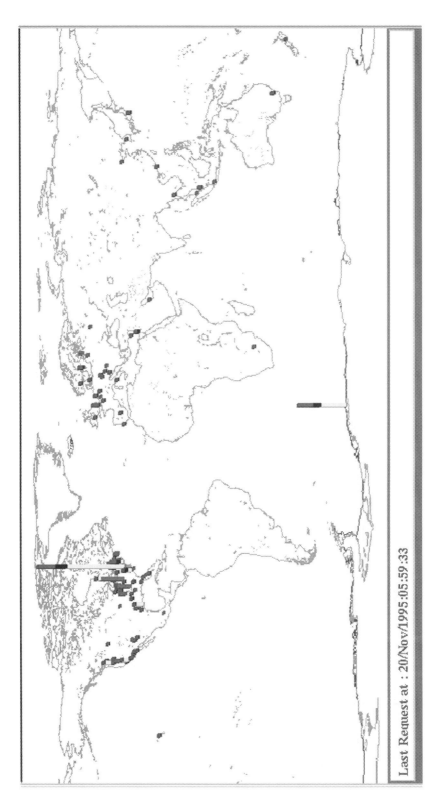

Last Request at : 20/Nov/1995:05:59:33

Figure 5.14 Palantir Web traffic visualisation

maps are one outcome of a larger project to monitor and analyse, in real time, the patterns of traffic on Web servers to help design better performing systems (see Scullin *et al.* 1995). The visualisations utilise what might be termed a 'skyscraper' metaphor mapped onto a globe to reveal Web traffic and type, originating at different locations in real time. As such, the height of the skyscrapers grow and shrink as the traffic flows vary. The skyscrapers are situated at the geographic location of traffic origin and are divided into segments, with each segment representing different types of Web traffic (text, images, video, data, etc.) that make up the total volume. Plates 2D and 2E show Web traffic in North America, although Europe can clearly be seen on the horizon. The data are Web requests to the National Center for Supercomputing Applications (NCSA) Web server in 1995. This server was the home of the Mosaic Web browser, at that time one of the most popular sites on the Web, on average receiving over 400,000 'hits' (requests for data) a day (Lamm *et al.* 1996).[46] As hits were received the display was updated, and the system could cope with a peak traffic load of thirty to fifty hits per second.

To enable the visualisation of traffic on the globe, Lamm and his team developed a rudimentary method for determining the geographic location of the origin of Web traffic. Their utility 'IP to Latitude/Longitude'[47] only produces approximate locations, with traffic originating outside the US simply aggregated to the level of country, with the capital city of a country assigned as the origin location. Within the US much greater spatial discrimination is achieved, with traffic being mapped to the city of origin.

It is important to note that these global 'skyscraper' maps were designed to be accessed in a sophisticated VR system that allowed the user to immerse him/herself in the visualisation. This required expensive virtual reality hardware including head-mounted displays and even a CAVE[48] to interact with the globe. In these VR-supported environments, the user's head movements are continuously tracked and the view constantly updated in response to where the user is looking. Moreover, it is possible to interact with the map by reaching out and 'touching' one of the skyscraper bars to initiate an enquiry and see full details of traffic flows. Control panels and a map key are also displayed (as shown in Plate 2D). The user has considerable power to change the data classification and the scale and rotation of the globe.

There are many commercial software packages that can perform statistical analyses on the log file of website visitors.[49] Some packages attempt limited geographical discrimination, breaking down visitor numbers by domain, but generally this is not the case. None use maps, instead they employ conventional statistical charts. An exception is a Web traffic analysis application called Palantir[50] (see Figure 5.14). Developed by Nekarios Papadakakis and his colleagues in the Computer Science Department, University of Crete, in Greece (Papadakakis *et al.* 1998). It maps the geographic patterns of Web traffic in real time and uses the same skyscraper metaphor as Lamm's maps, but unlike Lamm's it is written in Java and can be accessed over the Internet.[51] In addition, it also employs concentric circles as an alternative mode of visualisation. The traffic can be historical (mapping the data recorded in logs) or occurring in real time. It is also possible to filter the traffic, so that one can analyse a particular subset of interest, such as the visitors from .com domains or for a certain time period, or combine traffic from several sites into a single map. This is useful for a large organisation, such as a university, that may want to determine overall usage across a distributed set of servers. Like Lamm's maps, the geographical accuracy of the maps is best described as approximate. In most cases traffic is mapped to the nearest city, and in some instances allocated to the capital city of a country. Figure 5.14 displays the geography of Web traffic of a typical website.

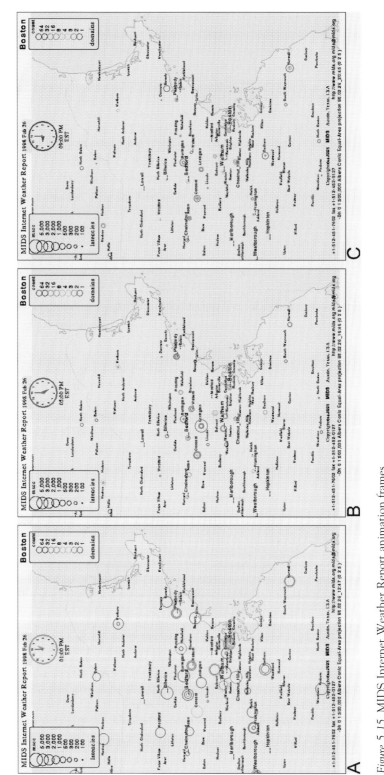

Figure 5.15 MIDS Internet Weather Report animation frames
Source: MIDS, http://www.mids.org/

Figure 5.15 (a, b and c) displays part of an 'Internet Weather Report' (IWR),[52] created by MIDS. Rather than showing traffic volumes, it displays the level of congestion and delays across the Internet. Forecasts are made six times a day, every day of the year, for over 4,000 Internet sample points around the world (Quarterman *et al.* 1994, 1995; Quarterman 1997). This measurement consists of 'pinging' (sending a tiny packet of data) sample computers from MIDS headquarters in Austin, Texas, and measuring the time the ping data takes to travel there and back. This round-trip time, measured in milli-seconds, gives an estimate of the latency for that sample point on the Internet. These ping measurements are turned into a map with proportional circle symbols to represent latency (the larger the circle the longer the delay). The colour of the circle represents the number of Internet hosts at each location that are being surveyed by the IWR. The individual maps are used to create daily animations of traffic congestion on the Internet.

Figure 5.15 shows three frames from the an IWR animation for 26 February 1998, focusing on the Boston area in the US. In basic terms, small circles on the map show a healthy Internet, while large circles are indicative of poor performance and possible prob-lems. The first map, (a), shows Internet conditions at 1 p.m. Eastern Standard Time (EST) and many large circles are apparent. This is not too much of a surprise as this is the peak time, when a maximum number of users are likely to be logged on. It is lunchtime on the Eastern seaboard of the US, mid-morning in California and early evening in Europe. The changing time and the number of people awake and logged on to the Internet obviously alters the traffic load that the networks experience (see below). The next map frame, (b), is four hours later, at 5 p.m. EST and many circles have shrunk back to an acceptable size, meaning latencies have fallen at the end of the work day on the East coast. The final, map, (c), is based on measurements taken at 9 p.m. EST, when traffic demand and therefore congestion is low. MIDS can produce these maps from the global to the local scale. Interpretation of IWR animations can be difficult, although it is claimed that you can see patterns of Internet latency ('storms' of congestion). A major limitation with IWR is that all the measurements are taken from a single origin point at their headquarters in Texas, which means that each map can never give a representative view of the whole of the Internet's weather.

Plate 1E displays one moment in the output of Web Hopper.[53] This application, de-veloped by Koichiro Eto,[54] visualises in real time the trails of Web surfing, plotting these as straight lines from origin point to destination. The Web trails are drawn in bright red, but gradually fade over time. In order to plot the data flows in real time, sample data packets of Web data traffic passing through a checkpoint are extracted and their destina-tion Internet address noted. This address is then translated into an approximate geogra-phical location using 'IP to latitude/longitude' (see p. 101). As the application is run, more and more Web trails are drawn, overlaying existing lines and building a multilayered map. After a space of twenty minutes or so, dense pathways are apparent from many over-plotted lines, showing the key patterns of Web traffic, just as footprints of people walking around a park combine to form clear paths of the most travelled routes. Whilst an effect-ive form of visualisation, because the traffic flow between the start and end point is drawn as a straight line connecting the two, the picture revealed is a considerable abstraction from the reality of the route taken as traffic 'hops' across the network from node to node. As such, Web Hopper only reveals information about the flow of traffic to destinations rather than the routes that data takes (as with some other applications we have discussed). Web Hopper also suffers in that it only samples traffic from a single point in the global Internet – a Japanese backbone network called WIDE.

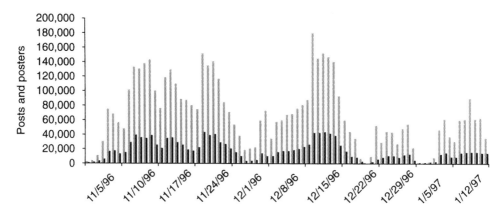

Figure 5.16 Daily rates of messages and participants in Usenet, 1 November 1996 to 12 January 1997
Source: Marc Smith

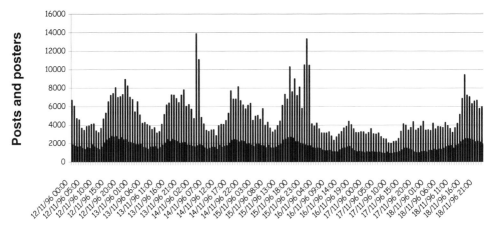

Figure 5.17 Hourly rates of messages and participants in Usenet, 12 November 1996 to 18 November 1996
Source: Marc Smith

Temporal aspects of ICTs

In this final section we consider the temporality of ICTs. As the 'Internet Weather Reports' demonstrate, there is a temporal aspect to the use of the Internet, both in terms of daily fluctuations in Web usage, but also in relation to delays in the transfer of traffic. Understanding this temporal dimension is important for two reasons. First, it helps network operators to plan for peak surges in the system use, and second, it reveals details about how time feeds into the social interactions that take place in cyberspace. The latter point is one that motivated Smith (1999) to produce charts of the daily and weekly cycles of posts and posters of Usenet (see Chapter 7). Figures 5.16 and 5.17 clearly demonstrate the temporality of posts, with interactions reaching peaks at certain times. As noted in the 'weather maps' these peaks coincide with the overlapping of key times in geographic space – for example, the coinciding of early morning on the East Coast of the US with midafternoon in Europe, or the coinciding of early morning on the West Coast of the US with early afternoon in the East Coast. Smith (1999) reports that these patterns form a consistent cyclic rhythm.

TeleGeography has also produced an interesting map of the global time geography of cyberspace (see Plate 3A), what it terms the 'circadian geography of the network' (Staple 1998: 12). This innovative and unconventional looking map shows the number of people who could potentially be networked in cyberspace during office hours (9 a.m. to 5 p.m.) in terms of telephones, cellular phones and Internet connections, throughout the twenty-four hours of a day. The map uses the visual style of a radar graph from statistical graphics, with three different colour-coded polygons representing the different communications technologies. The size of the polygons varies through the hours of the day, with the largest percentage of people online being shown when the polygon vertices stretch furthest from the centre. Time shown around the edge of the map is calibrated to Greenwich Mean Time, with three office days (in Shanghai, Paris and New York) highlighted by the circular strips surrounding the map. The area enclosed by the red line represents the percentage of Internet hosts live during office hours. This peaks in the afternoon, London time, as the Europeans are still in their offices and the North Americans are starting work. The green- and orange-lined polygons represent fixed and cellular telephone connections which peak at two different periods. First, in the morning, London time, when Europe goes to work and when Asia is still in the office. The next peak is in the afternoon when the North Americans and Europeans are together in network time–space. Interestingly, in the first European morning peak, the percentage of Internet hosts is relatively low (around 30 per cent) compared to the number of people connected by phone; this reflects the US dominance of the Internet. The world's population is shown by the blue polygon and its temporal pattern makes for an interesting contrast, in that for part of the day it is outside the other three polygons. This means that many more people could be online (i.e., they are awake during their office hours of 9 to 5) but are not connected to the global communications networks by the telephone or Internet. This is most apparent in early morning London time (3 to 8 a.m.), when Asia is awake, but the highly networked nations in Europe and North America are asleep. At the peak of the networked population in the afternoon, London time, the blue line is close to the centre of map showing that this represents less than half the world's population. This demonstrates graphically that cyberspace is accessible to the privileged minority of the world's population, who are located predominantly in the wealthy Western nations (ITU 1997, 1998). TeleGeography concludes that, 'Each day, while the demand for connections moves clockwise from east to west with the sun, available network capacity appears to move counterclockwise as network resources are idled during the night' (Staple 1998: 12).

The geography of personal communications using telecommunications and CMC technologies has been under-researched by academic geographers over the past twenty years (Hillis 1998). A resurgence of interest is being led, in part, by Paul Adams, a geographer at Texas A&M University, who is undertaking particularly innovative work examining personal social interactions at the micro-scale geography of human lives (Adams 1995, 1996, 1998, 1999). He examines how transportation and communications technologies can 'extend' the scope of the human body to reach out and interact across geographical space. Adams has visualised individual human extensibility using three-dimensional time-space models created with computer-aided design (CAD) software more usually associated with product design and architecture (Adams 1999). One of these models, showing one day in the lives of five interconnected people, is displayed in Figure 5.18. It shows the

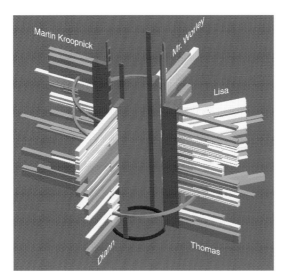

Figure 5.18 Three-dimensional CAD model of human extensibility
Source: Paul Adams

individual and social impact of space–time convergence through advances in transportation, communications and computer technology, enabling people to overcome the friction of distance and make connections with people at many scales.

For each person, a separate three-dimensional model of his/her daily routine was built in the CAD package, resembling a bar chart. The vertical axis represents time through the day and horizontal bars project out along the x-axis for different communications activities (such as making a phone call, watching television, talking face to face, or sending an email). The horizontal length of the bar from the axis shows the geographic distance of the activity, ranging from proximate face-to-face conversations to an international phone call. Thus the x-y dimensions of the bars represent the scale of the activity in time-space, with the length of the bar being distance and the width of the bar being temporal. These individual time–space activity 'bar charts' are combined into one model in Figure 5.18, by arranging them evenly on the circumference of a circle. This enables one to compare the shape and structure of the activities of the different people's daily lives as well showing the communications links between them. So the curving between the different individuals represents interactions between them. For example, at the front of the model, a link is drawn between Diann and Thomas which represents a face-to-face meeting between the two. The models are stored in a CAD system which allows them to be interactively analysed and displayed from any chosen viewing position.

The data for these models are gathered through detailed time diaries and interviews, recording the daily activities and social interactions (face-to-face meetings, phone calls, letters, watching television or listening to the radio, email communication) of a small group of connected people who live in the Albany, New York metropolitan area. Adams argues that the three-dimensional models 'reveal the existence of a kind of "commuting" between physical and virtual places, an oscillation that occurs much faster than the older form of home work commuting: every time one picks up a phone receiver, opens a book, or turns on the radio' (Adams 1999).

6 Spatialising cyberspace

> information visualization focuses on information, which is often abstract. In many cases information is not automatically mapped to the physical world (e.g. geographical space). This fundamental difference means that many interesting classes of information have no natural and obvious physical representation. A key research problem is to discover new visual metaphors for representing information and to understand what analytical tasks they support.
>
> (Gershon *et al.* 1998: 10)

In contrast to the mapping of the materiality of ICTs, in this chapter we discuss attempts to create spatialisations of cyberspace. Here, data with no inherent spatial properties is mapped onto a defined spatial framework so that it might be better understood. These spatialisations employ a number of graphical techniques and visual metaphors ranging from two-dimensional static representations to immersive three-dimensional landscapes. The best of these spatialisations are providing striking and powerful images that give people a unique sense of the space, arguably in a manner similar to the Apollo images, which gave people a new understanding of the Earth (Cosgrove 1994). In general, these spatialisations have been created by academic researchers from the disciplines of computer graphics, information design, human–computer interaction, virtual reality, information retrieval and scientific visualisation (see McCormick *et al.* 1987; Laurel 1990; Tufte 1990; Holtzman 1997; Johnson 1997; Shneiderman 1997; Wurman 1997; Jacobson 1999), although a number of companies have also experimented in the development of commercial spatialisation products (e.g., Visual Insights, Perspecta, Inxight Software and Cartia).[1] A consequence of this research has been the formation and growth of a new, distinct research field, *information visualisation*, that has particular focus on developing and improving the interface between user and the information spaces of the Internet (see Gershon and Eick 1995a and b; Card *et al.* 1999; Chen 1999).

We divide the discussion in this chapter into two distinct sections. In the first section, we detail spatialisations of the infrastructure of cyberspace, focusing on representations of computer networks. These are essentially topological maps of ICTs, where the infrastructure has been mapped into a form that does not correspond directly with geographic space. In the second, more substantial section, we detail spatialisations of cyberspace itself and in particular look at the creation of information spaces, that in many cases become the means by which the data is navigated. In order to keep our discussion manageable, we have been selective in our choice of examples, detailing a range of different kinds of spatialisation, once again utilising our distinction between static, animated, interactive and dynamic representations. Many of the examples we detail are experimental in nature, representing work in progress. As such, they provide a fragmentary, imperfect view of cyberspace, but they are none the less worth examining due to the fact that they open up new vistas of understanding.

Topological maps of networks and the Internet

In this section we discuss topological maps of the networks that compose the Internet. Whereas the infrastructure maps discussed in the last chapter mapped the networks onto geographic space, topological maps are more abstract in nature. Here, the absolute, geographical location of the infrastructure is not important, as the spatialisations are designed to reveal other kinds of information using a system of relative location. For example, the spatialisations might reveal connectivity and routing that may be indiscernible on a cluttered geographic representation. As with other maps of ICTs, these spatialisations take on a number of forms and are used by different groups of people; some spatialisations are technical blueprints used by those who manage the networks, others are aimed at the users of the networks, often employed as a marketing device. The spatialisations most often use graph-type representations as their basic visual metaphor, with a number of different layout styles.

Some of the earliest spatialisations date back to the pioneering development of computer networking. In addition to the series of maps of the ARPANET displayed in Figure 1.2 (see Chapter 1, p. 9), there were also a number of topological maps produced. These maps were drawn by the scientists and engineers who designed and built the networks. Some of the best surviving examples are the initial ARPANET network topology maps. Indeed, some fascinating sketches of preliminary designs for network topologies drawn by Larry Roberts[2] survive (reproduced in Hafner and Lyons 1996). Figure 6.1 displays an early sketch map of the nascent ARPANET,[3] with its first four operational nodes as of the end of 1969. The nodes of the network are represented by circular symbols which are numbered by the order they were installed and labelled with the site name. The square boxes

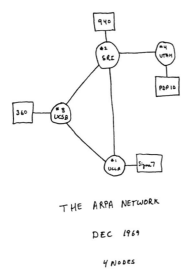

Figure 6.1 Topological map of ARPANET in 1969
Source: Salus 1995: 56

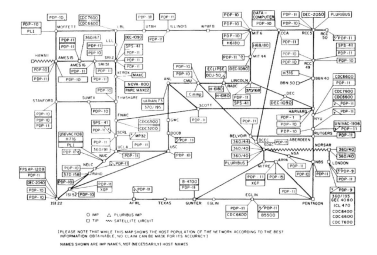

Figure 6.2 Topological map of ARPANET in 1977
Source: Heart *et al.* 1978

on the map represent the actual computers connected to the network, and they are labelled with the model name (PDP10 made by DEC and the 360 from IBM). Figure 6.2 presents a later version, dating from March 1977, when ARPANET had clearly grown considerably.[4]

These kind of maps are still produced today, but whereas Figure 6.2 represents the whole of ARPANET at that time, Figure 6.3, with similar levels of detail, only displays the University of Buffalo's[5] campus backbone in early 1999. This is a large internal network connecting many thousands of students and academics to the Internet. The six large squares represent the core nodes, connected together by high bandwidth fibre optic links. Fanning out around core nodes are numerous sub-networks for different buildings and departments across the university. Unlike the ARPANET map, individual end-node computers are not displayed.

It is also possible, however, to find topological maps at the national scale. For example, Figure 6.4 displays the Internet Initiative Japan (IIJ) backbone, one of the largest Internet networks in Japan.[6] The map uses a highly stylised graph to show a very generalised view of their network at the level of cities and the link capacities between them. Link capacity is represented by the varying thickness of lines, with the fattest ones being 155 Mbps. The map also gives prominence to external links from Tokyo to four major Internet hubs in the US (two in Silicon Valley and two in the New York region). Also shown is the connection to AIH, which is the Asia regional backbone (operated by Asia Internet Holding),[7] allowing interconnection with other nations in the region.

In some cases, researchers have attempted to provide a broader view of the infrastructure of the Internet. For example, Figure 6.5 displays a spatialisation of the MBone topology (see Chapter 5 for more details on MBone). The spatialisation takes on an almost astronomical character, with the MBone nodes and connections floating free in an abstract space, like stars in distant galactic clusters. The map was created by Elan Amir[8] while a graduate student in the Computer Science division at the University of California Berkeley. The

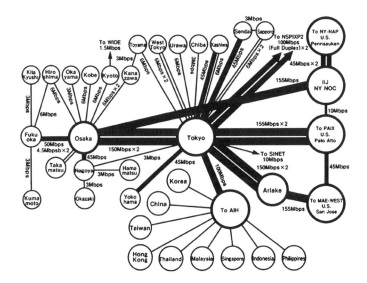

Figure 6.3 University of Buffalo, campus backbone
Source: Computing and Information Technology, University of Buffalo

Figure 6.4 Topological map of Internet Initiative Japan backbone, October 1999
Source: Internet Initiative Japan, Inc., http://www.iij.ad.jp/index-e.html

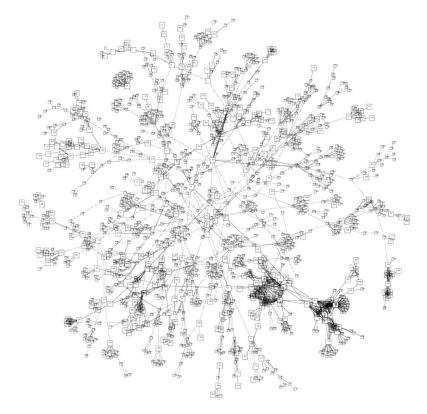

Figure 6.5 Topological map of the MBone, 1996
Source: Elan Amir

map shows some 1,377 routers on the MBone in August 1996 as an unruly graph, with dense clusters and many scattered outlying limbs. It was created using a network drawing tool developed by Amir called Carta (Amir 1993). MBone routers are represented by boxes labelled with an ID number.[9] Whilst having little practical use for navigation, the spatialisation is of use because it presents a holistic view of the MBone, providing a sense of the shape and interweaving structure of that part of cyberspace.

In a similar visual vein to Amir's MBone map, but on a much larger scale, are the richly detailed graphs of the Internet produced by Bill Cheswick and Hal Burch.[10] Rather than mapping a mere 1,000 nodes, their spatialisation seeks to visualise nearer 100,000 nodes. Their Internet Mapping Project, at Bell Labs-Lucent Technologies, uses the Internet itself to measure daily the routes to a large portion of end-points (usually Web servers) from a sample point in New Jersey in the United States (Burch and Cheswick 1999). The resulting spatialisations reveal how the hundreds of networks and many thousands of nodes connect together to form the core of the Internet.[11] Plate 4 displays an example of one spatialisation of the Internet in June 1999, representing over 88,000 nodes. The graph layout is an abstract, non-geographic space and it takes about 20 hours to generate a finished map on a typical Pentium PC. The layout algorithm uses simple rules, with forces of attraction and repulsion jostling the nodes into a stable, legible configuration. The end result is a static spatialisation, but there are many permutations in the algorithm to generate different layouts and colour codings of the links according to different criteria (such as network ownership, distance from the measurement point, status, traffic). In the example shown, links have been colour-coded according to the IP address, seeking to highlight communities of nodes that share common network addresses. One striking feature of this spatialisation is the large, dark blue cluster which represents a hub of Cable and Wireless (formerly MCI), an important node through which many networks connect; Cheswick describes this as 'the magnetic north of the Internet' (Weinberger 1998: 216).

Cheswick and Burch's measurement and mapping of Internet infrastructure is an ongoing project. The data is archived and is available for use by other researchers. Over time it is hoped that the data will be useful for monitoring growth and changes in the structure of the Internet. They are also working on animated time sequences and investigating ways to create a three-dimensional spatialisation. One limit to their current project is that all measurements are made from a single sample point. The Caida research group are running a similar Internet measurement project, called Skitter,[12] but with multiple sample points. The Skitter project involves a range of data spatialisations, including the graph layout techniques developed by Cheswick and Burch (Claffy *et al.* 1999).

A significant limit to the usefulness of the spatialisations that we have examined so far is that they are single, static snap shots of the Internet. A more useful approach is to provide users with interactive tools and the ability to dynamically generate data, so that they can explore and visualise the infrastructure for themselves. An attempt to provide such an approach is Plankton, a network mapping tool developed by Bradley Huffaker and Jaeyeon Jung at Caida (Huffaker *et al.* 1998).[13] Plankton provides a tool to create spatialisations of an important element of Internet infrastructure, the topology of international Web caches. Plate 3B displays a screen shot of Plankton in action, with each small square representing a cache node and the lines connected together to form hierarchies. The nodes and lines are coloured according to traffic demand with the darkest red being the most busy. Data on cache topology is gathered daily and the user is able to select which day he/she wants to examine. Plankton provides a number of interactive functions including graph layout (geographic or abstract), symbology (node size, link thickness, colour coding by traffic or domain), rotating, panning, zooming and the creation of time-series animation; in addition, individual nodes can be interrogated.

Our final example in this section, Plate 3C, uses animation and three-dimensional graphics to spatialise data on Internet infrastructure. The spatialisation technique envisions part of the Internet as a huge spinning cyber-globe using a metaphor that looks like a cross between the Death Star[14] and an eyeball, and was developed by Hans-Werner Braun of the Measurement and Operations Analysis Team (MOAT), at the National Laboratory for Applied Network Research in the US.[15] Plate 3C shows one frame from an animation sequence. The translucent blue sphere represents the Internet 'universe'. It is a high-level view of the Internet, with clusters of networks represented by small red balls embedded in the large sphere. Each red network ball, defined by a single, globally unique identifier known as an Autonomous Systems Number (ASN), contains hundreds or thousands of nodes and links. Braun's 'Death Star' movie presents how networks, defined by their ASN, connect together for one example route through the Internet. In this case, the starting point is at the University of Oregon and is shown as the green ball on the left-hand side of the image. The target network is that of Sun Microsystems represented by another green ball on the opposite side of the sphere. The Oregon network routes data to the Internet through nine different networks, which are shown by the purple balls that form a circular opening into the surface of the Internet sphere. Inside the sphere are various routes between different networks to reach the destination. The routing pathways between networks inside of the sphere are shown as thick yellow pipes. In essence, this spatialisation is an arc-node graph mapped into a three-dimensional space utilising a metaphor straight out of science-fiction imagination (see Chapter 10).

Mapping the information spaces of the Internet

In contrast to infrastructure topology spatialisations, the creation and mapping of information spaces involves the application of a wide array of spatial metaphors to data with no geographic referent. There are many different forms of information space and they have different structures, use different communications protocols and offer users different modes of navigation and interaction within the space. In Chapters 7 and 8, we examine the particular structures and mappings of information spaces used principally for communications between people, such as asynchronous email and Usenet and synchronous chat and virtual worlds. Consequently, here we only consider virtual spaces where users search and interact with inanimate information resources, such as webpages, documents, FTP archives, images, database and archives. In other words, we focus on the mapping of what has been termed the 'infoverse' (Card *et al.* 1999). This 'infoverse' is composed of different types of information exchange, and has varying degrees of synchronicity and levels of user interaction, thus requiring differing forms of spatialisation to appropriately model its nature.

Figure 6.6 shows a sketch map produced by John December, a leading Internet consultant and writer, showing the principal information spaces and some of the connections between them, as of the end of 1994 (December 1995). The spatialisation provides a good way of conceptualising the different information spaces of the Internet, not as distinct and self-contained domains, but as having fluid, complex boundaries and many interconnections and overlaps. Since its creation, the nature of the Internet has changed markedly, with certain information spaces all but disappearing as they fall out of favour with users

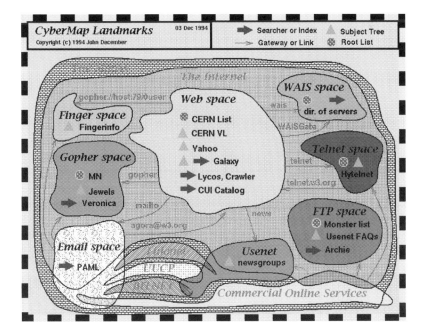

Figure 6.6 Conceptual map of cyberspace, 1994
Source: John December, http://www.december.com/

(e.g., WAIS and Gopher). For many end-users the Web is now the key information space of the Internet, although email is still the most widely used service (Clement 1998; and see Chapter 7). Important information spaces that have evolved and grown since 1994 include multi-user chat environments, virtual worlds and instant messaging. In addition, there are many large private networks and Intranets creating significant information spaces that are largely unseen and difficult to quantify.

Most spatialisations of the 'infoverse' are designed to improve navigation through an information space (to increase spatial legibility, see Chapter 9) and enable people to find the data they are searching for more easily. The problems of finding relevant information in a timely fashion using the conventional approaches of keyword searching and Web brows-ing are proving to be real impediments to gaining the maximum benefits from the informa-tion resources in cyberspace (Bowman *et al.* 1994; Brake 1997; Lawrence and Giles 1998, 1999). It is important to distinguish between *browsing* and *searching* for information in a large information space like the Web. They are very different activities which require dif-fering support tools. Browsing is largely an explorative activity, usually undertaken with no set plan or specific goals, with useful results dependent on serendipity (Marchionini 1997). At present, the Web supports two major forms of browsing – link-following and directories.

Browsing by link-following uses the utility at the core of Web space, namely hyperlinks. However, browsing using hyperlinks can often be frustrating and unproductive, as it is all too easy to get lost in the complex topologies of links. This situation has been termed 'lost in hyperspace' (Edwards and Hardman 1989; Cockburn and Jones 1996; Brake 1997). After a while, wandering, lost users are often forced to go back to the entrance point and start again (see Chapter 9 for a more detailed consideration of navigation in virtual space). Spatialisations in this context can provide 'maps' of the route taken or the route to be taken.

The second popular form of browsing is to use directories which group pages into easily 'browseable' categories, often organised in a hierarchical fashion.[16] The categories and the classification of pages is usually achieved using human judgement. One of the most well-known directories is Yahoo!, one of the key landmarks of the Web and the most visited site, attracting many millions of visitors a week.[17] There are thousands of other directories, many of which focus on a particular subject area and are run by commercial enterprises, academics or the efforts of dedicated individuals. Directories have proved to be a useful, con-venient and popular way of organising Web space into a structure that most people can utilise. However, there are problems with human-constructed directories relating to the granularity of their categories and maintaining currency in the rapidly growing Web (Chen *et al.* 1998). Perhaps the most important issues are, how meaningful are the categories created by the directory developers and do these categories match the conceptualisation of the user? Spatialisations in this second case can provide a graphical means by which directories can be understood and browsed.

'Flat' spatialisations of information space

As with maps of ICTs, spatialisations of cyberspace are both two-dimensional and three-dimensional in nature. In this section we concentrate on the former, examining 'flat' spatialisations. These employ either a continuous 'land use' map or an arc-node graph metaphor to provide users with a spatialisation of the content and structure of the in-formation space.

An example of the 'land use' metaphor is displayed in Figure 6.7, a hierarchical set of 'category maps', which are essentially visual directories (Chen *et al.* 1996). This prototype application, known as ET-Map,[18] is being developed by Hsinchun Chen and the research

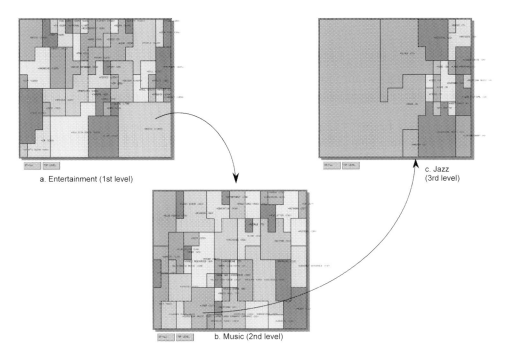

a. Entertainment (1st level)

c. Jazz
(3rd level)

b. Music (2nd level)

Figure 6.7 ET-Map, a hierarchical category map
Source: College of Business and Public Administration, University of Arizona

team at the University of Arizona's Artificial Intelligence (AI) Lab (Chen *et al.* 1998). Figure 6.7 displays the spatialisation of over 110,000 entertainment-related webpages listed by the Yahoo! directory (Chen *et al.* 1998), and is an example of how the spatialisations work to reveal information on jazz music. Each 'category map' displays groupings of associated webpages as regularly shaped, homogeneous 'subject regions', which can be thought of as virtual 'fields' all containing the same type of information 'crop'. The spatial extent of the 'subject regions' is directly related to the number of webpages in that category. For example, the 'MUSIC' subject area (Figure 6.7 (a)) contains over 11,000 pages and so has a much larger area than the neighbouring area of 'LIVE' which only has about 4,300 pages. Clicking on a 'subject region' with less than 200 pages takes one to a conventional text listing of the page titles. If a region has more than 200 pages, then a sub-map of greater resolution is created, with a finer degree of categorisation (Figure 6.7 (b and c)). In addition, a concept of neighbourhood proximity is applied so that 'subject regions' that are closely related in content are plotted close to each other. For example, 'FILM' and 'YEAR'S OSCARS', at the bottom left of Figure 6.7 (a), are neighbours.

ET-Map can be interactively browsed, explored and queried, using the familiar point and click navigation style of the Web, to find information of interest. Importantly, the spatialisation provides a 'big picture', an overview of the whole information space (Chen *et al.* 1998). ET-Map was created using a sophisticated AI technique called Kohonen self-organizing map (SOM), which is a neural network approach that has been used for automatic (i.e., no human supervision) analysis and classification of semantic content of text documents like webpages (Kohonen 1995). Chen *et al.* (1998: 592) believe 'that Kohonen SOM-based technique . . . can be used effectively and scaleably to browse a large information space such as the Internet'. However, it is also a challenge to automatically

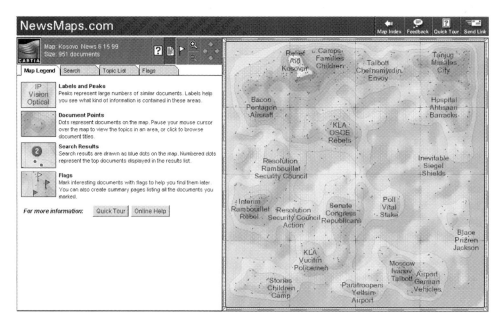

Plate 6.1 NewsMaps, June 1999
Source: Cartia, Inc., http://www.newsmaps.com/

classify pages from a very heterogeneous collection of webpages and it is not clear that the SOM categories will necessarily match the conceptions of a typical user. From the limited usability studies on category maps, it appears that they are good for conducting unstructured, 'window shopping' browsing, but are less useful for undertaking more directed searching.

Another example of this kind of approach are NewsMaps,[19] which employ a carto-graphic terrain metaphor. Plate 6.1 displays a typical NewsMaps mapping of over 900 online news reports from many sources, such as the *Washington Post* and Fox News, of the Kosovo crisis from mid-June 1999. NewsMaps are created by using a high-powered, soph-isticated information mapping system called ThemeScape,[20] developed by Cartia, Inc.[21] The visual metaphor of shaded terrain maps, lifted from topographic cartography, uses colour and density of shading, along with contour lines, to give the reader the impression of different elevations of the information landscape. The NewsMaps maps are the result of considerable offline data-crunching that analyses the actual content of the news reports, using proprietary algorithms and techniques that intelligently summarise the key topics and the relations between them. This is then spatialised so that topics which are popular and have generated many articles combine together to form hills and mountains. Using the same neighbourhood proximity concept employed by ET-Map, related topics are plotted near to each other on the terrain. The valleys between hills represent divisions and transitions between topics, where more isolated articles are likely to be found. The actual location of the news articles used to construct the spatialisation is indicated by small black dots.

Users can interactively browse the NewsMaps' spatialisation. By passing the mouse cursor over an area of interest, the top five topics within a small radius are displayed in a pop-up window. Clicking once on the terrain will cause a pop-up list of available articles in the area to be displayed. Clicking on an article title allows the full article to be opened

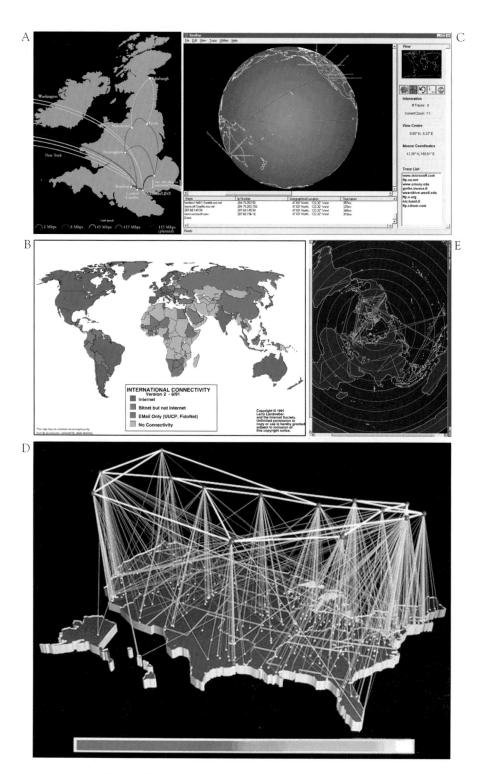

Plate 1 (A) UUNET UK backbone from 1998 (© UUNET UK, http://www.uunet.co.uk/); (B) Network connectivity by Larry Landweber; (C) GeoBoy three-dimensional traceroute; (D) NSFNET traffic flows by Donna Cox and Robert Patterson (© NCSA/UIUC); (E) Web Hopper dynamic traffic map.

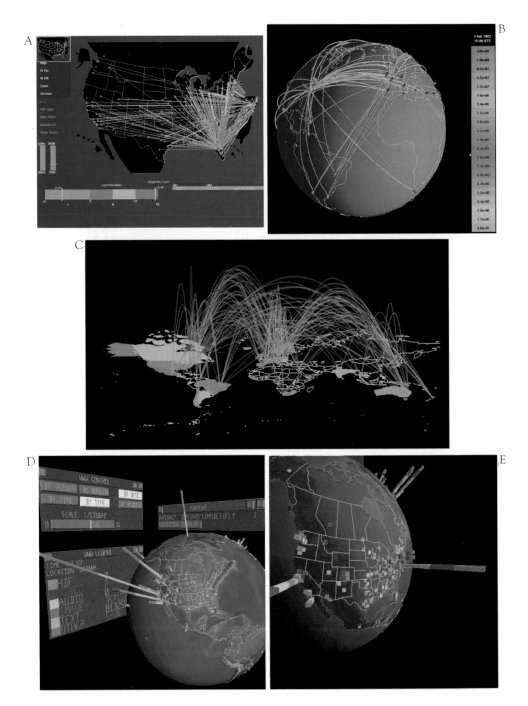

Plate 2 (A, B, C) Interactive visualisations of traffic flows by Stephen Eick and colleagues (© Stephen Eick, Bell Labs – Lucent Technologies); (D, E) Web traffic 'skyscrapers' by Stephen Lamm, Daniel Reed and Will Scullin (© Department of Computer Science, University of Illinois at Urbana-Champaign).

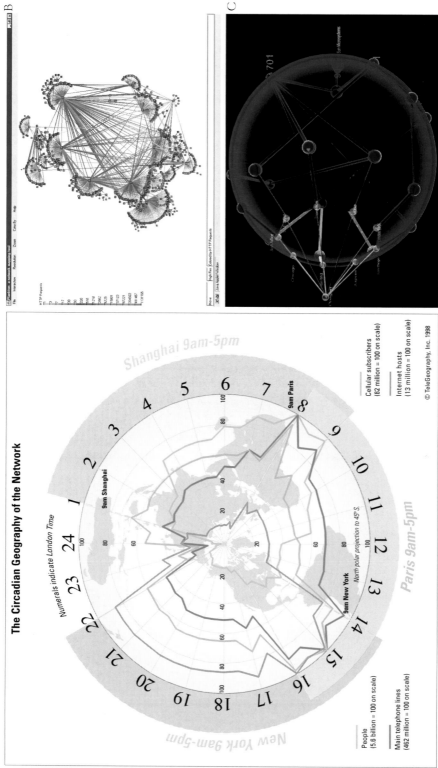

Plate 3 (A) Circadian geography of cyberspace (© TeleGeography, Inc., http://www.telegeography.com/); (B) Plankton interactive visualisation by Bradley Huffaker and Jaeyeon Jung (© Caida, http://www.caida.org/); (C) ISP interconnectivity by Hans-Werner Braun (http://moat.nlanr.net/).

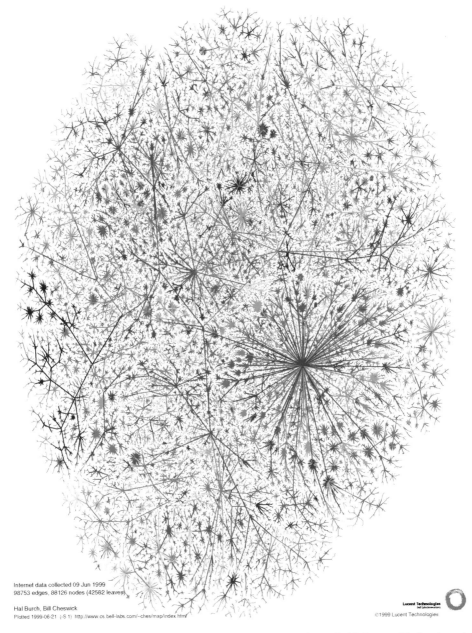

Internet data collected 09 Jun 1999
98753 edges, 88126 nodes (42582 leaves)

Hal Burch, Bill Cheswick
Plotted 1999-06-21 (-S 1) http://www.cs.bell-labs.com/~ches/map/index.htm

Lucent Technologies
Bell Labs Innovations

©1999 Lucent Technologies

Plate 4 Internet connectivity graph by Bill Cheswick and Hal Burch (© Lucent Technologies).

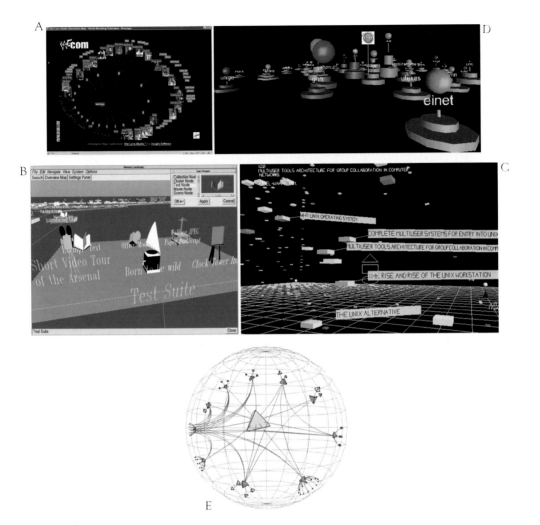

Plate 5 (A) Site Lens map of a website (© Inxight, http://www.inxight.com/); (B) Information landscape – Harmony hypermedia browser by Keith Andrews; (C) VR-VIBE datascape by the Communications Research Group, University of Nottingham; (D) Webspace landscape by Tim Bray (© Tim Bray, http://www.textuality.com/); (E) Hyperbolic visualisation of a website by Tamara Munzner and Paul Burchard (© Munzner and IEEE).

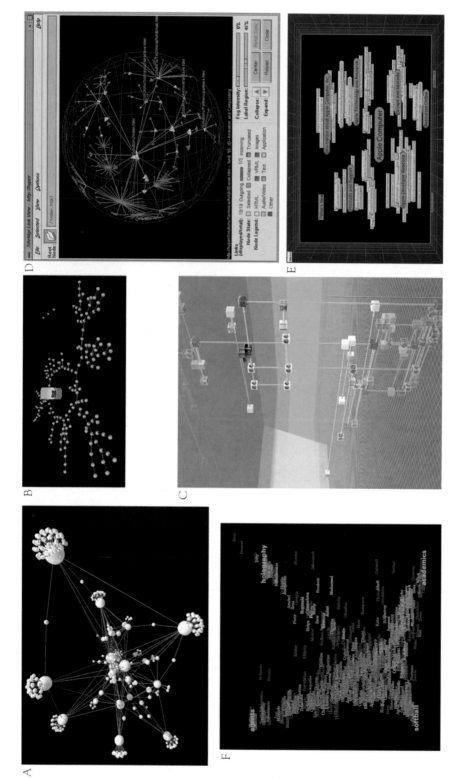

Plate 6 (A) Three-dimensional Graph Visualisation HyperSpace by Andrew Wood, Nick Drew, Russell Beale and Bob Hendley; (B) Semantic Constellation by Chaomei Chen; (C) WebPath browsing history by Emmanuel Frécon and Gareth Smith; (© IEEE); (D) SiteManager hyperbolic visualisation of a website; (E) HotSauce information fly-through; (F) Visual Who by Judith Donath.

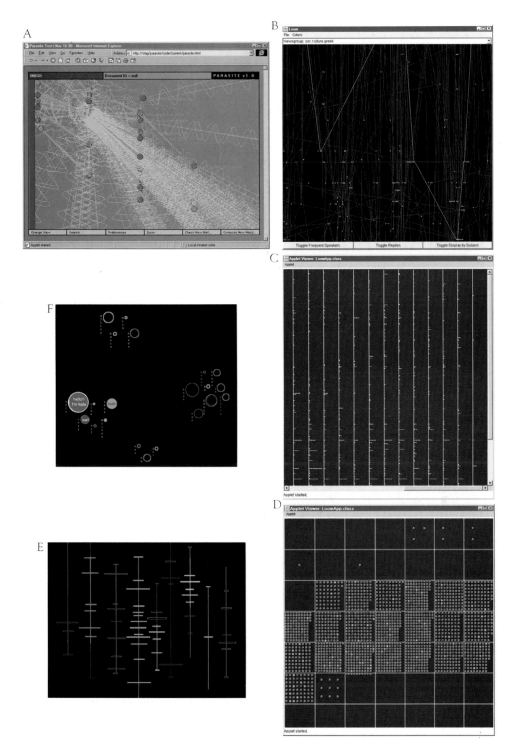

Plate 7 (A) Parasite experimental visual email application by Steve Cannon and Gong Szeto; (B, C, D) Loom maps of a Usenet newsgroup by Karrie Karahalios, Judith Donath and Todd Kamin (© Sociable Media Group, MIT Media Lab); (E, F) Chat Circles by Fernanda Viégas, Judith Donath, Joey Rozier, Rodrigo Leroux and Matt Lee (© Sociable Media Group, MIT Media Lab).

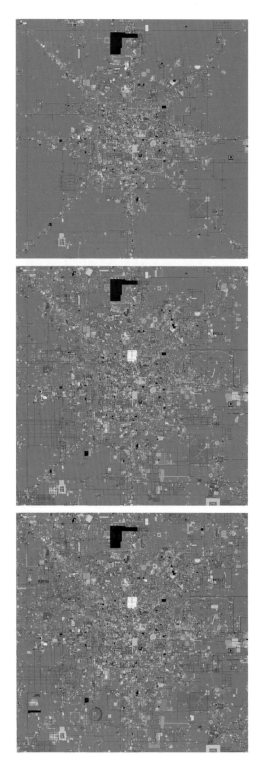

Plate 8 Satellite maps of the urban development of AlphaWorld by Roland Vilett: December 1996 (top), February 1998 (middle), August 1999 (bottom) (© Activeworlds.com, Inc., http://www.activeworlds.com/).

in a new browser window. Users can also perform a key word search for articles of interest or select articles from a topic list, the results of which are shown prominently by large blue dots on the spatialisation, numbered according to their relevancy ranking. It is also possible to stick small red marker flags into the terrain to identify documents of interests for future reference and to zoom-in and pan around the spatialisation to reveal more detail.

There are many other fascinating examples that employ the flat map metaphors utilised by ET-Map and NewsMaps. Plate 6.2 displays five of the more interesting ones, which we do not consider in detail. Plate 6.2 (a) displays an example of a category map created using Visual SiteMap,[22] developed by Xia Lin, based at the School of Information and Library Science, University of Kentucky (Lin 1992, 1997), which has many similarities to ET-Map. The second example, (b), is a more stylised and abstract terrain map of a region of the Web developed by Luc Girardin, at The Graduate Institute of International Studies, Switzerland, titled 'Cyberspace Geography Visualization'[23] (Girardin 1995). It is another SOM-based analysis of Web content. WEBSOM,[24] (c), similarly uses the technique of SOM analysis but rather than mapping the Web, it maps the thousands of articles posted on Usenet newsgroups (see Chapter 7 for more information on Usenet). The WEBSOM maps are multilayered and can be actively browsed. They are being developed by researchers at the Neural Networks Research Centre, Helsinki University of Technology, in Finland (Lagus *et al.* 1996; Honkela *et al.* 1998). The particular example shown is the top-level map of over 12,000 articles posted to the news group comp.ai.neural-nets from June 1995 to March 1997. The fourth example, (d), is a commercial information map titled 'Map of the Market', developed by SmartMoney.com.[25] It is an interactive map of the market performance of the stocks of major US corporations. Individual companies are represented by different sized plots of virtual land, scaled by their market capitalisation. The colour of the plot is based on recent movements in the company's stock price (red for falls, green for increases) and companies in the same sector (technology, health care, etc.) are grouped together on the map. The final example, (e), is a prototype spatialisation of a digital library,[26] developed by Sara Fabrikant as part of her graduate research in the Department of Geography, University of Colorado at Boulder (Fabrikant 1999, 2000), which similarly uses a terrain-style representation linked to other interface elements.

In contrast to land use maps, a number of attempts have been made to provide directory-style spatialisations using an arc-node metaphor. The most common of these are site maps designed to provide a visual overview of the key parts of a website in order to aid users in finding particular information (see Morville 1996; Gloor 1997; Cockburn and Greenberg 1999; Kahn 1999). Many examples are one-off, hand-crafted static spatialisations. There are also several more interactive systems. In addition to site maps aimed at end-users there are also applications designed for webmasters, the people who build and maintain websites. These are more comprehensive in their ability to determine and map the exact topological structure of a site required to find and fix problems.

The simplest site maps are the ones that provide a graphical table of the site's contents, and they are often structured hierarchically. They are interactive and users simply click on the part of the map that interests them and the corresponding page is loaded. A wide variety of designs are employed and Figure 6.8 (a) shows a typical example taken from the Hilton Hotels website[27] in September 1999. Another, more creative example, is the Yell Guides (the UK yellow (golden) pages) site map[28] (see Figure 6.8 (b)), which replicates many of the graphical conventions of the London Underground map (Garland 1994). The 'map' uses four different coloured subway lines to represent various sections of the site, and the standard symbol for an inter-change station to represent individual pages.

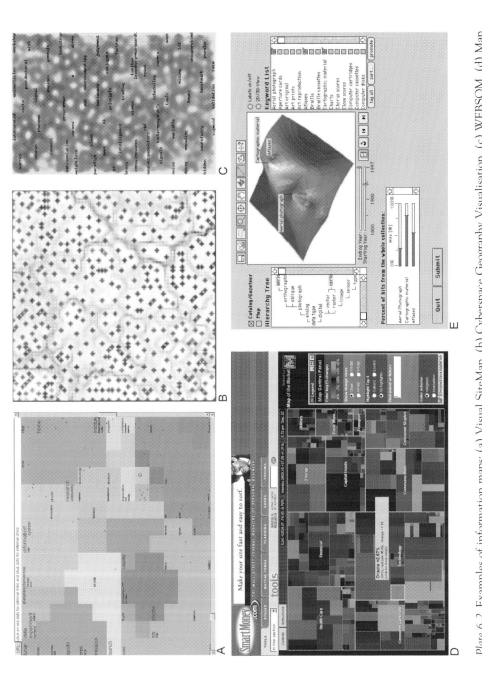

Plate 6.2 Examples of information maps: (a) Visual SiteMap, (b) Cyberspace Geography Visualisation, (c) WEBSOM, (d) Map of the Market, (e) Digital Library Access

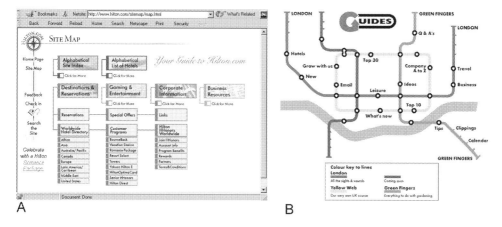

Figure 6.8 Hand-crafted, static maps of websites: (a) Hilton Hotels, (b) Yell Guides
Source: (a) Hilton Hotels Corporation http://www.hilton.com/; (b) British Telecommunications plc, http://www.yell.co.uk

A more interactive and powerful website map, Site Lens, has been developed by Inxight Software.[29] Plate 5A displays an example of the Site Lens map of part of the World Wrestling Federation website. The map is displayed using a fish-eye technique which distorts the spatial view of data under the 'lens' so that elements at the centre of the map appear larger than those at the periphery (Sarkar and Brown 1994). The user is able to 'grab' page objects and drag them to the centre of the map where they are enlarged. In this manner, the user can explore a large hierarchical graph, with local detail in the centre of the map, with wider context visible in the surrounding area. Double clicking on a page rectangle will load this page into the main browser window. Similar fish-eye techniques are used in another interesting interactive website mapping tool called Mapuccino from IBM.[30] It offers various graph styles, such as 'wheel-spoke' and horizontal/vertical trees to visualise the structure of a chosen website. It is presently being developed by researchers at IBM's Haifa Research Lab in Israel (Maarek *et al.* 1997).

Specialised website mapping applications that are aimed at those who manage large sites are generally analysis and management tools rather than navigational aids. They provide tools for detailed graph-based spatialisations of the structure of pages and hyperlinks. Figure 6.9 displays a screen shot of one example, Astra SiteManager,[31] developed by Mercury Interactive. The screen shot is of a mapping for the website for the Centre for Advanced Spatial Analysis. Astra SiteManager is a stand-alone application that can crawl a complete website and show the detailed structure of individual pages, links and images. For large graphs that can not be displayed in full on the screen, a small overview map is also available for panning. SiteManager identifies problems like broken links using colour coding and enables a Webmaster to fix them quickly. One powerful function of Astra SiteManager is its ability to overlay traffic patterns on the structure graph, showing how visitors travel through a site and where the 'hot spots' of interest are. There are several other products that compete in this market such as CLEARweb, Site Analyst, Visual Web and WebAnalyzer.[32]

The ability to show how a site is being used, as well as revealing its structure, is potentially a very interesting aspect of mapping websites. It can be instructive for those

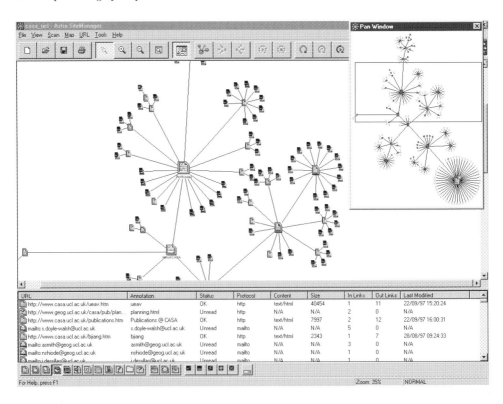

Figure 6.9 Astra SiteManager

who design, build and manage websites to see what people do within their site, where they start, where they go, how many pages they look at, how deep into the hierarchy they venture, and where they leave. One research project that attempts to spatialise this data for Web browsers is Footprints,[33] developed by Alan Wexelblat while a graduate student at the MIT Media Lab (Wexelblat and Maes 1999). Web browsing is currently a solitary activity with users unaware of others who may also be looking at the same page, and unaware of the many previous users. This kind of knowledge, however, can be used to improve navigation, with current users benefiting from the efforts of those who have explored the space before. The integration of this information would transform solitary surfing into a more collaborative, social activity like walking around a busy city centre (see Munro *et al.* 1999, for a review of current research). The Footprints system records the traces of individual users' travels through a website, and makes the aggregate patterns visible to other visitors on a map. A key component, shown in Figure 6.10, is a fish-eye site map showing through traffic. This 'map' may look quite disjointed, but this is because it only shows pages and links that have been visited and recorded by the Footprints system. In order to maintain the anonymity of individual traces, traffic is aggregated. Traffic is represented graphically by colour-coding the page symbols in the map. Similar work mapping the pathways people take through the Web is being carried out with the Recer system,[34] currently being developed by Matthew Chalmers at the Department of Computer Science, Glasgow University (Chalmers *et al.* 1998). A related idea to Footprints are browser add-on tools that provide graphical histories of user activities across

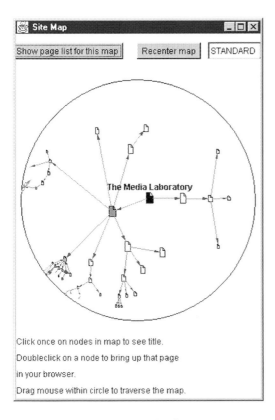

Figure 6.10 Footprints map of website activity
Source: Wexelblat (1999), http://wex.www.media.mit.edu/people/wex/

many websites which can be used to find previously visited sites. We look at some three-dimensional examples of these later in this chapter.

Landscapes of information

In contrast to 'flat' spatialisations, some developers have experimented with the application of landscape metaphors, extending visualisation into three dimensions. This gives one more 'space' in which to display information, but can often be harder for people to interpret and use. Most of these spatialisations are also interactive and many of them run within VR-style environments where the user can fly over and into the 'map' itself.

The first example we consider is the MAPA system developed by David Durand, Paul Kahn and colleagues at Dynamic Diagrams[35] (Durand and Kahn 1998). It is designed as a mapping tool to improve end-user navigation of large websites. Figure 6.11 displays a screen shot of the MAPA map of the Javasoft website. MAPA is a simple information landscape that stresses the hierarchical structure of a website. The visual display resembles note cards in a card index. Simple colour coding is also used to further identify hierarchical structure so that immediate child pages are shown in green and further grand-child pages are in blue. Below the location card, towards the bottom-left of the screen, ancestor pages are shown by orange-coloured cards which delineate the steps back to the

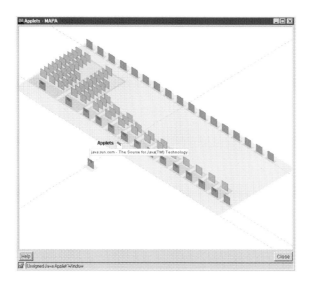

Figure 6.11 MAPA website landscape by Dynamic Diagrams

site's home page. The cards with dark bars across the top are section identifiers, and as such are the most significant in terms of navigation. Dynamic Diagrams call this graphical style the 'Z-Diagram', and each vertical card represents one webpage. They use Z-Diagram maps widely in their hand-crafted diagrams for planning the structure of new websites or redesigning existing sites for clients, with thumb-nail images of the actual pages pasted onto the front of cards (see Wurman 1997, for examples). MAPA can be invoked from any place in the website, producing a pop-up map window that shows the user location in relation to the rest of the site.

The information landscape metaphor can be extended by making the maps immersive, allowing users to 'get inside' the space. Two such systems are Harmony Information Landscape (Plate 5B) and VR-VIBE (Plate 5C). These are particularly interesting examples because, in many respects, they look most like the popular imagination of a Gibsonian datascape (see Chapter 10).

The first example is part of the Harmony browser, the client used to access an Internet hypermedia system called Hyperwave[36] (Maurer 1996). The Harmony browser provides an integrated two-dimensional graph map and three-dimensional landscape view of the structure of a hypermedia information space in addition to the conventional page view of the documents. Here, we are concerned with the three-dimensional information landscape component developed by Keith Andrews and colleagues at the Graz University of Technology in Austria (Andrews 1995; Andrews *et al.* 1996).

The Harmony Information Landscape provides a three-dimensional spatialisation of an information space for users to browse resources (such as documents, files, images, etc.) which are represented by blocks and icons laid out across an infinite flat plain (Plate 5B). Collections of resources (the Hyperwave equivalent of a website) are represented by flat slab blocks onto which the actual resources are placed as iconic glyphs, such as a book to represent a text file and an old-fashioned movie camera to identify a video clip. The spatial arrangement of the blocks encodes the hierarchical structure of the Hyperwave information space. The user is able to fly over the landscape and choose objects of interest, which are then displayed in the conventional browser window. As the user browses, new collections are added dynamically to the landscape.

The second example, VR-VIBE (Plate 5C) was developed by the Communications Research Group at the University of Nottingham, led by Steve Benford. The focus of their research is collaborative virtual environments (CVE), which provide virtual spaces that can be shared by many users (Benford *et al.* 1997). Internet virtual worlds, such as AlphaWorld, are commercial examples of CVE, and we consider them in detail in Chapter 8. VR-VIBE represents the application of CVE technologies for information searching and retrieval, creating a three-dimensional co-operative system that can be simultaneously shared by several users (Benford *et al.* 1995; Churchill *et al.* 1997). Matching documents from keyword enquiries are displayed as simple blocks floating in patterns above a flat landscape covered with a standard yellow grid. Keywords, represented as octahedrons, are positioned across the space and the document blocks are displayed in relation to strength of attraction to each keyword. As such, keywords act as virtual magnets, pulling documents towards them with differing strengths depending on their significance to the search. Plate 5C displays a screen shot of a VR-VIBE session where over 1,500 documents are spatialised according to five keyword 'magnets'. The size and colour of a document block encodes the relevance score of that document to the overall enquiry. This means that large, brighter blocks, that are visually the most prominent, are the best matches to the enquiry. Crucially, this data-space is a shared virtual environment, and Plate 5C is a static screen shot, from a first-person perspective. Other users present in the space are represented by simple avatars, which look like sticks with eyes. This shared aspect raises interesting possibilities for collaborative searching and exploration of a large information space.

Users are able to interact dynamically with the VR-VIBE dataspace in a number of ways. For example, a user can fly in to examine part of the document space in more detail and then quickly fly above to get an overall view of the configuration. Users can change the parameters of the search enquiry by adding, deleting and moving the keyword 'magnets'. Thus the user can select a keyword glyph and move it by dragging it, causing the spatial arrangement of the documents to adjust. It is also possible to change the thresholds of the enquiry using a three-dimensional scrollbar to limit the number of matching documents and add annotation to blocks, in the form of text 'flags', as can be seen in Plate 5C. Finally, documents of interest can be selected and these will be fetched and displayed, for example in a conventional Web browser.

There have also been attempts to spatialise the wider Web landscape with whole websites represented as singular, graphical objects. Plate 5D displays one such landscape created by Tim Bray (1996).[37] In order to answer the following four questions – how big is it? how wide is it? where is the centre? how interconnected is it?[38] – Bray used a large search engine index to calculate the key metrics on the structure of the known Web in 1995.[39] He examined the hyperlink structures of the Web and found that interlinking between sites was surprisingly sparse. Most links were local, within a site, and a few key sites (e.g., Yahoo!) acted as super-connectors tying sites together. He derived two intuitive measures of website character based on hyperlinks – visibility and luminosity. Visibility is a measure of incoming hyperlinks, the number of external websites that have a link to a particular site. In 1995, the most visible website was that of the University of Illinois at Urbana-Champaign (UIUC), the home of the Mosaic browser. The vast majority of sites had very low visibility and nearly 5 per cent had no incoming links. Measuring the reverse, the number of outgoing links, determines a site's luminosity. The most luminous sites carry a disproportional amount of navigational workload. In 1995, Yahoo! was the most luminous site and probably remains so today.[40]

Using these statistical characteristics, Bray spatialised the key landmarks of the Web in 1995, highlighting the largest, most visible and connected websites. The resulting information was shown as landscapes dotted with three-dimensional models which he called 'ziggurats'[41] (Plate 5D). Each ziggurat visualised the degree of luminosity and visibility of a single site, along with the size of the site and its primary domain (e.g., government, education, commercial, etc.). The basic graphic properties of the ziggurat – size, shape and colour – were used to encode these four dimensions. The overall height represented visibility, the width of the pole represented the size of the site (in terms of number of pages), the size of the globe atop the ziggurat indicated the site's luminosity, and colour coding displayed the primary domain (green for university, blue for commercial, red for government agencies). The ziggurats were also labelled with the domain name of the site for identification. The spatial layout of the ziggurats across the plane were based on the strength of the hyperlink ties between them. The model is constructed in VRML and can be 'flown around'. Plate 5D displays a field of ziggurats at the very core of the Web in 1995. Further from this core region there would be many thousands of other ziggurats, but most would be minuscule in relation to those at the heart.

A more photo-realistic, or literal, information landscape is the 3D Trading Floor (3DTF) of the New York Stock Exchange. In 3DTF information flows and real-time data are spatialised in a three-dimensional virtual environment modelled on a physical architectural space. Plate 6.3 presents two distinct views of the 3DTF which was unveiled in March 1999 (McLaren 1999; Scanlon 1999). The information environment is used as a real-time decision support system for operators who manage the stock exchange, keeping vital networks and computer systems running as well as monitoring the actual information flows of the market performance. It was designed by architects Lise Anne Couture and Hani Rashid at Asymptote Architecture.[42] Users can be immersed in the interactive environment. Various real-time data streams on the business systems, including stock performance of individual companies and user-defined aggregations, the underlying networks and computer servers, and news broadcasts can be spatialised. Particular failures, incidents and unusual activity can be highlighted and examined in detail.

3DTF is used by operators in a purpose-designed command centre, nicknamed The Ramp, in the heart of the actual stock exchange. It is displayed on a bank of large flat-screen panels. It is not accessible over the Internet. To power such a large and complex spatialisation, running with real-time data streams, requires considerable computer resources, employing six expensive graphic supercomputers from SGI. At present, it is still at the experimental stage and it remains to be seen how valuable and useable the spatialisation will be in real-time management of this information space. There are plans to develop the 3DTF further with extended interaction and users represented by avatars, as well as pro-viding wider and remote access to the environment.

Information spaces

In this final section we examine a range of spatialisations that use three-dimensional spaces and virtual reality-style interaction techniques to create environments in which the user becomes immersed. They differ from information landscapes in that data fully utilises all three dimensions and is not arranged on a planar construct.

The first two examples we discuss use what might be termed 'virtual molecular models' to represent arc-node topological structures. These molecular models are three-dimensional

A

B

Plate 6.3 Three-dimensional New York Stock Exchange trading floor
Source: Asymptote/New York Stock Exchange/SIAC (Securities Industry Automation Corporation)

models constructed inside virtual reality environments. In both cases, users can manipulate and view the models from any position and angle, flying inside and around them. Plate 6A shows HyperSpace Visualiser and Plate 6B displays Semantic Constellation.

The HyperSpace Visualiser (Plate 6A) was a prototype spatialisation of the local structure of a small portion of the Web (Wood *et al.* 1995). The goal of the system was to provide users with information on how their current location was connected to the neighbouring Web space to help overcome the symptoms of being 'lost in hyperspace', showing users where they were and where they could go to next. The system was developed by Andrew Wood, Nick Drew, Russell Beale and Bob Hendley, at the School of Computer Science, University of Birmingham, in the UK, and it was based on a general purpose information visualisation system called Narcissus (Hendley *et al.* 1995). Their spatialisation uses a conventional arc-node metaphor, with solid spheres representing individual webpages and the arcs the hyperlinks between them. The size of the sphere is scaled to the number of hyperlinks from the page. The layout of the arc-node model in three-dimensional space is achieved using a self-organising algorithm based on attraction/repulsion behaviour of individual lines and spheres – the Webpage spheres repulse each other and this is counteracted by the hyperlinks which attract each other. Starting from a random placement of a clump of interconnected webpages in the three-dimensional space, through a series of iterative steps, the spheres and arcs push and pull each other into a stable and coherent spatial arrangement. The result is self-organised equilibrium. Only pages that have been 'discovered' by the user's browsing are displayed in HyperSpace. The pages at the edge of the explored space are just single nodes, with a single arc leading back to the parent. This gives the edge spheres a pincushion appearance. The system is synchronised to the user's Web browser so that details of pages traversed are automatically transferred to HyperSpace. The actual spatialisation is displayed in a separate window on the desktop. Users can interact with the model as well as selecting what is displayed and nodes can be labelled with page titles and URLs.

The second example is the Semantic Constellation created by Chaomei Chen, in the Department of Information Systems and Computing, Brunel University, in the UK (Chen 1997). Plate 6B displays a view of Chen's Constellation which is used within a standard desktop VR environment, while the model itself is made from VRML.[43] It spatialises an information space of over one hundred and fifty conference papers from three years' worth of online proceedings. As with HyperSpace, an arc-node three-dimensional graph is employed, with spheres representing the individual papers and the arcs connecting them together based on the relatedness of their content. The spheres are colour-coded by year. Unlike HyperSpace, the arcs are not explicit, 'hard-coded' hyperlinks connections. Instead, in the Constellation they are based on a computed measure of semantic similarity between the papers. Thus, papers that discuss the same or related topics are semantically linked in the spatialisation. The more closely two papers are related in terms of their content, the nearer they are in the semantic space. The semantic linkages are spatially arranged and connected together into what is known as a PathFinder network. Pointing to a particular sphere causes the paper title to be displayed in a pop-up window and clicking on a sphere displays the paper abstract in a linked window in the Web browser.

Like HyperSpace, WebPath is an example of a 'surf map',[44] providing a graphical history of routes taken by a user through the Web. It was created by Emmanuel Frécon, at the Swedish Institute of Computer Science, and Gareth Smith, at Lancaster University, UK (Frécon and Smith 1998). Plate 6C shows a view of the three-dimensional spatialisation of browsing structures hanging, weightless, in a stylised purple cyber-world. WebPath aims to provide users with a 'flexibly tailorable real-time visualisation' of browsing history within a VR environment, working alongside a conventional browser. The principal advantage of WebPath, the authors contend, is that visual patterns of previously visited sites means easier retrieval. It employs a particularly angular arc-node spatialisation, with many straight lines and 90-degree angles. Individual webpages are represented by cubes. Cubes are used as their flat surfaces are easier to read from a distance. The page represented by the cubes is indicated by labelling (with the title or URL) and texture mapping on the faces. The texture can be the background image of the page, or an image on the page, or background colour of the page, depending on the user's choice. Clicking on a cube of interest will load that webpage into the browser.

The positioning of the cubes in the space is used to encode data about the webpage and when it was accessed. The three orthogonal dimensions of the space allow one to display three distinct parameters. First, the vertical axis is used exclusively for the time at which the webpage was accessed. The cubes at the top of the spatialisation are always the most recently visited. The x and y horizontal axes can be used to encode a variety of metrics, such as loading time of the page, page size or number of hyperlinks, which can be selected by the user. The user can change the meaning of the x and y dimensions at any time and the cube positions will be automatically recalculated. WebPath can also position the cube according to approximate real-world geography rather than using an abstract co-ordinate space. A base map is provided on the 'floor' of the information space and the cubes are positioned in the appropriate country based on the domain name of the website.

The links between webpage cubes show the paths the user has taken via hyperlinks. When a user visits a new webpage, a new cube is created, and an arc connects this back to the previous cube. The colour of the arc is used to indicate the pages that are from the same site. Repeat visits to the same website at different times are indicated by multiple cubes which are vertically separated but are connected by solid yellow columns. For the most popular webpages the column turns from yellow to red to indicate repeated accesses. Distinct browsing sessions are also separated visually using semi-transparent horizontal planes. This divides the space into separate layers.

In contrast to these Euclidean three-dimensional spaces, Tamara Munzner, a graduate student in the Computer Graphics Laboratory, Stanford University, has investigated the potential of constructing spatialisations in hyperbolic space (Munzner and Burchard 1995; Munzner 1998). Hyperbolic spaces have advantages for visualising the detailed structure of large graphs containing many thousands of nodes, such as the Web, as Munzner and Burchard comment, 'The felicitous property that hyperbolic space has "more room" than Euclidean space allows more information to be seen amid less clutter, and motion by hyperbolic isometries provide for mathematically elegant navigation' (1995: 33). Plates 5E and 6D display two examples of Munzner's three-dimensional, hyperbolic spaces.

The spatialisations provide a novel way for exploratory visual browsing of the page-hyperlink structures of large websites. The structure of nodes and links is projected in hyperbolic space inside a ball, known as the 'sphere at infinity'. The user is able to manipulate the graph, rotating and spinning it inside the sphere. Like the fish-eye distortion technique, discussed earlier in relation to Inxight's Site Lens spatialisation (see p. 119), hyperbolic space gives greater visual presence (in terms of screen-space) to elements at the centre of the space. As objects are moved to the periphery, they smoothly shrink in size. At the edge of the sphere the nodes are very small, but the user can easily drag them into the centre to enlarge them and see them in detail. In this manner, the hyperbolic spatialisation can provide a view of the detailed graph structure, whilst still showing the overall context.

An early example by Munzner and Burchard from 1995 is shown in Plate 5E which spatialises the structure of two layers of their department's website. The pyramid glyphs represent pages and the curving lines are the principal hierarchical hyperlinks. This spatialisation was part of a Web mapping system called Webviz which could gather the structure of a specified portion of the Web and then visualise it in a three-dimensional viewer called a Geomview (Munzner and Burchard 1995). Munzner further refined the underlying hyperbolic spatialisations, developing the H3 layout algorithm and a more powerful viewing system (H3Viewer) which enables interactive exploration of graphs of 100,000 or more nodes (Munzner 1998). Plate 6D displays an example of the H3Viewer incorporated into a product called Site Manager[45] from SGI. Site Manager is a tool for Webmasters that provides a fluid and scaleable view of a website structure.

The final example we detail is HotSauce, a three-dimensional fly-through interface for navigating information spaces.[46] It was developed largely as a one-man effort by Ramanathan V. Guha at Apple Research. Plate 6E shows one screen-grab of a fly-through of the Apple website from 1997. The fundamental concept behind HotSauce is, 'why just browse when you can fly?' HotSauce was a specific three-dimensional spatialisation of the Meta Content Framework (MCF), also developed by Guha. MCF was a schema for describing and organising the structure of an information space (Guha n.d.).[47] This is called metadata and is separate from the actual content. For example, a library catalogue is vital metadata that enables books to be found on the shelves.

HotSauce works as a plug-in to an existing browser. As such, when a hyperlink to a MCF website is selected, the user is dropped into a first-person perspective view of the information space, with pages floating like brightly coloured 'asteroids' in an infinite space. The view can be compared to that from a starship cockpit, the flying is smooth with the pages becoming larger as they are approached and then disappear behind. Webpages are represented by rectangular glyphs and are labelled with the page title. Topic areas are indicated by the round-cornered rectangles. Different hierarchical levels of information are denoted by different colours of glyphs as well as their spatial depth in the three-dimensional display. Thus, in Plate 6E, the top-level is the green 'Apple Computer' topic, which is then followed by major sections of the site represented by the bold red glyphs. Further back are yellow and then purple pages. The glyphs are also arranged spatially into distinct groups. Unfortunately, Apple ended its development of HotSauce in 1997 and Guha moved to Netscape (Andreessen 1999). Since then, a different company, Perspecta, has been, developing its own three-dimensional information fly-through called PerspectaView, similar in many respects to HotSauce, which is based on an underlying metadata structure called SmartContent.[48] Despite the practical difficulties in using HotSauce for information browsing and retrieval, it was an important development. As Steven Johnson in his book *Interface Culture* describes:

> But a day or two with HotSauce was enough to catch a glimpse of what a genuinely spatial system might feel like. At a few, enthralling moments, I found myself groping around for a familiar document and thinking: It's back there somewhere, up and to the left a little, about two or three planets deep. For a second or two I was thinking in purely spatial terms, zooming in and out of my own private dataspace. For those few moments, there was a hint of liberation in the air, the promise of things to come.
>
> (Johnson 1997: 80)

7   Mapping asynchronous media

In the preceding chapters we have considered, in some detail, the geography of various information spaces that are experienced primarily by oneself and used for accessing inanimate resources. However, there are those who would argue that the true power of cyberspace is that it provides a media which fosters social interaction *between* people. As discussed in detail in Chapter 3, to these commentators, one of the most basic human needs is to communicate and interact with others, and cyberspace – through email, newsgroups, mailing lists, bulletin boards, chat rooms, MUDs, networked games, and graphical virtual worlds – is providing new media through which this communication can occur and in many cases flourish. Everyday, these social media, largely based within the Internet, are used and inhabited by millions of people all talking, discussing, arguing, flirting and playing with one another. In this and the following chapter we consider the nature and spatiality of online social media and the social interactions that occur through and within them, focusing in particular on how they can be mapped and visualised.

Characterising the social media of cyberspace

In order to make sense of the many different social media available in cyberspace we have categorised them into a simple typology, demarcated by time and numbers of users. The time dimension divides social media into two groups: asynchronous (where participants communicate at different times) and synchronous (where participants are present at the same time). Outside cyberspace, letter writing is the archetypal asynchronous mode of social interaction and face-to-face communication is the archetypal synchronous mode. The number of users dimension divides social media in relation to how many people are participating using a particular social medium. Clearly this dimension is a continuum ranging from a minimum of two people, small conversations with a group of friends or family over the dinner table, up to large parties, seminars and concerts, and perhaps even the many millions who participate in large events like the football World Cup final via mass media broadcasting (Adams 1992). We impose a logical, simplifying break in this continuum, dividing social media into two groups. The first, one-to-one, is social media for interactions between two people, the second, one-to-many, is media where you can communicate with more than one other person simultaneously. Table 7.1 takes these two dimensions to create a typology of four categories which characterise the principal social spaces of cyberspace.

The most obvious and well-used example of asynchronous and one-to-one media is email, which is most often used for individual communication. Messages are sent to a named individual and stored in their mailbox, to be read at a later time. Users of email

Table 7.1 Typology of social media of cyberspace

Forms of media	Asynchronous	Synchronous
One-to-one	Email	Talk / instant messaging (ICQ) Private chat rooms 'Whispering' in MUDs / virtual worlds Internet telephony
One-to-many	Mailing lists and listservs Usenet Bulletin boards	Chat rooms / IRC MUDs Graphical virtual worlds Networked games

do not have to be online at the same time to successfully communicate. Email is the most widely used social media of cyberspace with many millions of messages flowing through the Internet every day.

The one-to-many asynchronous section contains important social media that are in some respects like the conventional mass media of newspapers and broadcasting. These media allow users to broadcast to a group of people who subscribe to a particular domain. Any member of a list can broadcast his/her views, and similarly any member can respond. The group may be a few colleagues on an office mailing list or many thousands on a popular listserv. The most popular of these media are the Usenet newsgroups which offer a huge, distributed system containing many thousands of distinct domains on different topics, daily attracting millions of readers and participants in a multitude of conversations.

Synchronous, one-to-one interaction generally involves private conversations that take place in real time, exactly like a telephone conversation, except they are on online. Historically these media have been provided by the 'talk'-type facilities commonly found on mainframes and Unix-based computers. Conversing via 'talk' is achieved by typing short sentences which are displayed in real time on the screen of the person you are conversing with. This 'talk' concept underlies the more recent development of instant messaging, the most prevalent commercial example of which is ICQ.[1] This category also contains private conversations which can take place in what are usually viewed as 'public' media. Here, users of a public space can open a private chat channel or room. This mode of private conversation can also be achieved in MUDs and graphical virtual worlds using the whisper mode.

The final section is social media with characteristics of synchronicity and one-to-many conversation. These are 'public' spaces where groups of people can converse in real time. The most well-known are the chat rooms and channels of IRC. These are usually simple textual media in which people can chat, gossip, argue and flirt using common language and a distinctive new Internet lexicon (see Table 7.2). MUDs, graphical virtual worlds and networked games provide similar spaces, but usually have more sophisticated inter-faces and spatial environments. In the case of games, there are often other objectives than mere communication: usually trying to kill the other participants!

In the remainder of this chapter we look at the principal asynchronous social media in more detail, starting with email, paying particular attention to their structure and how people have attempted to map them. In the following chapter we examine the spatialities and geometries of synchronous social spaces.

Table 7.2 Emoticons and the feelings they are used to express

Typical US/European emoticons		Typical Japanese emoticons	
:-)	regular smile	(^_^)	regular smile
:^)	happy	(^o^)	happy
;-)	wink/mischievous	(^ . ^)	girl's smile
:-o	wow	(*^o^*)	exciting
:-|	grim	(^o^;>)	excuse me!
:-||	anger	(_o_)	I'm sorry
:-(	sad	(;_;)	weeping
:^(	unhappy	(^ ^;)	cold sweat
.oO	thinking	(^_^;;)	awkward

Source: Aoki, cited in Kitchin (1998)

Email as social medium

Email is at the very heart of social cyberspace. It was the first social application, coming into effect in 1970, and quickly become the key reason to be on the network and arguably made it a success (Abbate 1999). Since then it has undoubtedly remained the most popular, well used and powerful of the virtual social media in existence. Although the Web may have garnered most of the headlines and plaudits when the Internet went mainstream in the mid-1990s, it is text-based email that is the fundamental core of what makes cyberspace such a useful media. For most Internet users, the first thing they do when they logon is check their email. As such, email is what is known in computer industry terms as the 'killer application'. Indeed, many people seek out an Internet account due to the lure of email communication with friends, family and colleagues, which when used appropriately can achieve many different tasks. For example, email is good for:

- instant, efficient, and cheap communication between geographically separated parties;
- maintaining multiple relationships through simple, short messages;
- fostering collaboration of work and consensus by allowing the distribution of digital documents;
- creating an archive of communication;
- reaching a number of parties with one message. It is just as simple to send an email message to ten people as it is to send it to one.

Email is also a low-cost, accessible Internet service, requiring minimal hardware and network connections, especially when compared to the resource-hungry, graphical Web. Moreover, because it is an asynchronous media one can send a message without worrying whether the recipient is online to receive it. Instead, the mail is stored centrally and downloaded when the user logs-on to the mail server. Mail can be sent and read on all computers regardless of platform or operating system and thanks to emergence of Web-based email services, such as Hotmail.com, it is free and readily available for those who are able to connect to the Internet. Indeed, many people now have multiple email accounts for different aspects of their lives. For example, it is not uncommon for employees to have a work account and a personal account. The reasons for this are various, but it is mainly so that personal and institutional views and time do not become confused. Essentially, email provides an effective and more efficient replacement for postal mail. Due to its relative simplicity, social conventions are learnt easily and using email quickly becomes second nature and an indispensable personal and business tool of communication to rival the telephone.

Email is most often used as a one-to-one, personal communications tool. As such, it receives much the same attention and consideration as ordinary postal mail. This is one of the reasons why spam (unsolicited email) causes such annoyance, as it is an invasion of the personal space of the mailbox by an uninvited stranger. Consequently, email does not exist in a communications vacuum, people develop their email usage within their existing social context and in conjunction with other channels. Email is most often used to communicate with family, friends and colleagues who one already has social ties with. As Esther Dyson succinctly says:

> It extends and maintains relationships more than it creates them. Indeed, it's not *the* web, but it's part of *a* web of relationships that includes meetings, phone calls, purchase orders, voice mails, third-party gossip – all the ways people communicate with one another.
>
> (Dyson 1999: 11)

Email is a social media that is predominately an extension of, or supplement to, offline social interaction. Its main influence on socio-spatial relations is to provide a new media of communication, rather than to foster a new spatiality online. Email thus differs to other online social media where the development of new socio-spatial relations is more prevalent (see below, and Chapters 3 and 8). Email, it can be argued, was at the vanguard of these developments because it: 'laid the groundwork for creating virtual communities through the network. Increasingly, people within and outside the ARPA community would come to see the ARPANET not as a computer system but rather as a communications system' (Abbate 1999: 111).

Given its popularity, email's usage has been studied by a number of social scientists (e.g., Sproull and Kiesler 1992; Hiltz and Turoff 1993). In these studies, particular attention has been paid to the social and organisational impact of email in the workplace. It is hypothesised that email has the advantage of being a more informal mode of communication than the telephone or face-to-face meetings, so it can be used more freely to communicate across conventional social hierarchies. It is this quality that has the potential to undermine conventional power structures in the workplace.

Despite these few studies, there has been surprisingly little work done on analysing email usage from a geographic perspective (exceptions are Anderson *et al.* 1995; O'Lear 1997). This leads us to ask: what is the structure of email's social media like? The architectural structure of email, from the perspective of the typical user, is simple, consisting of the message itself and the software application to receive, compose, send and store the message. The message is normally composed of text, and consists of a header and a main body. The header contains certain standard components on separate lines – from, to, date and subject – while the body contains the text of the message and can be of any length. Part of the power of email is that these messages can be easily stored, filtered and sorted by the header details. However, there are drawbacks such as the ease by which messages can be forwarded or replied to, which can result in embarrassing mistakes, like accidentally sending an inappropriate message to the wrong person. In addition, the limits of expression of tone and emotion, due to the restricted range of communication cues available in simple text, can sometimes lead to the misinterpretation of a message.

As befits a predominately text-based form of communication, messages are stored in sorted lists and arranged in folders. There have, to date, been few attempts to provide a more visual interface using spatial structures to represent messages. This is surprising, but perhaps current interface metaphors are functional and sufficient for the job at hand. One notable experiment in designing a visual email application was Parasite, developed by Steve Cannon and Gong Szeto at the New York-based, digital design company, i/o 360. Plate 7A displays a screen shot of the prototype interface. The visual metaphor used to map out the storage of messages in mailboxes is an adaptation of Feynman diagrams normally used in the field of particle physics. Unfortunately, development of Parasite ceased before a finished version of the software was released, so it will remain an interesting but untested approach. However, email remains an area ripe for future development of visual, map-based interfaces.

Mailing lists

The power of email for one-to-one communication can be extended to allow for one-to-many interactions and many-to-many discussions. Mailing lists allow a single mail message to be redistributed by a central server to all members of a list. A single message can therefore be delivered to hundreds of subscribers at no extra cost to the sender. Whilst this has benefits in that information and questions can be distributed to a wide range of people with a specific interest in the mailing list focus, it also has disadvantages, such as the distribution of spam.

List space is a very important online social media as many people subscribe to one or many mailing lists for purposes of work and personal interests. It is not known how many lists exist, but there may well be a quarter of a million public ones (Bennahum 1998b), and many more smaller private ones. The public mailing lists and listserv discussions cover all manner of topics.[2]

We classify list space into two types. First, there are what we term mailing lists which are used for one-to-many communication. Here, the creator/owner of the list is the only person who can send messages to that list. This form of mailing list is easy to create from a simple list of addresses using a standard email application. A classic example is a class mailing list of a professor used for easy communication with all her/his students. Another example is a commercial daily news bulletin which is distributed to many thousands of people (e.g., Wired News[3]). There are many motivations for the people who run public mailing lists, but most are not run for direct financial rewards. Instead, they offer a chance for someone to publish and share their ideas at a very low cost of entry. As Bennahum (1998b) states, 'list publishing is a real example of our idealized picture of the Net as a place where all can speak and ideas flow freely'.

A different kind of list is the two-way communications of discussion lists, known as listservs.[4] This allows many people to post to the list which opens up the potential for many-to-many social interactions and discussions. An initial analysis of the geography of listserv discussions has been undertaken by Stanley Brunn, a Professor of Geography at the University of Kentucky (Brunn 1998). He examined the temporal sequencing and spatial patterns of discussions on the GEOGED listserv, a long-standing, well attended list for those interested in all aspects of geographical education. Given the predominately US origin of posters, not surprisingly most post originated in the US. The location of these posters was scattered, with no discernible patterns evident as one might expect at a conventional meeting where a distance-decay function would normally delineate the origin location of contributors. In addition, he found that there were various kinds of community members ranging from the complete number of subscribers, to the 'lurkers' who read but never post, the occasional posters, and, finally, a core community of regular posters. The importance of listservs like GEOGED is that they can be accessed by geographically disparate individuals who could not meet face-to-face on a daily basis.

The lists that people subscribe to can reveal much about their interests and the form of the wider electronic social communities in which they participate. Judith Donath, a professor at the MIT Media Lab, has created an interesting interactive mapping tool called Visual Who to dynamically spatialise the social patterns (based on affinities) of an electronic community using lists (Donath 1995). Plate 6F shows a screen shot of the Visual Who mapping system showing the affinity of a large number of people to four different mailing lists – softball, academics, agents and holography. The lists are positioned as anchor points around the edge of the map window and the people are represented by their names. The position of a person's name is relative to the strength of his/her affinity to each of the anchors. People with a strong affinity to a certain list will be drawn close to it on the map, whereas someone with approximately equal affinity to two lists will be drawn midway between each list. As a case study, Donath used the 700 people affiliated to the Media Lab. The colour coding of people's names is based on their work status (faculty are yellow, staff are purple, graduate students are red, and undergraduates are green). Visual Who is also an interactive map in that the user of the system can move anchors, delete and add new lists, with the mapping being rearranged dynamically to take account of the new forces. The system can also show presence by only mapping the people who actively logged-on to the computer system at any one time. In this way, the user can explore quantitatively the social patterns and uses of list space.

From badgers to body art: newsgroups on anything and everything

Usenet is another important social media that has many structural features in common with email and listservs. It is a vast, distributed, global network, running on top of the Internet, that provides a complex mesh of interrelated domains known as newsgroups. As detailed below, there are thousands of different newsgroups covering a huge range of topics. Each newsgroup has its own particular social characteristics and together they comprise perhaps the most diverse portion of social cyberspace, in which hundreds of thousands of people carry on discussions, share technical tips, provide support, and hold heated arguments and debates. Many more millions silently read these daily interactions without actively contributing. Usenet stands for 'User's Network', but is known by various names, including 'netnews', 'usenet news', 'newsgroups', and 'plain news' (Quarterman 1990). Unlike mailing lists, submissions to Usenet are not redistributed to all members of the list, instead users check news central servers to follow a thread (theme of a discussion). Contributions to the discussion are sent to the central news server which then distributes to other Usenet hosts.

The idea of Usenet was conceived by Jim Ellis and Tom Truscott, graduate students at Duke University, in 1979, with the first software to implement the network written by Steve Bellovin, also a graduate student, at the nearby University of North Carolina (UNC).[5] Not surprisingly, the first link in the Usenet network was between Duke and UNC. The initial aim of Usenet was for technical discussion relating to Unix for those not fortunate enough to be on the ARPANET. Usenet had no official funding and the software was distributed in the public domain and promoted by word of mouth. The growth of Usenet,

Table 7.3 The hierarchical structure of newsgroups

alt.drugs	—	General
alt.drugs.abuse		
alt.drugs.culture		
alt.drugs.hard		
alt.drugs.mushrooms	—	Specific
alt.drugs.pot		
alt.drugs.pot.cultivation		
alt.drugs.pot.cultivation.hydroponics		
alt.drugs.pot.cultivation.hydroponics.nospooks	—	Very specific

in terms of the number of users (both active participants and passive readers), messages posted per day, and newsgroups, was rapid throughout the 1980s. Usenet was low-cost and effectively free to access (you just had to find someone willing to supply you a 'news feed' and pay the cost of transmission) which enabled it to rapidly diffuse around the globe (see the geographic map of Usenet traffic flows by Brian Reid in Chapter 5, Figure 5.12).

Usenet is perhaps the third most commonly used element of the Internet (after email and the Web). In many ways Usenet can be thought of as the archetype of uncontrolled cyberspace, since it is highly distributed, free-wheeling, and has no official funding, external quality control or censorship. It was a 'bottom-up' experiment built by a small group of graduate students, and its success and exponential growth was fuelled by the user community. The users also drove the structural development of Usenet, in both technical and social terms, helping develop software and protocols and the rules of conduct and netiquette.

The spatial form of Usenet is based on two fundamental structural components – newsgroups and articles. There are tens of thousands of newsgroups, although many are only distributed locally (within large organisations or country-based), and many others are rather lifeless, with little user participation. Newsgroups are arranged in a number of large hierarchies based on a very broad classification. The seven of the oldest hierarchies are comp, sci, soc, talk, news, misc, rec, later joined by the largest and most controversial of all – the alt (alternative) hierarchy. The comp hierarchy contains the newsgroups for computer-related discussion, rec is for hobbies and leisure activities, while alt holds a miscellaneous bunch of groups including what some people consider offensive groups like alt.drugs, alt.sex and the alt.binaries.pictures.erotica.[6] There are also hierarchies for major organisations, commercial software vendors, countries, and local site specific ones.

The names of the individual newsgroups under each of these hierarchies are structured in a hierarchical fashion and are read from left to right, with the topic narrowing the further right one goes. For example, Table 7.3 displays a sub-sample of the alt.drugs newsgroups and its 'siblings', illustrating the narrowing of newsgroup focus. This narrowing indicates the evolution of a topic as people continually subdivide the social space to make it more manageable and relevant to particular interests. From this table it should be clear that each of the thousands of newsgroups available has a different theme or topic enabling users to find a particular social niche that best matches their interests. For example, if you were keen on badgers then the group alt.animals.badgers might be your scene, or maybe you have an interest in body art, so check out the rec.arts.bodyart group. And if a newsgroup does not exist that matches your interests, it is quite easy to create one (providing you can find enough supporters). The creation of new newsgroups is again the result of grass-roots activism, with anyone free to propose the creation of a new one providing they can gain sufficient support from the broader community in a democratic election. It is much easier to create new newsgroups in the alt hierarchy, hence the proliferation of many with strange names, many of which are often created as a prank and are never used.[7]

How many newsgroups are there? There is no definitive answer because it depends where on the Usenet network you look. Different organisations, ISPs, and news servers carry different portions of Usenet space. For example, in June 1999, the news server at University College London carried around 13,000 groups while the news at Demon Internet (one of the largest UK ISPs) carried over 32,000, of which just over 11,000 were in the alt hierarchy and around 900 were comp groups. Not all these are active (i.e., have a participant community of regular posters and readers), so there are probably around 20,000 widely-distributed, active newsgroups. This is quite a number and the breadth of topics covered is considerable.

The second structural component of Usenet is the individual news article. Active newsgroups can contain anything from a few posted articles a day to several hundred in the most popular ones. Articles are short textual messages, common in appearance to email messages, containing a header and a main body. The volume of articles posted each day varies greatly between different groups and also over time within a particular group (particularly if it is related to real-world topical events). It is important to realise that each group has particular internal social dynamics and 'life-force' created by the users of that list. Some groups, especially the technical ones, maintain a distinctly 'question and answer' style of social interaction, where one person posts a question and others reply with answers. Other groups hold much more involved and long-running discussions and some are plagued with spam messages that can render the group almost uninhabitable for meaningful dialogue. Many articles posted are part of an ongoing, sequential discussion and as such are said to be threaded. Threads can vary in length from a couple messages to many messages when discussions become heated. Articles can also be posted to several newsgroups at the same time, a phenomenon known as 'cross-posting'.

The underlying forces that determine the structural form of newsgroups and articles are the Usenet network protocols.[8] From the perspective of the typical Usenet participant, the spatial structure is also determined to a significant degree by the user interface of the client program used to access the groups. These clients are called newsreaders and are now built into popular Web browsers. A commonly used interface structure is the mapping of Usenet into three separate windows, as shown in Figure 7.1. This is a screen shot of the popular PC newsreader WinWN[9] which employs the '3 window' approach. Window one shows the available newsgroups as a simple alphabetical scrolling list. Double clicking on a newsgroup of interest (in this case, rec.pets.cats.anecdotes) opens window two which gives a list of the currently available articles in that group. Again, clicking on a article that is of interest will open a third window where the text of the article is displayed.

Usenet also has an important temporal dimension in that all articles have a limited life span and are set with an expiry date. Generally, Usenet articles persist in the network for only a couple of weeks (the exact duration depends on the particular settings of the news server one uses to access Usenet).[10] The fact that all social activity using Usenet is transitory strongly influences the tone of the newsgroups making them informal, spontaneous and free-flowing. It also means the communities that exist in different newsgroups are dynamic and it requires almost daily participation in order to be accepted and treated as a fully active member. Clearly, this can be demanding on time, particularly if one participates

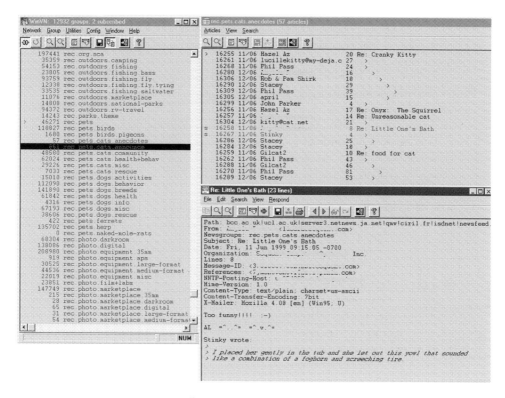

Figure 7.1 Typical Usenet interface[11]

in several groups. Rheingold provides a vivid description of the multifaceted social form of Usenet as:

> a place for conversation or publication, like a giant coffeehouse with a thousand rooms; it is also a worldwide digital version of the Speaker's Corner in London's Hyde Park, an unedited collection of letters to the editor, a floating flea market, a huge vanity publisher, and a collection of every odd special-interest group in the world.
>
> (Rheingold 1993: 130)

In general, anyone with full Usenet access is free to post to any newsgroup of interest, except for the few that are moderated. However, one needs to be careful to post articles in the appropriate group as posting an 'off-topic' article is likely to annoy people and generate flaming and spamming. People in the body art group do not want to hear about the latest badger news!

Social discourse, interaction, and the oft-claimed sense of community (see Chapter 3) is solely reliant on the text distributed. Judith Donath, responsible for the Visual Who mapping we examined in the previous section, has undertaken a detailed dissection of individual Usenet articles revealing the hidden depths of language, identity construction and deception that is being played out in this form of communication (Donath 1999). Donath is one of a growing number of academics, particularly from the social sciences, who have studied Usenet over the past decade. In some senses, this research has used

Usenet as a live and lively social 'laboratory', which can be studied and analysed with relative ease, to explore wider questions concerning community formation, identity and gender issues, and social conduct (Baym 1995; McLaughlin *et al.* 1995; Tepper 1997). There have been detailed studies of the content of particular newsgroups, for example pornography (Mehta and Plaza 1997) and soap operas (Baym 1997). A particularly interesting analysis of Usenet, relating the virtual community to social activity in the real world, is the work of Wesley Roehl (1999). He examined the posting traffic in the newsgroups dedicated to different US baseball teams (under the hierarchy of alt.sport.baseball) in relation to the success of the teams on the field and their attendance figures by fans. Not surprisingly, he found strong, positive relationships between activities of baseball fans in real and virtual activities.

Usenet has also been widely reported and misreported by the popular media. It has received considerable critical attention as the delivery channel for the worst of the Net's pornographers, hate-mongers and bomb-making malcontents. This was particularly so in the sensationalist scare stories when the Internet broke into mainstream culture in the mid-1990s, and there was widespread concern about the dangers, particularly for children, on the Internet. These stories spoke of the Internet in general terms, but were really concerned with only a few of the most 'shocking' Usenet groups that carry pornographic material.[12] While offensive material does exist on Usenet, there is also many more progressive and positive aspects that are ignored, with many groups providing technical help, emotional support or frivolous fun.

The most notable research into the virtual geography of Usenet has been undertaken by the Netscan project,[13] directed by Marc A. Smith (Smith 1997, 1999). Netscan began as part of Smith's graduate research in the Center for the Study of Online Communities at UCLA and has continued at Microsoft Research, where Smith works as a research sociologist. The Netscan system collects and analyses the daily social activity of the majority of active newsgroups, the goal being to answer some of the fundamental descriptive questions about the structures and dynamics of this social medium – 'How big is the Usenet? How many people post? Where are they from? When and where do they post? How do groups vary from one another and over time?' (Smith 1999: 195). Netscan provides relatively representative and consistent base-line statistics to answer these questions, finding patterns in the social interactions in terms of average daily posts to a group, the average thread length, the domain and geographic location of posters, and cross-posting structures. Much of this is visualised using tables and statistical charts. It also aims to provide the means for deeper, ethnographic studies of social groups and dynamics.

Smith (1999) reports some interesting statistics on the nature of Usenet derived from Netscan. Participants in Usenet are geographically diffused, with postings recorded from 205 of the 238 Internet domains (as measured by the email addresses of the posters), although 41 per cent of posters are from the US (Smith 1999). In 1997, the 14,000-odd active newsgroups received an average of 300,000 articles a day from approximately 23,000 different people. The alt hierarchy was by far the largest, and Smith notes that it represents a distinct 'political sub-culture'. Netscan also discerns significant variation in activity between newsgroups; Smith reports that one-fifth of newsgroups were empty (i.e., dead) and 43 per cent were only partially inhabited (with less than one hundred messages recorded over a ten-week period). At the opposite extreme, there are a small percentage of very active groups, which can be measured in terms of number of articles or number of posters. Table 7.4 shows the top ten groups as measured by Netscan in a ten-week period from November 1997 to January 1998, revealing that seven of the ten, in terms of traffic,

Table 7.4 Top ten newsgroups by number of articles and number of posters

Rank	(a) *Top ten by message volume* Group name	Number of articles
1	Misc.jobs.offered	232,612
2	biz.jobs.offered	217,472
3	ba.jobs.offered	210,562
4	Misc.jobs.contract	98,803
5	alt.jobs	76,605
6	ba.jobs.contract	29,077
7	News.newusers.questions	27,332
8	Comp.sys.ibm.pc.hardware.video	25,293
9	tx.jobs	20,741
10	Comp.sys.ibm.pc.games.strategic	20,482

Rank	(b) *Top ten by poster population* Group name	Number of posters
1	News.newusers.questions	12,012
2	Comp.os.ms-windows.win95.misc	7,189
3	Comp.sys.ibm.pc.hardware.video	5,576
4	mis.jobs.offered	5,184
5	Comp.sys.ibm.pc.games.strategic	5,055
6	Comp.sys.ibm.pc.hardware.chips	4,394
7	Comp.os.ms-windows.nt.misc	4,377
8	Comp.sys.ibm.pc-games.action	4,329
9	Comp.os.ms-windows.win95.setup	4,202
10	Comp.sys.ibm.pc.games.rpg	4,201

Source: Smith 1999: 203, 205

were related to job-seeking, while the most active in terms of the number of posters were groups related to discussions on PC hardware and games.

Importantly, Netscan also analyses the cross-posting between groups by constructing an interactive map of the connections. A screen shot of this graph-based spatialisation, shown in Figure 7.2, maps the groups connected to comp.infosystems.gis. In total, nine groups are connected with the majority located in the 'sci' hierarchy and are concerned with physical environment applications of GIS (e.g., sci.image.processing, sci.environment. geology). The system enables one to view the cross-post graph for any of the thousands of newsgroups covered by the Netscan system, quickly revealing hidden clusters of related newsgroups. The visualisation is also interactive, enabling the user to explore the graph with various settings. The size and colour of the boxes represents the newsgroup characteristics with the size of the box proportional to the number of posts and the colour showing the posts to poster ratio. The graph is also fluid, so that the user can move and position the boxes as desired. The lines between boxes show the ties between groups and can be labelled to indicate the strength of the relationship in terms of the number of cross-posted articles.

The visualisation of cross-posting is important because it is a key structural component of Usenet, which Smith argues forges small neighbourhoods of interconnected groups. He found that only 6 per cent of groups analysed by Netscan stood completely alone with no cross-posting at all. On average, a newsgroup has some measure of connectivity to fifty other groups. Smith argues that cross-posting is important because the 'dense level of

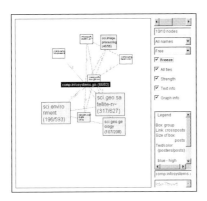

Figure 7.2 Crosspost map of a Usenet newsgroup using Netscan

interconnection give Usenet the ability to act as a powerful social information switch. Questions that appear in one newsgroup are likely to be seen by someone who has a connection with a more appropriate newsgroup, who then forwards the message or redirects the questioner to a proper newsgroup' (Smith 1999: 208).

Another project mapping Usenet is Loom[14] by Karrie Karaholios and Judith Donath, at the MIT Media Lab, which visualises the patterns of article posting within individual newsgroups over time (Donath *et al.* 1999). Loom can be considered a shift down to a finer scale of mapping compared to Netscan, with a focus on the participants and the article threads within a single group. Plates 7B, 7C and 7D display three screen shots of different maps of the social terrain of a newsgroup produced by the Loom system. The reasoning underlying the naming of this application makes reference 'both to the "threads" of a Usenet group and to the appearance of the visualization; the patterns and texture of the events within the group are reflected in the patterns and texture of this digital fabric' (Donath *et al.* 1999).

The key goal of Loom is to create an intuitive mapping tool so that people can browse newsgroup archives in visual terms, something which can be difficult with current text-based systems. In addition, the system provides a kind of newsgroup finger-print from which people can tell at a glance what type of group it is. Loom maps use a simple spatial structure based on a two-dimensional grid. The horizontal axis represents time and the virtual axis is divided into rows that represent individual participants in the group. A variety of different maps can be produced to show the different social structures underlying group interaction.

In Plate 7B, each message is represented by a single colour glyph in the map; the colour can be used to represent some characteristic of the article (such as subject, or domain of the poster). The lines connect articles of a thread together. At first glance, the map looks like a confusing mess, making it difficult to read. In interpreting the map there are three key patterns to look for. First, a strong vertical patterning of glyphs represents an intensity of activity at a particular time slice. Second, prominent horizontal rows of glyphs show the most active participants. Third, the structure of connecting lines between articles provides a visual indication of the conversational atmosphere of the group; where there are long, complex overlapping lines one can infer a group of intense discussions involving many participants with many replies and long threads. Where the lines are much shorter and form a more disconnected mesh, this indicates a more 'question and answer' style of interaction.

The second map, Plate 7C, concentrates on the temporal patterns revealed by Loom. The solid virtual lines are user-specified time units (such as days or weeks) and the dots at different heights are the articles posted by different people. Plate 7D is an attempt to classify and map the actual content of the articles rather than just the structure of the group. Automatic content classification is a major challenge facing those wishing to map social cyberspace. Loom uses a heuristic device to classify articles into four categories: angry (red), peaceful (green), news-based (yellow), and all others (blue). The map is divided into a cellular structure based on the boxes of a calendar. Each day's articles are represented by appropriately coloured disks in the cell. Clicking on a particular disk will cause the text of an article to be displayed in a pop-up window. The map displays an entire month's worth of newsgroup posting and it is possible to see the daily intensity and tone of conversations from the density and colour of the disks. In this way, this particular type of Loom map acts as a kind of visual overview, and also generates a unique signature of the group.

8 Mapping synchronous social spaces

In the previous chapter we introduced the geographies of asynchronous media, focusing on the various forms and initial attempts to reveal their spatialities and geometries. We now extend this analysis, turning our attention to synchronous social media where the social interactions between people take place in real time. Here, although the inhabitants of cyberspace are geographically dislocated, they share the same media at the same time. Synchronous social media are the virtual realm that have received perhaps the most academic and public attention, particularly by sociologists. As discussed in Chapter 3, it is felt that synchronous social media provide new spaces in which to explore identity, as well as creating and redefining notions of community. Whilst the broader arguments concerning the implications of synchronous social media are discussed in Chapter 3, in this chapter we discuss empirical work relating to a number of different kinds of synchronous media and specific attempts to chart their spatialities and geometries.

Live conversations in chat

Real-time conversation via IRC (Internet Relay Chat) and numerous other proprietary 'chat room' media are very popular cyberspace activities. The conversation between participants is conducted by typing short text sentences. As participants type the text, it appears simultaneously (or nearly so depending on network lag) on the screen of their own computer, the screens of people with whom they are conversing, and on the screen of anybody else logged-in to that 'room'. The chat 'space' can be private between two or more invited and authorised participants, or they may be public spaces open to all. The IRC system is probably the largest chat media, although it is itself broken into several major networks, each of which may have hundreds and even thousands of channels (individual, distinct 'spaces'). These self-contained channels each hold a group of participants and a conversation, somewhat like a radio channel. One can join and leave public IRC channels at will. Users are also free to create new channels of their own. Rheingold (1993) aptly terms the participants of chat space the 'real-time tribes'.

Connecting to an IRC media involves using a client program.[1] The client provides the spatial structure of chat through its interface design. Figure 8.1 displays a typical IRC client interface, in this instance a mIRC32 for Windows.[2] It is a fairly simple interface, befitting the simplicity of chat. It consists of one main window where the stream of sentences that makes up the conversation flow, plus a side bar showing all the participants logged-in to the space. Along the bottom is a dialogue box where the user types his/her dialogue. Pressing the return key sends this dialogue to the chat server and it will then appear at the bottom of the chat window, prefixed with the user's name, known as a 'nick' (as in nickname).

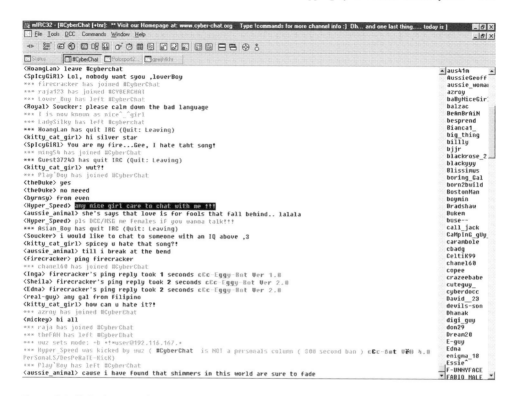

Figure 8.1 IRC chat interface

As IRC is a multi-user experience, other people's dialogue also appears in the chat window, intermingling with the other participants. Each sentence usually takes up one line in the chat window. The conversation scrolls upward and off the screen as the conversation develops. As such, chat is a very dynamic media, comprising continuously flowing text that can be difficult to follow. Certainly, trying to read the conversation in Figure 8.1 is difficult, as it does not make much sense. Many first-time users can become easily confused, as it is difficult to follow multiple conversations that overlap and interweave, with sentences appearing out of sequence as people 'talk' over the top of each other. Unlike email messages or news articles, conversation in chat mode are rarely archived, so once the conversation has disappeared from the window it is lost.[3] As such, the interactions in chat are highly ephemeral, and like the conversations at a party, unknown to those not intimately involved and often quickly forgotten.

Unlike a party in the real world, however, the participants in chat are in the main anonymous, only identifiable by their nickname. This anonymity is an important component in reducing the social inhibitions and constraints of face-to-face conversation (Reid 1991). The lack of inhibition means that chat has a large degree of freedom of expression, with people saying things they would not say in the real world; this can, obviously, have negative as well as positive aspects. On the positive side, it encourages the sharing of emotions, openness and intimacy which facilitates the formation of deep friendships and relationships. Chat rooms are famed for their atmosphere of overt flirtation, with sexually-orientated chat perhaps the single most important reason for its popularity. It may enable the shy and the less social adept to speak their mind. On the negative side, it offers a space where people can express views of hate and harass other users under the veil of anonymity. This is aided by the fact that chat is largely unregulated, although individual channels are owned and controlled by the users that created them. The owner is known as the 'chanop' (channel operator) and he/she has considerable powers to eject and ban people from the channel. There are no means of appeal, and it is expected that one simply finds a more congenial place to hang out in IRC. As Elizabeth Reid explains:

> Disinhibition and the lack of sanction encouraging self-regulation lead to extremes of behaviour on IRC. Users express hate, love, intimacy and anger, employing the freedom of the electronic medium to air views and engage in relationships that would in other circumstances be deemed unacceptable in relating to strangers. This 'freedom' does not imply that IRC is an idyllic environment.
>
> (Reid 1991: 13)

Because participants can only identify other participants by their nickname, nicks become important social markers. As such, a basic convention is that nicks must be unique within a particular domain of IRC and the deliberate stealing and misuse of someone else's nick is considered a serious violation of the minimal self-made social rules of the IRC community. Of course, individuals can and do create multiple characters for IRC, each with a different nick. This encourages the fluidity and instability of personality in chat as people play out elements of their own personality that they would not necessarily feel comfortable doing in real life (Turkle 1995). This is most obviously manifest in gender shifting. Indeed, Elizabeth Reid (1991), one of the leading researchers of social spaces on the Internet, argues that chat is a playground for the exploration of identity, self-representation and forms of self-expression (see Chapter 3). For some, the freedoms of chat can be addictive. Indeed, concern has been raised about the psychological impact of the time and effort serious chat users expend in cyberspace, and the damage it may cause to interactions in geographic space (Slouka 1995).

To Reid, and other social scientists, one of the most interesting aspects of chat is 'what kind of culture emerges when you remove from human discourse all cultural artefacts except written words?' (Reid 1991); when the visual and aural cues and the context of geographic place are removed? In her work, Reid explores how IRC participants use the power of written language and human creativity to construct a form of community. Two of the key elements she identifies are the creative extension of language to encode expressiveness usually conveyed by other channels in face-to-face conversation, and the development of communal sanctions to punish those who flout the common laws (known as 'netiquette'). The language of IRC is a strange, highly informal speech, but with defined styles and conventions. Conversing using short-typed messages is necessarily direct and 'chatty' in nature. This has given rise to new conventions of language unique to real-time text-chat, with a prevalence of abbreviations and acronyms (like 'LOL' for 'laugh out loud' or 'BRB' for 'be right back') to represent frequently used phrases, minimising the amount of repetitive typing and helping to maintain the speed of conversation (Menges 1996; Suler 1997). In addition, these abbreviations act as a shared dialect which helps to define the particular character and distinctiveness of the virtual community (Reid 1991; Danet *et al.* 1998). Other distinctive features of this mode of communication include the use of emoticons (e.g., :-)), action phrases (*Martin smiles*), heavy punctuation ('. . . . ????'), capitalisation ('SCREAM!'), and onomatopoeia ('hehehe'), to express feelings and emotions that are normally conveyed by body language and tone of voice in spoken conversation. Those who are the most creative wordsmiths at using this language hold centre stage. As Reid notes, 'speed of response and wit are the stuff of popularity and community in IRC' (Reid 1991: 16).

One new way to explore the social relations and social interactions between participants, that extends beyond ethnographic work, is to try and map them. One such project, Chat Circles,[4] is being developed by Fernanda Viégas and Judith Donath, assisted by Joey Rozier, Rodrigo Leroux and Matt Lee. They describe Chat Circles as 'an abstract graphical interface for synchronous conversation' found in chat media IRC (Viégas and Donath 1999). Essentially, Chat Circles seeks to map the scrolling reams of text, creating a graphical interface that is able to convey important factual information concerning the chat as well as some of the unspoken nuances of face-to-face communication. The system is made up of two distinct components, the first is the dynamic map of Chat Circles and the second is an archival map which enables one to see and browse an overview of conversations over time. Their work employs abstract two-dimensional graphics, encoding the key chat characteristics of participant identity and activity with fundamental properties of shape, size and colour.

Plate 7F displays an example of the prototype Chat Circles interface. The participants of chat rooms are represented as different coloured circles. All participants are represented, regardless of how much and how open they speak, enabling one to see at a glance all those in the chat room. A major problem with conventional chat interfaces is that one is often only aware of the active speakers, as the text acts as a central indicator of presence. This makes people feel as if they must continually write to maintain their presence otherwise people will forget they are there. It also means one is not aware of 'lurkers', who may be monitoring the conversation.

Another problem experienced by chat users is that chat space can become divided into a number of separate rooms or channels, each with small groups that are unaware of each other. Chat Circles overcomes this problem by displaying all the different rooms on a chat server on a single screen, dividing the chats into distinct conversational groups based on the spatial clustering of circles. However, it is only possible to 'hear' the conversation in the limited geographic proximity of one's circle, so one is not overwhelmed by a cacophony of chatter. Yet, crucially, it is possible to see the number and strength of the other conversations, indicated by the number and size of the circles. It also enables a user to easily move to another conversation by moving his/her circle to that cluster. Chat Circles thus employs spatial location in the interface as a tool for 'geographic' (chat room) filtering. In Plate 7F there are four conversational groupings in different areas of the screen, but we can only 'hear' the one in the bottom left, hence the text of the conversation is displayed in these circles.

The size and brightness of the circles is dependent on how much and how often the people talk. A person's circle grows to accommodate the text of his/her speech. The circle fades over time, after each active sentence. So, the most active participants in the conversation are visually prominent on the map with large, bright circles, while the lurkers are represented by small, faded dots. The circle is also labelled with the person's name for easy identification and an individual's own circle is drawn with a white outline for enhanced distinctiveness. Over time, the dynamic of the conversation can be seen as the circles grow and shrink, and drift to different groups; and the creators note that, 'The use of these dynamic graphics creates a sequence of bright splashes of colours and fading circles in a pulsating rhythm that reflects the turn-taking of regular conversations' (Viégas and Donath 1999).

Viégas and Donath are also developing a history function for chat using a time-line graphical metaphor. They call this the 'conversational landscape', and an example is shown in Plate 7E. This tool enables the user to construct a visual log of chat. The maps from the 'conversational landscape' represent each participant as a coloured vertical line. The z-dimension of the screen is used to show time and horizontal bars are individual sentences. Taking all the lines together shows the threads of the conversation. Browsing the threads is possible, with one simply pointing the mouse cursor over a bar leading to the display of the actual dialogue. For conversations that took place outside the 'hearing' range of the particular participant, the sentences are shown by hollow horizontal bars. As with the main Chat Circles interface, it is a simple, minimalist approach to cyberspace spatialisation, but it is potentially a powerful mapping of chat space history, as it 'allows for a visualization of both group and individual patterns at the same time as it creates, by its mere shape and colors, a snapshot of an entire conversation in one image' (Viégas and Donath 1999).

Worlds of writing

MUDs are chat rooms with an explicit geography. MUDs (Multi-User Dungeons, or more recently, Domains), like chat, are synchronous media of social interaction, but they differ in that they employ explicit spatial metaphors to create stages or scenes, in which inter-actions are situated. Thus, instead of a conversation taking place within a chat room, interactions take place within a designated room, such as a bar. Generally, stages operate at an architectural scale, consisting of inter-linked rooms, although they can consist of open spaces and landscapes. Indeed, some of the largest MUDs have hundred of rooms linked together into complex topologies. Each room is known to its occupants by its *textual* description, and can contain any number of objects, for example furniture. A typical room description, taken from DragonMud, is given here:

> Narthat Street
> You stand surrounded by ornate buildings of gothic design. Carriages rush
> by you, carrying elders and townsfolk through the fog. Above you staring out
> into the mist are gargoyles perched on rooftops, and the barely visible glow
> from TinyBen's luminous clockface.
> [Exits: South to the town square, NorthEast into the Lawyer's Guild, cab,
> West into the builders guild, East into the Town Hall, North along the street]
> Contents:
> A window box of daisies
> The Town Hall

Rooms are linked together by exits which enable MUD participants (generally known as players) to traverse from room to room. The linkages of the rooms into larger topologies of space, and the ability of players to move purposefully through it, travelling in distinct directions, all serve to create an approximate sense of spatiality. In this, and the following section, we detail attempts to map the geography of MUDs. First, however, we provide some background details for those unfamiliar with this often weird and wonderful realm of cyberspace.

The experience of MUDding (playing a MUD) is often described as like being inside a literary novel (rather than 'outside' as a reader). Rather than turn the pages, the narrative of the MUD unfolds on the computer screen in the form of scrolling text. Also, unlike a novel, the MUD is often considered a living novel, written in real-time by the players, who are a disparate band of independent, co-operating (sometimes competing) authors. In some cases the geography of the MUD is predefined and fixed, but in other cases players can change their spatial surrounds. Like chat, MUDs encourage participants to play with their identities. As such they promote the creation of distinct player characters whose virtual identities may be very different from the person's real persona. Sherry Turkle, a clinical psychologist at MIT who has studied MUD players, notes, 'as players participate, they become authors not only of text but of themselves, constructing new selves through social interaction' (Turkle 1995: 12).

As might be expected, MUDs work in much the same way as chat rooms. Participants connect to a specific MUD using a client program.[5] On initial joining of a MUD, a participant creates a character which involves choosing a name (usually a short, single word), a gender (male, female, or neutral), and a verbal description of how they look. The character should be unique to a specific MUD. All interaction with the MUD and other players is via textual dialogue. In general, three major activities are available to players: socialising (chatting) with other players; exploring the rooms of MUD; and building new rooms and objects.

Socialising can happen via chance encounters with friends and strangers or by actively seeking out friends. There is evidence to suggest that it is generally easier to strike up a conversation with a stranger in a MUD because social inhibitions are lowered by the anonymity of the technology, and the fact that other players are similarly seeking social interaction. Public conversations are conducted using two major commands – 'say' and 'emote' (also known as pose). The 'say' command formats a sentence so it appears spoken. For example, if the player Dodgy types, 'say Hi Smithee, what's happening?', then the player would see, 'You say, "Hi Smithee, what's happening?"', while the rest of the players in the room would see 'Dodgy says, "Hi Smithee, what's happening?"'. The emote command is used to convey verb actions like smiling, hugging, and sighing to other players. For example, if Dodgy typed, 'emote grins', then everyone in the room would see, 'Dodgy grins'. For private communications between two players, the whisper and page commands are used. Whispering enables one to speak only to a named player in the room and can not be heard by any others present. Paging is used to send a message to a player on the MUD but not within direct 'hearing' range in the room.

Today, MUDs are seen by many people to be oddities, systems left over from the 1980s and early 1990s, seemingly outdated and lacking visual glamour, left behind by the rise of the Web, e-commerce and networked computer games. A newspaper article from spring 1999 noted that in the mid-1990s, MUDs accounted for perhaps 10 per cent of Internet traffic, but 'then the Web took off, and MUDs drifted off the mainstream cultural radar screen' (McClellan 1999: 2). Despite the declining fortunes of MUDs, they have a vital place in the history of cyberspace. MUDs were pioneering, virtual spaces and many have referred to them as the first consensual virtual realities, even though they lack three-dimensional graphics (Curtis 1996). In some ways MUDs were a product of their era, flourishing at a time when most computers could only manage text interfaces and when network capacities were a fraction of those available on the Internet today.

The history of MUDs, like many of the other social media of cyberspace, can be traced to the experimental hacking of university students, long before the Net hype of the 1990s.[6] The first MUD was created by two students in Britain at the end of the 1970s. The two undergraduate students were Richard Bartle and Roy Trubshaw, studying at the

University of Essex. Their first MUD was written on the shared university mainframe and in spring 1980 it became available for networked users outside the university, including users from the US (Bartle 1990).[7] Of course, the concepts that lay behind Bartle and Trubshaw's first MUD had a much longer genealogy, owing a great deal to the dungeon and dragons computer games Zork and Adventure of the mid-1970s (Dibbell 1999). However, these were only single player games, lacking the real-time social interaction of MUDs.

In turn, these computer games were inspired by the Dungeons and Dragons (D&D for short) board game created by Gary Gygax and Dave Arneson in 1973. D&D was set in a Tolkienesque landscape of myth and fantasy, inhabited by heroes, elves, dragons and wizards. Its popularity spawned a new genre of role-playing games. In terms of MUD genealogy, Dibbell (1999) argues that D&D made four significant contributions. First, the games were social, with the players co-operating and undertaking quests together. Second, the games were open-ended, with no fixed end point. As such, the game was not so much about a final victory but rather the creation by the players of an ongoing adventure story. Third, players had to take on a distinct character and play the game in that role. Last, the conventional game board and map was taken away from the players who instead relied on the verbal descriptions of the dungeon master to conjure up the locales in their minds. Dibbell (1999) argues that these innovative features of D&D meant that it was:

> more than just a new kind of game. They made it, frankly, a whole new mode of representation – an undomesticated crossbreed, combining the structured interactively of the broad game with the psychological density of literary fiction, yet eluding the ability of either medium to fully embody it.
>
> (Dibbell 1999: 55)

Bartle and Trubshaw copied this formula, and in turn their original MUD was developed, adapted and extended by many others throughout the 1980s. In MUD history, the next critical juncture came in summer 1989 with the release of TinyMUD. TinyMUD marked the emergence of purely social MUDding, eliminating the element of fantasy adventure. Like many other developments, TinyMUD was created by a graduate student, James Aspnes, based at Carnegie-Mellon University. Some crucial design decisions, made unintentionally by Aspnes, had the effect of empowering a whole new form of MUD. His aim was to create a very open, free-wheeling MUD, so he eliminated the rigid player ranking system and questing goals which were an integral feature of the adventure-style MUDs. He also removed the restrictions on who could build new rooms in a MUD, traditionally the preserve of the wizard elites, giving the creative freedom to all players to extend and construct the virtual environment to suit their needs, wishes and fantasies. As Amy Bruckman, a pioneering MUD researcher states:

> James Aspnes decided to see what would happen if you took away the monsters and the magic swords but instead let people extend the virtual world. People's main activity went from trying to conquer the virtual world to trying to build it, collaboratively.
>
> (Bruckman 1996)

The social MUD enjoyed a boom in the early 1990s as the number of MUDs mushroomed, with a whole panoply of acronyms to describe particular variants, such as MOOs, MUSHes, MUCKes, Tiny and Teeny.

In many ways, the MUD community developed their own multi-user virtual reality. In stark contrast to the high-end, graphical VR where a few users struggled to pick up a virtual cup with their cumbersome datagloves, headsets and expensive computers, the MUDders were 'already cranking out one lucidly believable digital microcosm after another, more or less just for the fun of it' (Dibbell 1999: 59). Consequently, the MUDding community attracted the attention of academics from a variety of social sciences who wished to study online social relations or use MUDs as a lens to examine changes in 'real-life' (Reid 1994, 1995; Schwartz 1996), examining issues like identity (Turkle 1995), gender swapping (Bruckman 1993), language (Cherny 1995), and social control and power structures (Reid 1999). There was also concern expressed over the potentially harmful impact of MUD addiction, experienced by some users (Kelly and Rheingold 1993; Turkle 1995). The apparent strangeness of the social life inside MUDs helped them burst into the popular media in the 1990s as one of the 'dangerous realms' of the cyberspace, with the most famous article being Julian Dibbell's 'Rape in Cyberspace', originally published in the *Village Voice* in 1993.[8] The article recounted the story of a virtual rape and the consequences on social relations in LambdaMOO, one of the most well-known and studied MUDs, started by Pavel Curtis at Xerox PARC (Curtis 1996; Dibbell 1999). The MUD community is still alive and well in the year 2000, although with a much lower public profile.[9]

Mapping MUDs/MUDs as maps

MUDs are particularly suited to socio-spatial analysis because of their explicit spatial qualities. Indeed, the spatial metaphors at the heart of MUDs are richer and more complex than in any of the other social media we have so far examined. The discrete but interconnected rooms of a MUD provide a geographical geometry that is topologically correct. For example, the topological relations between two connected spaces generally holds, so if heading north out of the hall takes you into the lounge, then taking the south exit from the lounge will take you back to the hall. In this sense, MUDs adopt a recognisable geographic quality; they are logically navigable. It should be noted, however, that the logical topology of room connections can be broken, in particular by disconnected private spaces that do not directly link to the main MUD structure. That is, they can not be reached by conventional 'walking' movement because they do not have a programmed entrance/exit. In these cases, the only way to reach these disconnected spaces is by teleportation to the specific locations.

Given the basic geographic characteristics, it is possible to construct a schematic of the generic types of space found inside a MUD (see Figure 8.2). First, there is core public space, accessible and used by all players. Second, there are two types of private, individual space. One type is connected to the core by conventional walking pathways and the other type is disconnected and can only be reached by direct teleportation. Third, there are what might be termed multi-scale spaces. These spaces are those that break the rules of Cartesian space. For example, a 'Tardis'[10] effect might exist where a larger space (e.g., a casino) exists inside a smaller space (e.g., a broom cupboard). Dibbell states that the most 'dizzying trick of scale' in LambdaMOO was the world that existed inside a spinning globe in the main entrance hall (Dibbell 1999: 49).[11]

In many cases MUDs are based on a real-world template or use an identifiable fictional setting on which the core of the world can be based. For example, the central areas of LambdaMOO is based on the founder's home, and MediaMOO on the building that

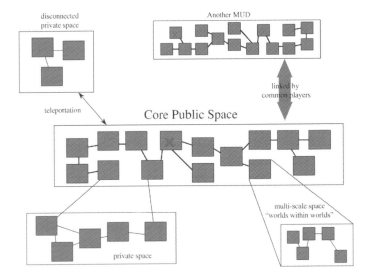

Figure 8.2 Schematic of the generic spaces found in a MUD

houses the MIT Media Lab. Discworld MUD[12] is based on the world described in Terry Pratchett's novels. This spatial grounding of central areas provides a firm and familiar setting for players. However, this grounding often breaks down further from the core, particularly as more players build and extend the MUD in different directions.

Given the explicit spatial topography of MUDs, one logical, analytical approach is to map their spatial form and extent. In one sense, this is an unnecessary task as the MUD is literally a map of itself. This conflation of map and territory is one we explored earlier in relation to webpages, and is an idea that has been most famously expressed in the 'Tale of Cartography' by the Argentinean writer Jorge Luis Borges.[13] Julian Dibbell, in his MUD travelogue *My Tiny Life* (1999), recounts the following realisation, 'It occurred to me that there was in fact one map that represented the width, breadth, and depth of the MOO with absolute and unapologetic reliability – and that map was the MOO itself' (Dibbell 1999: 50). This realisation was brought about by his search for a new home in LambdaMOO, as he explored the MUD seeking to gain a cognitive sense of its geography (see Chapter 9).

While the territory of the MUD as a perfect map is conceptually true, it is of little use to people who want a more simplified, generalised view of the world. One-to-one maps do not enable one to comprehend geography at a larger scale, and are not practical for navigation, as the Borges fable makes clear. As noted in Chapter 4, a significant element in the art of cartography has been the skill to produce maps at a generalised scale, that hide superfluous detail and highlight salient, interesting features of the landscape. Producing such maps for MUDs is a difficult task, given their complex geography that can generally only be plotted by exploration, their geographical peculiarities such as black-hole rooms,[14] and their dynamic nature (active geography) with rooms being structurally altered and new locations added on a daily basis. Despite these difficulties, a number of people have made attempts to map specific MUDs.

In all cases, these attempts have been manual exercises involving laborious surveys. Consequently, most tend to cover only a small portion of the centre of the MUD, usually

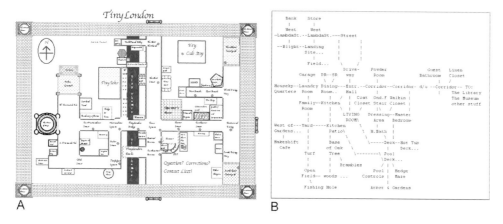

Figure 8.3 Hand-drawn maps of the centre of two MUDs: (a) TinyLondon, (b) LambdaMOO

surrounding the start-point. In some senses, the maps are like city centre tourist guides – useful for the visitor, but largely redundant for the resident who knows the space well. Figure 8.3 shows two interesting examples of this kind of map. The first, (a), is a hand-drawn map of TinyLondon, the capital of DragonMUD, created by the user Liszt. TinyLondon is imaginatively described as a 1800s walled city that is 'perpetually pre-industrial and heavily sprinkled with magic, . . . a city of cobblestones, carriages, gaslamps, stone bridges and ancient towers'.[15] The second map, (b), is an ASCII line-art map of LambdaHouse, the core of LambdaMOO. This room layout is based on the apartment of the founder of the MUD, Pavel Curtis. The map is very simple, with name labels to represent rooms and dashes to show connections between rooms.

In a few cases, very dedicated MUD cartographers have undertaken more ambitious mapping projects. One of the most notable is the atlas of Discworld MUD created by Choppy.[16] The atlas contains many detailed maps that represent hours of manual work. They provide a detailed survey of the geography of Discworld MUD at various scales from regional, to city, to street, and individual public buildings. Figure 8.4 displays one of the regional maps. Each room is represented by a small square which is colour coded according to use: a white square represents normal, green is a shop, blue is a temple, etc. Black lines between the squares indicate possible pathways. Streets and notable shops/buildings are labelled. Inset maps are available to show the internal complexity of key buildings such as the Wizards Guild and the Mansion.

Peter Anders, an architect at the University of Michigan, and his students have also attempted to chart the topological structures of MUD space (Anders 1996, 1998). Working in pairs, students conducted a detailed field survey of specific MUDs, noting and sketching the geography in a notebook as they explored. Next, they constructed what Anders terms 'logical adjacency models' – physical three-dimensional models built with plexiglas cubes (rooms) and rods (connections), that visually graph the MUD structure.[17] As far as possible, the arrangement of cubes and rods are positioned spatially congruent to the MUD structure. Rooms that are connected, but not by Euclidean geometry (e.g., they are accessed by teleporting), are represented by spheres positioned arbitrarily. Only the publicly accessible geography of the MUD was mapped (e.g., private spaces were omitted). In total, ten MUDs were mapped and Plate 8.1 displays the logical adjacency model of

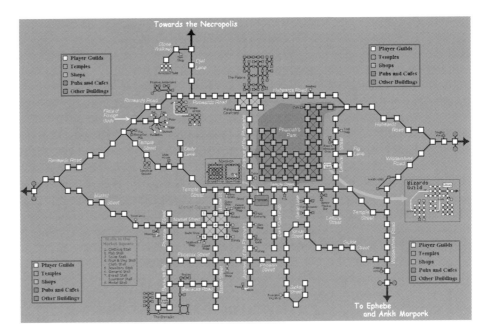

Figure 8.4 Map of part of Discworld MUD

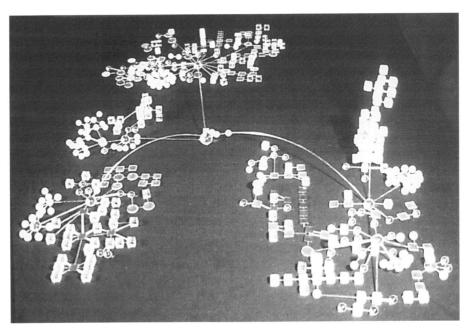

Plate 8.1 Model of the room topology of BayMOO by Thomas Vollaro and Susan Seeler
Source: Peter Anders, http://www.mindspace.net/

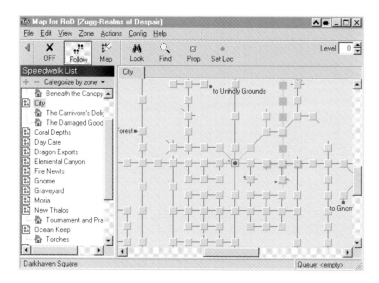

Figure 8.5 Automatic mapping of MUD room topology using zMUD

BayMOO[18] produced by Thomas Vollaro and Susan Seeler. This model highlights that BayMOO is divided into three distinct zones (The Bay Area, Netspace, and Other Worlds), linked by a central node known as the Aquatic Dome.

Mapping MUDs using field surveying techniques and hand-crafted maps is an inefficient process, especially given that most MUDs have many hundreds of rooms and a dynamic geography. Therefore, what is needed is some means of automatically surveying the MUD. One interesting approach that attempts to achieve this, is one which automatically records movement through MUD space, using this information to dynamically map the spaces visited. Such an approach has been adopted by the zMUD client, from Zugg Software,[19] which includes the automapping tool shown in Figure 8.5. zMUD can be configured to decode the room descriptions, and to record the standard cardinal walking directions, teleports and one-way links. The map is similar in style to Choppy's Discworld map, with rooms drawn as small squares and the topological links as thin lines connecting them. Rooms can be labelled and colour-coded to identify them. The zMUD map can also be used as a powerful navigation tool as double-clicking on a room of interest will 'speedwalk' you to that location using the shortest path.

Virtual worlds

Virtual worlds are MUDs with a graphical interface that reveals visually the physical landscape and the avatars of the participants – it explicitly shows the visual geography and topological structure of the MUD and its spatiality. Whilst we refer to them as virtual worlds, they are also known as 'multi-user worlds'; 'networked virtual reality' (Schroeder 1997); 'metaworlds' (Rossney 1996); 'avatar worlds' (Damer 1997); 'inhabited digital space' (Damer 1996); and 'shared worlds' (Roehl 1997).[20] At present, they are the closest form of online interaction to the shared, immersive, VR worlds envisaged by cyberpunk writers (see Chapter 10). In this section, we explore the spatialities and geometries of one of these worlds, AlphaWorld.

Plate 8.2 Three different virtual world interfaces: (a) The Palace (http://www.thepalace.com/),
(b) Blaxxun Cybergate (http://www.blaxxun.com/), (c) OnLive! Traveler (http://www.onlive.com/)

AlphaWorld[21] is one of a number of virtual worlds, accessible through the Internet and run as commercial ventures. Other worlds include V-Chat, InterSpace, Worlds Chat, WorldsAway, The Palace, Deuxième Monde, CyberGate and Online Traveller,[22] and examples of their interfaces are displayed in Plate 8.2. Each system aims to provide a visual, spatial domain (using two-dimensional, 2.5-dimensional or fully three-dimensional graphics) that can be shared by many people using a specialised browser application.

We have chosen to focus on AlphaWorld for a number of reasons. First, AlphaWorld is one of the most technically sophisticated virtual worlds, with a realistic three-dimensional environment, through which users can move in any direction. Moreover, whilst the geography of many virtual worlds are hard-coded, AlphaWorld extends the MUD facility of place-creation through its building capabilities. Inhabitants are able to own land and, using a tool-kit of standard objects, build homesteads. This visual quality, and the ability to build, it is argued, populates AlphaWorld with *places* that have meaning and foster a sense of belonging. This further is enhanced by its use of realistic human avatars[23] (a visual character identifier) to represent participants.

Avatars allow the inhabitants of AlphaWorld to present a visual appearance and project a tangible sense of self.[24] Avatars are able to move, they can manipulate objects, talk to each other and make gestures. Reid (1997: 197) describes them as a '"real" person's proxy, puppet or delegate to an online environment'. Avatars attempt to model some bodily movement with articulated arms, legs and head, along with some limited facial expressions. Movement is presented as a simulated walking motion and a number of pre-programmed actions are possible such as 'dance', 'wave' and 'anger'. Given that they visually portray an inhabitant and allow basic visual communication, Suler (1997) and Jeffrey and Mark (1998) contend that avatar appearance is important for identity formation in virtual worlds. In AlphaWorld, a first-person perspective is adopted so that a participant views the landscape through the eyes of his/her avatar. As such, as the avatar moves, the view of the world changes (although it is also possible to view an avatar in third-person perspective). Plate 8.3 displays a typical AlphaWorld landscape with other users represented by their avatars. This landscape is framed by the client interface, which is divided into three main sections: the viewing window, the text-chat interface, and control panels. Social interaction is mediated through visual cues performed by the avatars and, more commonly, typed chat.

We have also chosen to focus on AlphaWorld because it has a detailed recorded history that is easily accessible to other researchers,[25] a wide array of community information is available on the Web and via active newsgroup discussions, it is host to all forms of social interaction including tuition, tours, contests and religion (Schroeder *et al.* 1998), it has been studied by several social scientists using participant observation (e.g., Schroeder 1997; Schroeder *et al.* 1998; Jeffrey and Mark 1998), and its urban geography has been mapped in detail. Moreover, the world is free to explore.[26]

AlphaWorld is just one of the worlds in a large 'universe' run by Active Worlds, Inc. In total, there are over 500 different worlds, of which AlphaWorld is the largest, many of which are licensed to other individuals, universities and companies. Each world offers the same basic modes of interaction, but is characterised by its focus and visual style. Some are virtual versions of real places like Yellowstone National Park or the planet Mars, while others are for marketing purposes, for example promoting major movies (such as *Godzilla*, *The X-Files*). There are also worlds for different languages and religions, of which the Russian and Scandinavian language worlds are particularly popular (see Schroeder 1997; Schroeder *et al.* 1998). The appeal of AlphaWorld has meant that since its inception in summer 1995,[27] these worlds have been visited by over 800,000 unique users, many of

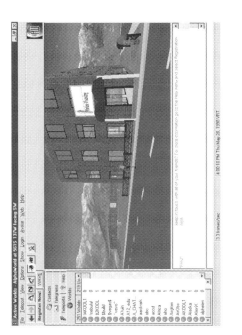
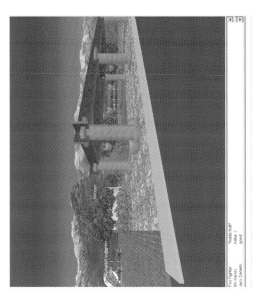

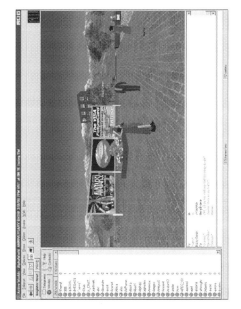

Plate 8.3 Views of AlphaWorld

whom have added to their built form. Whilst these worlds in themselves are of interest, we focus our discussion on AlphaWorld, the largest and most populated world in the Active Worlds 'universe'.

The spatial forms and geometries of AlphaWorld

The original spatial form of AlphaWorld was a flat, featureless and empty isotropic plain, with no natural features, that stretched for hundreds of virtual kilometres in every direction. This plain was coloured a uniform shade of green to signify that it was virgin territory, waiting to be claimed. The total area of the flat plain is exactly 429,038 square kilometres, which is some 43 per cent larger than the United Kingdom or 4.4 per cent larger than California (Vevo 1998). Unlike the UK, the borders of AlphaWorld are straight, forming an exact square of land 655 kilometres across. A Cartesian co-ordinate system is used to delineate this space, with an origin point (0,0) located at the dead centre of the world. This centre point is known as 'Ground Zero' (GZ) and acts as a focal point, the point at which most people enter AlphaWorld. Consequently, the area around Ground Zero is always the most densely populated with avatars, and contains the area of oldest and most developed urban forms. The co-ordinate system is important because it divides the plain into a series of 10×10 metre cells, and allows people to navigate through teleportation to any point in AlphaWorld using an x-y coordinate address. For example, co-ordinates '67N, 42W' translates to 670 metres north and 420 metres west of GZ.

Since AlphaWorld came into existence, inhabitants have been busy claiming land and building all manner of structures from modest suburban-style homes to grand castles. The number of visitors, however, far outweighs the number of builders, of which there have only been about 30,000. As of August 1999, 39.5 million objects had been placed by the inhabitants (Vevo 1999). All the objects used in construction, such as windows, doors, stairs and furniture, are appropriately scaled in relation to avatar size, which is limited to a constant size.

Despite this seemingly large number of objects, so far the inhabitants have made little impact on the vast expanse of AlphaWorld. Only a tiny percentage of the AlphaWorld's land contains any building, and if we take builders to represent overall population, then the population density is just 0.07 people per kilometre. Plate 8.4 is a map of the whole of AlphaWorld, showing the density of urban development as of August 1999, produced by Vevo.[28] The most heavily built-up areas on the map are represented by the brightest pixels. It is clear from this image that the most developed area of AlphaWorld is the densely built city around GZ, which sprawls out in all directions for about 15 kilometres. Ribbons of urban growth project out from this city along the principal compass axes to form a distinctive star shape. Towns and other small settlements lie along these axes, looking like bright beads strung along a necklace. The spatial structure of urban development is largely the result of the single entry point and co-ordinate system, with people choosing regular and memorable co-ordinates, such as 50N, 50W or 1555E, 1555S, as the location for their homestead. Once a pioneer has started building, other citizens will build alongside either by invitation or just to be close to other people.

The series of maps in Plate 8.5 were also produced by the Vevo project, which has developed a sophisticated mapping system, using a quadtree structure, that is capable of producing maps of AlphaWorld at twelve different resolutions. These maps can be browsed interactively,[29] and the highest resolution map, from which one can clearly discern the detail of individual buildings and roads, can be used as a teleportation tool. Similarly,

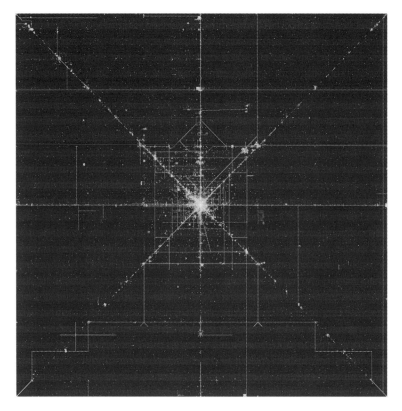

Plate 8.4 Urban density of AlphaWorld, August 1999, by Greg Roelofs and Pieter van der
        Meulen
Source: Philips Research Silicon Valley, http://awmap.vevo.com

Roland Vilett, one of the AlphaWorld programmers, has produced colour satellite-style
maps of the city (Vilett 1999). Plate 8 displays his three maps, portraying the development
of the GZ city (1000N, 1000W to 1000S, 1000E, a 400 km² tract of land) between
December 1996 and August 1999. These images vividly reveal the organic complexity of
the urbanisation caused by the unplanned action of thousands of real users.

   This unplanned action is facilitated by the lack of rules in AlphaWorld. There are no
planning laws, urban policies, or zones of development and inhabitants can lay claim to
and build on any empty plot of land that has not been built on by others. Moreover, there
is no limit to how much land that can be claimed. Finding an empty plot anywhere near
the centre of GZ is now difficult due to the density of existing urban development. This
density is compounded because every plot that is not claimed by inhabitants building on
it can be claimed by somebody else, meaning that a backyard, unless built over, could
be appropriated by another inhabitant. It also leads to one of the most common sources of
conflict, building disputes.

   Building is undertaken using a 'construction kit' of predefined objects, much like virtual
Lego bricks, such as road sections, walls panels, doors, windows, flowers, and furniture.
In total, there are over 1,000 different objects available, that can be combined together
into any structure imaginable. This means that the built environment has a somewhat

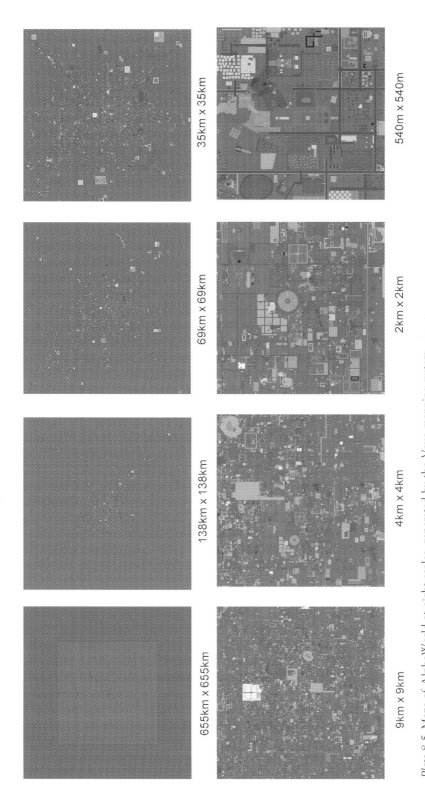

Plate 8.5 Maps of AlphaWorld at eight scales, generated by the Vevo mapping system

Source: Greg Roelofs and Pieter van der Meulen Philips Research Silicon Valley, http://awmap.vevo.com

35km x 35km

540m x 540m

69km x 69km

2km x 2km

138km x 138km

4km x 4km

655km x 655km

9km x 9km

homogeneous appearance as all inhabitants are using the same basic materials. Despite this limitation, individual creativity has flourished, with the building of all manner of structures. Some are well-designed and aesthetically pleasing, but there are also ugly, half-finished, and poorly built structures, none of which can be destroyed except by the original builders. This means that AlphaWorld cannot be redeveloped, large areas are derelict, and some spaces have been vandalised, which permanently damages the plot. In general, structures copy those in geographic space, matching familiar architectural forms and lay-outs (particularly from US suburbia), despite the freedom to stretch and warp the conventional architectural norms. As such, whilst it possible to build abstract structures that float in mid-air, they are rare. This is partly enforced by the range of building blocks provided, but also because inhabitants seemingly wish to create a world which might engender a familiar or yearned for sense of place. Moreover, unconventional structures are difficult to navigate through and thus discourage visitors.

As well as having spatial forms and geometries, AlphaWorld also exists within in its own space–time geometry, adopting the time zone called AlphaWorld Standard Time (AWST). The need for a special time zone arose because of the difficulty of scheduling meetings and events for inhabitants geographically located around the globe, and the confusion this caused as inhabitants tried to convert times into their local time zone. In November 1996, a group of AlphaWorld activists designated AWST as Greenwich Mean Time minus two hours. This time zone, known as mid-Atlantic, is not used by any country in the real world. Subsequently, it has been adopted by other synchronous virtual communities, and has recently been renamed Virtual Reality Time (VRT).

Spatial mobility in AlphaWorld

Although the spatial form and geometry of AlphaWorld seems to conform largely to those experienced in geographic space, it does differ in a number of ways, particularly in relation to spatial mobility. For example, avatars do not need to enter buildings through doors, as they can be guided through a wall by simply holding down the shift key. The default for avatar movement is walking. However, it is just as easy to fly, a mode that is often adopted because it is simpler to navigate across terrain and it is less tedious than walking. As yet, there are no other means to travel along the ground as there are no cars, trains or planes.

One mode of travel that breaks the Cartesian geometry is that of teleportation. Teleportation warps the concept of distance and geographical accessibility, as any location in the 429,038 km^2 expanse of AlphaWorld can be reached instantaneously from any other point, at no cost in terms of time or money. Consequently, every point in AlphaWorld is equally accessible. The ability to teleport is a relatively late addition to AlphaWorld, and has been progressively introduced for fear of its affects on the world – it was initially feared 'that teleportation will ruin the simulation of reality in AlphaWorld' (*New World Times*, no. 4, p. 2). There is no doubt that teleportation does have disadvantages. First, it means that inhabitants tend not to explore other parts of the world and therefore have little understanding of the overall geography, that is, how one place relates to another (Anders 1998). Second, it reduces the opportunities for chance encounters and discoveries as citizens tend to teleport directly to their homesteads and restrict their exploration to its immediate locale.

Beyond the tedium of walking, navigation is made extremely difficult because visibility is limited by the browser. At present, it is only possible to see a maximum of 120 metres in any direction, and the default setting is half this distance. As a consequence, despite the vast expanse of space which composes AlphaWorld, it is difficult to gain any real sense of scale (except by flying) or to build a cognitive map (see Chapter 9) as distant landmarks, for example, are not visible. This constraint is a necessary requirement in order to keep information processing to a manageable level, but it creates the effect of walking around inside an opaque bubble. This is a problem because streets and buildings appear to end with a sharp cut-off line at the edge of the visibility bubble. In practical terms, it is very hard to find buildings and features of interest unless your know their exact x and y co-ordinates.

Spatialities of AlphaWorld

Because of the combination of visual datascape and the use of avatars, one can think of AlphaWorld as consisting of hybrid places – lacking the materiality of geographic space but having a mimetic quality, containing enough geographical referents and structure to make them tangible. This, we suggest, engenders a level of spatiality beyond that found in other virtual spaces, with social interaction explicitly situated and grounded in a geo-graphic context. As with textual MUDs, the place-like qualities of AlphaWorld provides a context in which specific forms of social interaction and experiments with identity are played out (Schroeder 1997; also see Chapter 3). From a geographical point of view, the study of AlphaWorld then is interesting because it provides a social laboratory in which to observe the socio-spatial construction of place (Donath 1997; Huxor 1997; Schroeder 1997).

In AlphaWorld the 'sense of place' is centred around the activity of claiming land and building homesteads, the means by which the space is transformed into meaningful *places*, and by social interaction between the inhabitants. Both lead to specific forms of socio-spatial practice: the playing with identity, the creation of community, land disputes, virtual vandalism, and policing. These in turn are framed within a regulatory structure centred on citizenship.

As with other MUDs, AlphaWorld allows inhabitants to play with their identity. This play is both enhanced and hindered by the use of the avatar. An avatar provides a visual appearance to support a nickname and situate dialogue, but also removes imaginative and ambiguous elements. Thus inhabitants choose, rather than construct, their appearance. Avatars potentially augment characters by providing a set of gestures, although they are not widely used in AlphaWorld except to simulate dancing at parties (Jeffrey and Mark 1998; Damer 1997).

Avatars play an important role in structuring social interactions in AlphaWorld, as it is apparent that inhabitants both consciously and unconsciously use them in ways similar to their material body. The best example of this is the convention of facing an avatar when conversing in order to 'look' at the avatar. When people talk in groups they tend to arrange their avatars in a loose circle, all facing each other. Turning away from an avatar signals the end of an encounter. Moreover, inhabitants use avatars to maintain the same sense of personal space that bodies do in the real world. As such, avatars tend to keep a polite distance between one another, and inhabitants walk their avatars around others, rather than proceeding straight through them. Indeed, the sanctity of personal space around an avatar means unwarranted and deliberate attempts to invade it often feel threatening, and this is known as 'avabuse' (Damer 1997). This leads to new forms of virtual abuses that extend beyond the verbal to include physical blocking, shadowing and stalking (Suler 1997).

Whilst it is clear that AlphaWorld supports many virtual communities, at present a key problem is severe under-population. Generally, there is a lack of inhabitants beyond the immediate surroundings of Ground Zero, and much of the landscape remains a series of ghost-towns. As a consequence, inhabitants tend to congregate around GZ in order to meet other inhabitants, thus perpetuating the barreness of other locales. This problem of under-population is particularly prevalent in other worlds in the Active Worlds universe, most of which are empty and derelict. Put simply, at present, there are simply not enough people using most of the virtual worlds to make them interesting, engaging and self-sustaining social environments.

To counter this problem, there have been several attempts to form specific communities in AlphaWorld through the communal planning and building of townships. The most well-documented of these has been the Sherwood Forest[30] community project run by the Contact Consortium (Damer 1997; Contact Consortium 1998). The project was started in early 1996, with a formal charter and designated town plan. This project aimed to recreate a utopian, Californian-style, suburban township (Kling and Lamb 1996). Another example is the Pink Village, a gay and lesbian community, which has bars, cafés, a night club, a town hall, remembrance gardens, a museum, galleries, as well as 'private' homes (Pink Village 1998). The community is active, with a local newspaper, a calendar of social events including a Pride Festival, and an elected village council.

The uncontrolled and unplanned nature of building also provides a context for a new form of virtual, aberrant behaviour: virtual graffiti and vandalism. Even though the software prevents anyone but the owner of a piece of land changing a building, vandalism is possible by deliberately placing objects as close as possible to other people's homesteads. A small number of users appear to take pleasure from this, using annoying objects, like flames, bogus teleports, and even large billboards with offensive pornographic pictures on them, and placing them in front of the entrance to a homestead. Interestingly, the flame object appears to be used as a 'physical' equivalent of a flame email. This type of vandalism is viewed as a serious 'crime' in AlphaWorld because of the importance citizens attach to their homesteads and it has led to one of the earliest forms of community action in 1995, with the formation of the AlphaWorld Police Department (Damer 1997). In reality, however, there is little a victim can do and it is very difficult to get vandalism removed. As yet, it is not clear how widespread vandalism is, or whether they are random acts or a more concerted campaign against certain properties and individuals.

Community building is framed within a set of regulatory structures. Although anybody can enter AlphaWorld as tourists, several important rights are restricted to citizens (who, since September 1997, pay a fee of $20 a year). These rights include the ability to choose an avatar, to own land permanently and build on it, and also to send telegrams to other users. A tourist is free to wander and engage in conversation, but is stuck with the default 'tourist' avatar and cannot build permanent structures. Schroeder (1997) contends that these differential rights between tourists and citizens has given rise to a two-tier social structure of 'insiders' and 'outsiders', with the tourist often being treated differently and unfavourably by citizens. Even between citizens there is a subtle social hierarchy delineated by the length of citizenship, as indicated by the ordering of ID numbers, and spatial location, with 'pioneers' holding territory in key co-ordinate locations, organising meetings and events at specific locations, as well as exploring more widely, and 'newbies' tending to cluster at GZ (Schroeder 1997).

The constitutional position of citizens at best can be described as vague, and ultimately they should be considered as consumers, despite the rhetoric of immigration/citizenship.

AlphaWorld is a privately-owned space and citizens inhabit the space under a licence agreement. This provides the owners of AlphaWorld, Activeworlds.com, Inc., the right to regulate the space as they see fit, with no means for citizens to independently challenge the management's actions. As with other MUDs, there is evidence that some citizens are unhappy with the nature of their rights under this agreement and how the world is managed. There have been accusations levelled against the management with regard to the arbitrary use of their powers, resulting in the ejection and ban of people from AlphaWorld. These rumblings of discontent are reported in various online newsgroups for discussion of AlphaWorld. In part, the frustration of citizens is that they have no control over a space in which they have invested considerable time and effort, and which has undergone a number of changes, such as the citizen fee and increasing commercialisation.

AlphaWorld, then, can be thought of as another example of a privately owned and operated, semi-public space designed for consumption, just like shopping malls and theme parks in the real world (Sorkin 1992; Graham and Aurigi 1997) – a Disney-like FrontierLand. Just like theme parks and shopping malls, AlphaWorld has its own private security guards, called peacekeepers, along with a more informal set of policing. A peacekeeper's[31] main job is to patrol the world, maintain order, and to try and prevent and counter disruption to other customers, such as verbal abuse, avatar assaults and virtual vandalism. They are organised with a duty roster to provide continuous police cover and their powers include the ability to instantly eject tourists and citizens and to ban them from returning to the world for varying lengths of time. Whilst most citizens see the need for policing, some users have expressed serious concerns over how the peacekeeper's role is executed, with accusations of heavy-handed policing with summary expulsions and an inadequate appeals systems. In addition to peacekeepers there are other volunteer 'security guards' who operate under the direction of the Active Worlds.com, Inc. Generally, these people are employed to welcome new visitors to AlphaWorld, and to provide assistance, but they also act as a form of neighbourhood watch. Active Worlds.com, Inc. maintains that policing ensures that citizens have 'the levels of decency that you and I would expect in the real world', although it is clear that it appreciates that offensive behaviour discourages casual visitors from registering as citizens and alienates existing customers.

The socio-spatial construction of space in AlphaWorld

The ability of AlphaWorld inhabitants to build virtual architecture and communities led Andrew Smith (n.d.) to empirically examine the processes through which place-making occurs. He created a new world called the Collaborative Virtual Design Studio in which users could build a new community. He then monitored in detail the building of urban structures and the social interaction of inhabitants over a thirty-day period (start date, 30 November 1998). The plot of land was 3 million square metres in size, and capable of supporting thirty-two simultaneous users. Both registered and tourist inhabitants were able to enter the world and build structures. No specific guidelines were provided, although inhabitants were encouraged to visit a website[32] which detailed the experiment and a prize was offered for the best structure built within the thirty-day period. Inhabitants entered the world in a town square surrounded by message billboards. Nearby, a builder's yard provided 368 objects from which to build structures.

The experiment revealed a number of interesting results. Most important, users built a diverse range of structures and developed a strong core community who conversed

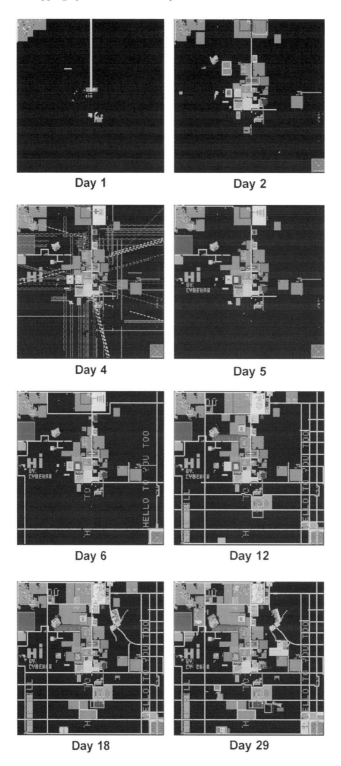

Plate 8.6 30 Days in ActiveWorlds
Source: Andy Smith, http://www.casa.ucl.ac.uk/30days

regularly. The extent of the building is evident in Plate 8.6, which shows satellite images of the world revealing the urban growth over the thirty days. The first 24-hour period in particular experienced considerable development, with 7,219 objects placed. In total, 27,699 objects were placed by forty-nine registered users and an unknown number of tourists, and 49 per cent of all available land was built on. Smith reports that a recognisable community of 8 to 10 users had already developed by day four, appearing much sooner than he had predicted. This group used the same nicknames and avatar appearances over the course of the thirty days. This community developed throughout the experiment, and produced a number of communal structures such as a temple and undertook several communal events (such as all of them adopting Smith's avatar for a day).

In addition, the world experienced some of the more anti-social phenomena of AlphaWorld, as described above. For example, on day four it was subjected to attack from the 'ActiveWorlds Terrorist Group'. On this occasion, over 85,000 objects were added to the world, as evidenced by the patterns of dashed lines in Plate 8.6. Also, some inhabitants took to 'sky writing' – claiming sizeable tracks of land to spell out a message that can only be seen when the world is viewed from the air. The first of these appeared on day four ('Hi'), and was subsequently followed by more. Smith is still in the process of analysing the data, but his findings to date reveal interesting insights into social production of space and communities in AlphaWorld. His thirty-day experiment has now been extended to 360 days.

9   Spatial cognition of cyberspace

He was thoroughly lost now; spatial disorientation held a peculiar horror for cowboys.

(Gibson 1984: 249)

It does not matter if your space is psychological or physical: it is good for the ass to know where it is in space.

(Foy 1997: 153)

Over the past forty years, since the publication of Kevin Lynch's (1960) seminal text, *The Image of the City*, researchers from a number of disciplines, including psychology, geography, planning, anthropology, computer science, cognitive science, and neuro-psychology, have engaged in studies that seek to identify and understand how we think about and behave in geographic space; to determine how we can remember routes, learn new routes, make distance and direction estimates, and know where places are in relation to one another (Golledge and Stimson 1997; Kitchin and Freundschuh 2000). This research is termed cognitive mapping, and the focus of study has been on our cognitive map knowledge and on our ability to effectively process this knowledge; how we consciously, and more commonly, subconsciously, acquire, learn, think about and store data relating to our everyday geographic environment (Downs and Stea 1973). Cognitive map knowledge thus consists of information concerning spatial relations and data on environmental characteristics which reside within a space–time context, and allows people to successfully operate within a complex geographic environment and process environmental and geographic data.

As most users will attest, it is easy to get lost or to be unable to locate a site or person in cyberspace; system interfaces are often not intuitive and data are organised in such a way as to make navigation difficult and circuitous. As we have discussed, maps and the spatialisation of cyberspace seek to facilitate understanding and more efficient usage of different virtual media – either hypermedia, textual or three-dimensional graphical spaces – by providing information spaces that are familiar in appearance and mode of interaction, and are cognitively accessible. The extent to which these spatialisations succeed is largely unknown. In this chapter we explore the relevance of cognitive mapping studies of geographic space to the construction of spatialisations of cyberspace and the spatial behaviour they engender, drawing parallels between the findings of geographic studies and the small number of studies which have explicitly examined the cognitive understanding of spatial relations in cyberspace. Such studies are important because they provide designers with information concerning how different virtual geometries are understood. Cognitive mapping data can thus suggest ways in which to improve the spatial legibility of virtual spaces, and the usability and usefulness (e.g., increase user understanding, learning speed, and navigation) of different spatialisations of cyberspace. Such improvements are necessary if cyberspace inhabitants are to fully exploit the full diversity and vastness of cyberspace.

Cognising geographic space

It seems logical that an examination of how we comprehend geographic space might reveal insights and lines of enquiry into how we spatially comprehend cyberspace (Shum 1990). As such, in this section we detail current thought in the field of cognitive mapping, focusing on theories of spatial learning. In the next section, we examine their application to cyberspace.

As researchers have discovered, the processes underlying the learning and understanding of geographic space are complex (see Golledge and Stimson 1997; Kitchin and Freundschuh 2000, for a full review). Studies have focused on identifying a number of things: the constituent elements of cognitive map knowledge; the knowledge structures used in storing and processing those data; the form in which cognitive map knowledge is stored within the brain; where different facets of knowledge are stored within the brain; how spatial information in different mediums and at different scales is learned; the factors that affect learning; and the development of knowledge across the lifespan. Here, we detail how spatial information from primary and secondary experience is learnt, structured and used.

Primary-based learning is navigation-based, with the collection and processing of spatial information explicitly linked to an individual's interaction with an environment through spatial activity (e.g., wayfinding). There are three main theories about how we learn an environment through primary interaction. In the first theory, it is hypothesised that environmental cues, such as landmarks, form the fundamental building blocks and framework of knowledge on which subsequent information, such as paths, are added. The second theory hypothesises that path-based information forms the initial framework of knowledge and that landmarks and other information are then placed in relation to this. The third theory suggests that because wayfinding only necessitates ordered productions (remembering the order of route segments), with initial learning consisting not of building a spatial database but on memorising vistas, views or scenes rather than the memory of landmarks or paths form the basis of early learning.

Siegel and White (1975) suggest that cognitive map knowledge is acquired in a three stage process: landmark learning, route learning and configurational learning. These stages are strongly related to the development of parallel knowledge structures (see below). The acquisition of landmark knowledge is based on recognition-in-context memory, followed by route knowledge which develops through the association of bearings between landmarks. Configurational knowledge occurs with the scaling and interrelating of routes into network structures. Siegel (1977) suggests that there is a set pattern of development. Initially an individual notes and remembers landmarks. Once landmarks are established as a frame of reference, an individual's acts are registered and assessed in relation to them. Next, an individual forms clusters of landmarks or 'mini-maps' which are internally coherent clusters of elements. Clusters might not, however, be relatively coherent with other clusters. This results in the formation of an objective frame of reference which organises separate perspectives into a spatial system. Full configurational knowledge is achieved only after the establishment both of routes and of an objective frame of reference.

Closely related to Siegel and White's theory of learning is Golledge's (1978) anchor-point theory. Golledge's theory emphasises the role of landmarks in the learning process.

He suggests that different places have different salience to individuals and therefore become hierarchically structured, with more important landmarks acting as mental 'anchors' to less important information. These cues are the anchors on which other information is 'hung' and act as spatial mnemonics, increasing the probability of recognising, or knowing the position of, an associated target cue. Other landmarks have lower levels of use and recognisability (Golledge *et al.* 1987), each acting as minor anchors to the levels below. Secondary nodes identify places of decision-making, recreation and entertainment, such as major road junctions, parks and cinemas. Tertiary nodes are usually places of minor decision-making, such as minor or little used junctions or little used or known landmarks. Minor order nodes are places that are specifically known, but which do not act as decision-making points. Often these are unique to the individual. A number of studies have found evidence that landmarks are learnt first (e.g., Brewster and Blades 1989) and that landmark knowledge is itself hierarchically organised (e.g., Golledge *et al.* 1985; Ferguson and Hegarty 1994).

Golledge (1991) argues that because of their importance, the choice and availability of suitable landmarks within an environment may become critical to both learning and structuring cognitive map knowledge. Areas with differentiated landmarks and landscapes are thus thought to be easier to remember. It may be the case that areas with few landmark cues are less easily cognised than those with many. However, a proliferation of landmarks may lead to errors in our knowledge because of over-processing. Golledge (1991) explains that this is because of a glut of landmarks can lead to confusion. As such, areas of 'optimal' cognitive map knowledge are likely to be those with a balanced number of landmarks, with suitable amounts in each of Golledge's (1978) anchor point categories. Hardwick *et al.* (1983) report that it is not just the availability of suitable landmarks that it is important but also the ability of people to select optimal landmarks. They found that poorer levels of cognitive map knowledge were correlated to selecting ambiguous landmarks.

In contrast to the primacy of landmarks, Gärling *et al.* (1981) argues that it is paths that need to be learnt before landmarks: paths act as a framework for the accurate recall of landmarks. Paths, it is argued, have primacy as they need to be traversed in order to encounter landmarks. MacEachren (1992b) claims that the evidence for the development of routes as the initial framework comes from sketch map data. For example, Appleyard (1970) and Devlin (1976) both found that short-term residents produced sketch maps that were path dominated. Residents of long-standing produced more integrated maps that contained more landmarks. Piche (1977) suggests that people progress from a knowledge of particular routes, to then linking the routes, and eventually building up a neighbourhood map of these routes in a continuous cognitive map of the area. Allen and colleagues have found evidence to support this hypothesis (Allen *et al.* 1978; Allen 1988; Allen and Kirasic 1985). They discovered that routes are learnt through a segmentation process which, over time and frequent encounters, are integrated into a complete cognitive structure.

Couclelis' (Couclelis *et al.* 1987) theory joins together aspects of landmark and route learning. She and her colleagues expand Golledge's anchor point theory, suggesting that nodes within the hierarchy may not necessarily represent landmarks but may include any feature that acts as a cue or anchor. For example, a stretch of main road may act as a primary anchor. Similarly, Gärling *et al.* (1986) suggest that it might be more accurate to think of anchor nodes not as points, but as areal extents. These areas then act as the anchors for the rest of the knowledge base. As such, key landmarks, linear and areal features 'individually and jointly "anchor" subregions of space and hierarchically link together environmental information' (Lee and Schmidt 1988: 340).

In contrast to both landmark-based and path-based learning, Cornell and Hay (1984) have hypothesised that the initial learning of an environment consists not of learning specific landmarks and paths which then form the basis of a complex spatial database but, rather, of recognising vistas; the scenes along a route. Here, initial navigation is not based on an internalised spatial knowledge system but on being able to order productions of vistas; remembering the order of views. Such a process has salience because it mirrors how sighted people encounter environments as they navigate through them. Using higher levels of cognition the information gathered from ordered productions can be converted to accurate mapping and configurational knowledge.

People can also learn about the spatial relations of an environment through other mediums such as maps and textual descriptions. Since we are concerned with spatialisations of cyberspace, and many virtual environments are textual, how people learn and understand maps and narrative has added significance. The use of maps is thought to have two effects on cognitive map knowledge. First, training in map use provides specific guidance in how to process and comprehend spatial information. Most maps are complex models of spatial information that require the use of specific skills in order to process them (Liben 1991). This implies that a novice will not learn from a professionally-produced map unless he/she knows how the map represents an area. For example, Butler *et al.* (1993) found that respondents with little map training needed several minutes to discover and memorise from a map a path in a complex building, and even then some respondents still had trouble navigating. Liben and Downs (1989) found that children could generally understand the holistic aspect of the map, but failed to understand specific features, suggesting that children have a 'fragile and limited understanding of the symbolic nature of the representation' (Liben 1991: 267). This is not to say that we do not possess an ability to understand basic maps from an early age. As Uttal and Wellman (1994) have demonstrated, very young children do understand basic maps, but an ability to understand complex maps requires acquired skills.

Second, studying a map can lead to greater knowledge of an area by revealing 'true' spatial relations. Many studies have demonstrated that cognitive map knowledge derived from maps is different and significantly more accurate than that derived by direct experience (Evans and Pedzek 1980; Thorndyke and Hayes-Roth 1982; Lloyd 1991; 1993; MacEachren 1992a). It is hypothesised that this is because the spatial relations of an area are revealed in entirety on a map while an area is generally experienced in several partial encounters. Maps are also viewed in one perspective while the environment is experienced from many. As such, configurational knowledge is revealed by a map whereas it has to be constructed from various encounters from direct experience. Presson and Hazelrigg (1984) thus argue that information from maps and direct experience are treated in distinct ways, with map information coded into a figural representation that is orientation specific (tied to an alignment). Indeed, Lloyd (1989b) has argued that maps are suitable for storage as an imagery code that is itself accessed like a paper map. This, he hypothesises, will make map-learnt configurational knowledge more spatially accurate as it has not been constructed from route knowledge but, rather, has been scanned as a whole (Lloyd 1989a). As a consequence, distance and direction estimates may be more accurate as well because the map's structure reveals this information within a set frame of reference.

Using map-learnt spatial information for navigating, however, is not a simple task. In order to navigate configurational knowledge gained from a map, it must be transformed into route knowledge. This involves a navigator being able to align the orientation-specificity held in the map to the real world as he/she is positioned within it, and to maintain that alignment as he/she moves through the environment. Peruch and Lapin (1993) and May *et al.* (1995), in VR-based experiments, found that there were significant alignment effects when the map learnt was at a different alignment to the environment experienced. For example, imagine learning a map where position A is at the top of the page and B at the bottom. Here, the map is learnt with an orientation-specificity of A at the top. Now, you are located at the doorway of a large room where you have to navigate between several objects. You are standing at point A and need to move to point B. To complete the task the memorised map has to be rotated 180-degrees so that the objects in the room maintain their left-right locations as displayed on the objective map. Indeed, many of us when using a map to navigate turn the actual paper map so that it is aligned with the features in the environment so as to avoid complex spatial processing.

Several studies (e.g., Taylor and Tversky 1992a; Franklin *et al.* 1992; Denis and Zimmer 1992; Ferguson and Hegarty 1994) have investigated whether cognitive map knowledge can be generated from written texts. Taylor and Tversky (1992a) note that while respondents could produce maps from text descriptions, they found that those constructed from survey text (which described a map) were indistinguishable from route texts. They further stress that narratives from which these maps are constructed must be coherent, organised and unambiguous or respondents will struggle with the task. In a second study (Taylor and Tversky 1992b), they found that map information can be successfully converted to text form. Ferguson and Hegarty (1994) found that cognitive map knowledge constructed from texts has properties similar to those constructed from travelling in real or simulated environments. Denis and Zimmer (1992: 297) found 'very suggestive indications' that such cognitive map knowledge exhibited genuine map-like metric properties and was isomorphic to that learnt from maps (i.e. imaginal). Franklin *et al.* (1992) reported that spatial information in texts are generally categorical in nature (e.g., 'above', 'near to', 'North of'), but that cognitive map knowledge formed from them are rapidly transformed into an ordinal or interval space to provide information concerning locations and orientation. These studies indicate that text-based information can be used effectively to create and update cognitive map knowledge. However, its role in relation to other variables is as yet unknown.

It is generally accepted that, whatever the means, spatial learning leads to progressively more sophisticated spatial knowledge, with knowledge progressing from being declarative in nature to procedural through to configurational. Liben (1981) describes declarative knowledge as the mental database of specific spatial features. Procedural knowledge consists of the rules used to synthesise declarative knowledge into practical action. These rules are essentially wayfinding knowledge that directs movement between places, an example of which is the transformation of path elements into a navigable route. This transformation, however, does not include the ability to make inferences about routes never experienced. Thorndyke (1983) hypothesises that there are two different types of procedural knowledge. The first of these is unordered productions. Here, spatial behaviour along a route is dependent on a series of independent pieces of spatial knowledge. Rather than knowledge being combined into an understanding of the whole route, the individual relies on taking actions at decision points along the route. Thorndyke substantiates his claims by referring to the story of a colleague who stated that she could take him to a place but not direct him to it. The second type of procedural knowledge is ordered productions. Here, order information is also known allowing whole routes to be remembered and described without having to physically traverse them. Hart and Berzok (1983) suggest that ordered productions are divided into three stages. Ordinal mapping is when an individual is almost certain of the sequence of places but is less sure of the relative distances that separate places. Interval mapping includes relative distances and times in the sequence but not direction. The third stage is accurate mapping that contains these missing directional components to provide a complete set of procedural knowledge for a particular route. Butler *et al.* (1993) suggest that people prefer to encode and use ordinal or interval procedural knowledge when wayfinding because it is easier to do so. They found that newcomers to a complex building preferred using ordered signs rather than interpreting 'you-are-here' maps, where distance and direction had to be remembered. Their evidence suggests that accurate mapping only occurs when a route seems particularly difficult to learn or remember. Similarly, Cornell *et al.* (1994) and Annoshian and Siebert (1996) suggest that wayfinding is predominately based on ordered productions with people making navigation choices based on recognising familiar vistas and places along a route. Cornell and Hay (1984) thus suggest that children (and adults) develop the ability to recognise and order vistas before they can construct maps.

The highest level of cognitive map knowledge is called configurational knowledge. Configurational knowledge surpasses procedural knowledge in depth and complexity by incorporating information such as angles, directions, orientation, location and distances between places (Golledge *et al.* 1987). It forms the basis of a comprehensive spatial knowledge system with the possessor having a detailed knowledge of the associations between, and the relative positions of, places (Golledge 1992), the ability to connect different and independent routes, and to make spatial inferences and propositions (Allen 1985). Thorndyke (1983) hypothesises that there are two types of configurational knowledge. Schematised knowledge consists of simple, prototypical configurations of elements that form basic representations of an area. Hart and Berzok (1983) call this 'loose topological mappings', and suggest that locations are mapped on the basis of ordinal or categorical strategies, such as, 'near to', 'parallel to', and 'in front of'. Alternatively, detailed knowledge is well developed, hierarchically organised and is nearly perfect in nature, being built using co-ordinate strategies through extensive navigational experience. Thorndyke (1983) also suggests that procedural and configurational knowledge are linked by symbolic abstractions so that navigation based on landmark and icon recognition is supplanted with semantic knowledge about the names of locations and routes.

Directly related to levels of knowledge are frames of references. Frames of reference (or systems of reference) are heuristics used to place knowledge of locations in relation to one another (Tversky 1981; Moar and Bower 1983). They help us to orientate ourselves and give us a 'sense of direction' (Kuipers 1978). An egocentric frame of reference refers to knowledge which is only known in relation to oneself. Here, a person knows where places exist in relationship to his/her own position through a system of self-reference. A fixed frame of reference refers to knowledge that is known in relation to oneself but also in relation to external features within the environment. Here, knowledge of spatial relations is organised into a natural co-ordinate system. Fixed frames of reference are important because they allow an individual to use cognitive map knowledge to orient him/herself in a known area regardless of his/her position in the environment (Gärling *et al.* 1986). Tversky (1981) discusses three types of fixed frames of reference: axes of symmetry, main-line axes and landmark axes, in which cognitive map knowledge is aligned and rotated to features in the environment. Axes of symmetry refer to natural features that bisect areas into two, such as a river. Linear features may well be aligned so that they run parallel to these features, thus distorting cognitive map knowledge (Tversky 1981). The same effect will happen when knowledge is organised with reference to main-line axes, such as, roads, railways and rivers, or axes that link well-known landmarks.

Global frames of reference extend beyond fixed frames, allowing individuals to orientate themselves in unknown environments, regardless of the direction faced (Gärling *et al.* 1986). With a global frame of reference it is possible to locate all other known places within the same frame of reference and examples of such systems include cardinal directions, mapping co-ordinates, and latitude/longitude values.

In general, there is a belief that knowledge progresses through these three frames (egocentric, fixed, global) with age and experience, and that they tie well to the three knowledge structures, declarative, procedural and configurational. That is, egocentric frames are tied to the declarative stage; fixed frames to procedural; and abstract to configurational. An alternate argument is that these frames of reference are a sub-section of the rules of procedural knowledge and heuristics, and essentially concern the spatial sophistication of declarative knowledge. As such, different frames of reference can be used in different situations (Pick 1976). For example, an abstract co-ordinate system is of little use in trying to remember the location of an item within a room, or wayfinding around a building, whereas a fixed frame may be of more use. An abstract system might be of more use navigating in an area devoid of landmarks, for instance at sea.

This discussion, we believe, reveals that those who spatially produce cyberspace (e.g., construct the virtual worlds) and those who produce spatialisations aimed at facilitating navigating, searching and understanding of information spaces, need to think carefully about how they structure their spaces if they are to facilitate optimal use. As we have discussed throughout the book, many of the media of cyberspace are spatially complex and difficult to navigate through. In the following section, we discuss studies that have sought to understand how we spatially cognise different virtual spaces of cyberspace. In the final section, we consider how spatial cognition research could improve the spatial legibility of cyberspace.

Cognising cyberspace

> This is very important for the users of cyberspace – with a unified layout, people can remember where they are and what's around them. Without this, people will find cyberspace rather disorientating and discontinuous – something the real world is not. In a unified cyberspace, you can make maps, or stop somewhere and ask directions.
>
> (Pesce 1995: 317, cited in Imken 1999)

To date, only a few studies have investigated how people spatially cognise cyberspace and the utility of the methods (e.g., spatialisation) used to improve its spatial legibility. Generally, these studies have fallen into two categories: those that focus on wayfinding in virtual, graphic worlds such as AlphaWorld, and those that concentrate on navigating through hypertextual environments like the Web.

Cognising virtual worlds

> Even in the physical world, the natural navigational abilities of humans and other animals are not completely understood. When the world is a virtual one, the problem is exacerbated by the degradation of sensory cues resulting from poor resolution, device latencies, and other shortcomings of current technologies. As virtual spaces become larger, more abstract, and more dynamic, the cues and stimuli associated with the physical world may be lacking altogether.
>
> (Darken and Sibert 1993)

A handful of studies have now investigated the ability of people to learn and navigate in a visual, virtual environment of three-dimensional graphics using screen or immersive virtual reality (VR), most compare the findings to a control study where the same/similar task was replicated in geographic space. Nearly all the studies have found differences in the ability of respondents to navigate in geographic and virtual spaces, with respondents finding it more difficult to navigate in virtual space. Despite these differences, all research suggests that the same cognitive processes are employed in virtual space, but that these processes are compromised by the interface and other technological factors.

Witmer *et al.* (1996) compared the spatial knowledge of respondents who had learned an environment through a virtual medium with those who had not. They reported that their respondents could successfully learn a virtual model of a real building and were able to transfer this knowledge when tested in the building, although they made significantly more route-finding errors than participants who were trained and tested in the building. Respondents in a study by Wilson *et al.* (1996), who learned the spatial configuration of a three-story building in a virtual environment, similarly performed significantly worse at estimating direction estimates than a control group who learnt the real building.

Richardson *et al.* (1999) compared the ability of sixty-one respondents to learn the layout of two floors in a complex building from a map, from direct experience, or by traversing through a 'desktop' virtual representation of the building. The group using the virtual environment performed the poorest, although similar levels of performance were displayed for learning the layout of landmarks on a single floor. They also displayed orientation specific representations defined by their initial orientation in the environment, and were particularly susceptible to disorientation after rotation. The authors conclude that, in general, learning a virtual environment is similar to learning geographic space, and uses the same cognitive processes, although respondents are more likely to become disorientated and have difficulty integrating layouts of other floors.

Other studies, rather than compare spatial performance in virtual and geographic spaces, compare spatial performance in virtual space with learning a map of the same space. Satalich (1995) explored the wayfinding ability of sixty-five respondents in a visual, virtual environment that comprised of a U-shaped building measuring 100 feet by 100 feet. The building contained thirty-nine separate rooms and over 500 objects. Collision detection was incorporated into the environment so that respondents could not walk through the walls, although it was not incorporated for the objects. Respondents moved in the direction their head pointed with motion directed by a three-dimensional joystick. Respondents learned the environment in one of three conditions: self-exploration (free to explore the building as they wished); active guided (follow a pre-determined path using the joystick); and passive guided (the respondent could move through the environment at a constant speed with no interaction, although he/she could move his/her head to look around). In addition, some respondents were allowed to view a map for five minutes before entering the environment and some were allowed to 'carry' a map while exploring which also showed their position. A control group learnt only the map, without experiencing the environment. After exploring the environment, respondents were asked to complete direction and distance estimation tasks and two wayfinding tasks using the most efficient route. Satalich found that regardless of the measure used, the control group performed either equivalently or better than the group that experienced the virtual environment.

Ruddle *et al.* (1997) tested the spatial knowledge of two groups of respondents to complete distance, direction and route-finding tasks. The first group learnt a building layout (135 rooms of which 126 were empty and nine contained landmarks) by studying a floor plan, and the second group learnt the same layout in a screen-based virtual environment.

Both groups were then tested in the virtual environment. They found no significant differences in the route-finding ability of respondents who had learnt a building layout within a virtual environment or through map learning. They attributed this fact to members of the second group regularly using an environment (nine training sessions, of approximately four hours each) with high fidelity (more photo-realistic). Time in the environment seems of particular importance. In their initial trials respondents were disorientated. However, there was a steep learning curve across trials with the route through the building becoming progressively more accurate so that by the ninth trial, seven out of twelve respondents had near-perfect route knowledge (i.e., took the shortest route between the nine test locations). In a second experiment, the effect of local 'abstract' landmarks (coloured cubes) sited in corridors on the rate of learning within a virtual environment (152 rooms of which 141 were empty and ten contained furniture) were tested. They found that although respondents changed their learning strategy to make use of the landmarks, their use did not lead to significant improvements in route-finding ability. In a third study, experiment two was repeated, but this time a familiar item (e.g., a clock, a cup, etc.) was placed on top of the coloured cube. This led to significant improvements in route-finding ability but not in distance and direction estimates. They speculate that this is because the objects aided within-environment location but did not affect configurational knowledge.

Tlauka and Wilson (1996) compared the spatial cognition of objects located within a room learnt through either virtual navigation or viewing a map. Respondents were required to point to objects that were not directly visible from both aligned and contraligned perspectives and to draw a map. No differences between conditions were found (navigation versus map). This led them to conclude that navigation in cyberspace and 'geographic' space lead to similar kinds of cognitive map knowledge.

Regian and Shebilske (1990) conducted studies of the use of VR as a training medium for visual-spatial tasks. One of their experiments involved wayfinding through a virtual maze which consisted of four rooms in each of three stories. Each room was the same size and each contained a unique colour-coded object, either a star, cube sphere, or pyramid. The colour could be red, green or blue. Every room was connected to at least one adjoining room by a hallway, or a passageway leading to a room above. The walls were coloured grey, the floors red and the hallways yellow. Respondents were guided on three tours and then given an hour to explore the environment. They were then tested on their ability to find the shortest route linking a set of rooms. The authors performed a comparison between these data and data from a random walk algorithm and found significant differences. The results of this study showed that the mean number of rooms for each tour was 10.1 (had to visit eight), 4.8 (had to visit four) and 6.7 (had to visit six). These results, Regian and Shebilske contend, demonstrate that virtual environments can be successfully navigated when designed appropriately.

Turner and Turner (1997) concluded from their study of distance estimates within a small five-room virtual environment that their participants' spatial knowledge was similar to that gained from exploring the real world, but is best described as being most like that gained from exploring a restricted real-world environment, such as a cave, or exploring with a restricted field of vision (e.g., wearing a helmet).

The authors of the above studies suggest a number of reasons as to why learning virtual space is more taxing than learning geographic space. For example, Ruddle *et al.* (1997) suggest that differences could be attributable to the restriction of the field of vision to 60–100 degrees, and movements within a virtual environment being guided through an abstract and limiting interface (e.g., mouse, keyboard, head-tracking helmet, three-dimensional joystick) rather than by eye, head and body movements. Richardson *et al.* (in press) develop this later point, explaining that the lack of proprioceptive (muscle movement) cues during navigation causes an optic/vestibular (eye movement/leg muscles stationary) mismatch. They further note that in non-immersive environments there is an element of scale translation needed to compute spatial dimensions. Virtual environments are also less visually complex, with fewer subtle landmark cues (notices, marks on walls, worn steps, etc.) and the inclusion of sound is restricted. Satalich (n.d.) also suggests that unfamiliarity with the medium may be a factor (a theory that is also advanced by the Ruddle study).

In contrast to studies which compare spatial learning in geographic and virtual space, some studies have investigated the extent and success of specific wayfinding strategies in visual, three-dimensional graphical environments. For example, Darken and Sibert (1995, 1996a, 1996b) have tested whether people use wayfinding strategies common in geographic space in large, visual, VR worlds. They did so by examining participants' performance on a complex searching task in five different virtual worlds, each containing a differing set of environmental cues. Each virtual world was equivalent to approximately 12,000 square kilometres in real-world dimensions and consisted of a mix of land and sea. In each environment, the viewpoint was restricted to movement above the terrain but below a maximum altitude of 400 metres. In each environment, participants had to locate a number of ships. Four conditions were tested: (1) control treatment – no wayfinding assistance provided; (2) grid treatment – the virtual world was draped in a colour-coded grid; (3) map treatment – a map of the world presenting the precise location of the ship was present at the bottom of the view; (4) map/grid treatment – both grid and map were present. They concluded that respondent wayfinding behaviour and level of orientation strongly correlated to the number and type of guiding aid. Other significant conclusions included:

(1) When not given an adequate source of directional cues, disorientation will inhibit both wayfinding performance and spatial knowledge acquisition.
(2) A large world with no explicit structure is difficult, if not impossible, to search exhaustively. This was shown by repeated reacquisition behaviour in the control treatment.
(3) A conceptual co-ordinate system is often imposed on the world to act as a divider. This is a side-effect of not being able to divide the world explicitly. A structure must be imposed on the world if an organised exhaustive search is to be attempted.
(4) Observations support the notion that path-following is a natural spatial behaviour. Subjects frequently used such features as coastlines or grid lines as if they were paths.
(5) A map allows for optimisations to be made to search strategies. This is because it can be considered a supplement to survey knowledge.
(6) Dead reckoning was observed to be an intuitive and natural part of navigation; all subjects exhibited this behaviour even though frequently they were unaware of it. The ability to infer position from a past location and constant velocity over time, while sometimes complex in reality, appears to be more easily understood and implemented in virtual spaces.

Darken and Sibert note that the users of virtual worlds tend to wander aimlessly when visiting a place for the first time. They also have difficulty relocating a place and are often unable to grasp the overall topological structure of the space. These problems are compounded because sites are often not revisited, offering the navigator no opportunity to develop a usable cognitive map of the environment. As such, naive searches, although rare in the real world, are common in virtual worlds, especially among first time explorers.

In addition to research concerning the navigation and the acquisition of spatial knowledge in a virtual environment, some research projects have used virtual environments as general laboratories for studying spatial cognition. For example, May *et al.* (1995) used a virtual environment to test the orient-specificity of knowledge gained from studying a map. In this experiment, participants were shown a map of a room (40 × 40 metres) containing twenty objects with a route plotted through it which they redrew or described from memory, and then followed through the virtual environment. Each participant completed eight trials. The virtual environment was used as a test environment only because it provided more freedom in constructing and manipulating spatial layouts and because it secured easy data recording of spatial and temporal aspects of movement. In another study, Peruch *et al.* (1995) tested respondents' ability to learn the location of four cubes within a virtual room containing sight screens. Some respondents could freely explore the space, another group were taken on a tour, and a third group were shown slide shots from certain locations within the room. Again, the VR medium was chosen because it could be carefully controlled. Despite this degree of control, it is clear that the cognition of geographic and virtual spaces does differ. Therefore, using virtual space as a controlled laboratory to make conclusions about spatial cognition *per se* may be a flawed exercise.

Cognising hypertext

> all users of the web are faced with a common task that will determine their success or failure. That is the task of finding your way along the information superhighway, or wayfinding.
> (Bachiochi *et al.* 1997)

Dieberger (1996) notes that the hypertext spaces of the World Wide Web often appear to have little or no structure. This is because, as Murray (1997) has pointed out, hypertext is structured in a rhizomic fashion, in which any point may be connected to any other point. This rhizomic structure can make webpages difficult to navigate between because they demand a high cognitive load (Kim and Hirtle 1995). Since the 1980s, researchers have sought to understand how people navigate hypertext, often with the aim of improving interface design to facilitate navigation. In many cases, the use of spatial metaphors, particularly mapping and wayfinding, are explored along with more traditional browsing strategies: scanning (covering a large area without depth), browsing (following a path until a goal is achieved), searching (explicit goal search), exploring (finding the extent of information), and wandering (unstructured search) (Canter *et al.* 1985, cited in Kim and Hirtle 1995).

Bachiochi *et al.* (1997) examined user-navigation through a hierarchically structured website, testing the ability of forty-five participants to navigate and search a website. They concluded that current browser tools are not sufficient for effective and efficient searching, particularly for novices who often became confused and 'lost' within the many pages of a single site. They suggest that browsers must be augmented with:

(1) a logical design structure;
(2) structure buttons and their associated webpages;

(3) a website (rather than log-in) home button labelled appropriately;
(4) fixed navigational aids at the top, beneath the browser tools.

Khan and Locatis (1998) found that one way to improve navigation efficiency was to pare down the links within a site. This can focus searching and exploration, and reduce cognitive load on users. They also found that listing, rather than embedding, links improved navigability. Cockburn and Jones (1996) suggest that browsers need to be user-centred in design, considering user models of navigation across the Web in order to carefully craft navigation features. An alternative is to make the browser's current navigation models more transparent. They note that, at present, browsers do not provide feedback concerning the current context of the page displayed, nor do they allow alternative representations of the visited sub-space. One problem of 'spatial' browsers, especially those that are dynamic in nature, is that constant changing will destroy any incremental ability to build a mental map of how to navigate between sites (Andrews 1995). Thus, the most effective sites to learn are those that remain stable.

Spertus (1997) has identified seven heuristics that can be used to customise a search through Web-based hypermedia, each one providing a strategy that can be used to guide information searches:

(1) Taxonomy closure: starting at a page within a hierarchical index, following downward or crosswise links leads to another page in the hierarchical index whose topic is a specialisation of the original page's topic.
(2) Index link: starting at an index, any page reached by following a single outward link is likely to be on the same topic.
(3) Repeated index link: starting at an index P and following an outward link to index P′, a page reached through a further outward link is likely to be on the same topic (or a specialisation of it) as the original page P.
(4) Authorship location: if P is a homepage and file P/ is in a directory below that of P, then P/ is likely to be authored by the person identified on page P.
(5) Directory/hyperlink correlation: if page P is above page P/ in the file hierarchy, one can probably follow hyperlinks from P to P/ and vice versa. This is especially likely if P is a home page.
(6) Spatial locality of references: if URLs U1 and U2 appear 'near' each other on a page, they are likely to have similar topics or some other shared characteristic. Nearness can be defined as proximity in the structural hierarchy of the page.
(7) Temporal locality of references: if a page R referenced a page P in the past, it is a good place to look for a reference to P, even if the URL of P has changed.

The extent to which these heuristics are used by people searching for specific webpages is, however, unknown, although Tauscher and Greenberg (1997) identified three predominant browsing search patterns used by their subjects.

(1) Hub-and-spoke. People visit a central page (hub) and navigate the many links to a new page (spoke) and back again.
(2) Guided tour. Some page sets include structured links (e.g., next page), and people can choose to follow these.
(3) Depth-first search. People follow links deeply before returning to a central page, if at all.

They note that 58 per cent of all pages visited are re-visits, with most access limited to a few highly related pages accessed through short sequences of repeated URL paths. They suggest that because most behaviour is habitual (same places visited regularly) one way to make the Web more efficient to navigate through is to create more efficient history mechanisms that extend beyond stack-based lists of the last sites visited.

Shum (1990), Kim and Hirtle (1995) and Dieberger (1996, 1997) suggest that one way to make hypermedia structures more navigable is through the use of appropriate navigation metaphors. Shum (1990) suggests the development of physical hypermedia structures which reflect conceptual structure, so that our ability to learn and navigate through space is fully utilised. Kim and Hirtle (1995) contend that because navigating in hypermedia involves many of the same tasks as wayfinding in geographic spaces (knowing where one is, keeping track of a route, planning a route), one way to improve navigability is to employ spatialisations, translating the hypermedia into a visual space, or by providing site maps or guided tours. Dieberger has implemented this idea and has empirically explored it utilising both architectural and city-based spatial navigation metaphors to improve interface design. In Juggler he linked architectural metaphors to webpages, so that the environment was organised into an architectural form with added functionality such as teleporting. He also has an alternative idea to develop a conceptual user interface modelled on the structure of a city. He suggests that such an interface metaphor would have utility because, although complex, people are familiar with navigating around cities; they know how to get information, how to reach particular destinations, and how to make use of the infrastructure. Moreover, cities possess a unique set of navigational tools (such as landmarks, buildings, transport systems) that lend themselves to creating sub-metaphors. This approach is also being developed by the Benford's research team at the University of Nottingham (CRG 1997). Whilst not employing the architectural or city metaphors, many of the spatialisations discussed in Chapter 6 are an explicit attempt to provide a more cognitively accessible structure to hypermedia systems. However, most have not been rigorously evaluated, so it is unclear how effective they are. As none of them have become widely used, one might conclude that they are not yet sufficiently useful to facilitate navigation or search.

Increasing spatial legibility

From the above discussion, it is clear that the primary method to increase the usability of spatialisations of cyberspace is to increase their spatial legibility. Our discussion of the cognition of geographic space suggests that several ingredients are needed in order to improve spatial legibility: that environments are differentiated, not too complex, and easily ordered into hierarchies; that they facilitate the development of a fixed frame of reference; that primary navigation experience should be accompanied by secondary aids such as maps; that paths should have recognisable vistas that can easily be placed into ordered productions. The empirical findings we report in this chapter tend to concur with these suggestions.

Charitos (1997) and Benford *et al.* (1997), drawing on the geographic-based work of Lynch (1960) and Passini (1992), suggest that virtual environments do not need to mimic geographic space in terms of their geometrical form, but that they do need to be legible in the same way as geographic space in order to facilitate navigation. In other words, virtual spaces need to draw on pre-existing mental models of spatial information (Chen *et al.* 1998). As such, virtual environments should include landmarks that are prominent physical features, contrast with the background in terms of form, colour and other characteristics (meaning, style, etc.), be visible from many locations, and perform a symbolic function. Benford *et al.* (1997) suggest structuring the space using the Landmark, Node, Edge, Path, and District categories as used by Lynch to demarcate areas of a city into legible forms. In addition, virtual environments should also include a number of signs, including directional signs which designate a direction towards a place, identification signs which identify locations, and reassurance signs which act as check-points and reassure the user that he/she is on the right route. Boundaries that demarcate rooms and zones should be displayed and collision detection enforced. Spaces within an environment, then, should be clearly defined and the routes between places expressed explicitly through signs or identifiable paths or implicitly through landmarks. Chen *et al.* (1998) explain that a space that does not conform to a user's mental model of an information space suffers from an 'embedded digression problem', resulting in the user spending a great deal of time wandering the space while learning nothing of interest, a condition known as the 'art museum phenomenon'.

Darken and Sibert (1996b) suggest that virtual wayfinding aids must support both exhaustive and non-exhaustive searches and must facilitate configurational knowledge acquisition. As such, virtual design principles need to be enhanced to facilitate expert-like navigation performance in novice users. This enhancement, they suggest, should consist of:

(1) dividing virtual spaces into distinct smaller spaces in order to create a sense of 'place';
(2) organising the smaller spaces using a simple organisational principle;
(3) providing frequent directional cues by inserting cues and signs into an environment to more effectively guide users through it.

In addition, they suggest that accompanying maps should be generated so that users can situate their virtual movements in relation to a fixed-referenced guide. These maps they contend need to:

(1) show all organisational elements (paths, landmarks, districts, etc.) and the organisational principle (how the space is structured);
(2) always show the observer's position;
(3) be orientated with respect to the direction faced by the observer so that the top of the map always corresponds to the direction faced, reducing tasks such as mental rotation.

In order to facilitate wayfinding in visual, virtual space, Darken and Sibert (1993) suggest the adoption of a navigation tool kit that consists of a sub-set of the navigation techniques used in geographic space by both humans and birds (avian) (see Table 9.1). In their study they tested the effectiveness of each of these tools to aid navigation through a large virtual environment overlaid by a grid and sparsely populated with objects such as ships and rectangles. Respondents learnt the environment under three different conditions: free exploration; naive search (respondent knew what the object looked like but did not know its location); and informed search (respondent knew both the description and the location of the object). When a target was located, respondents were instructed to return to the starting position as efficiently as possible. While each tool improved navigation and spatial search, they found that different tools did lead to significant differences in spatial behaviour. Of particular success in their environment was the use of a synthetic sun, which improved performance in both the search and return phases. Prior to the sun being added, all respondents experienced difficulty returning during the homing phase.

Table 9.1 Navigation tool kit

Technique	Real-world analog
Flying	Avian navigation
Spatial audio	Avian landmarking (sound cues)
Breadcrumb markers	Trailblazing (leaving a trail)
Co-ordinate feedback	Global positioning indicator (GPS)
Districting	Urban environmental cues (areal)
Landmarks	Urban environmental cues (point)
Grid navigation	Contour map orientation (terrain)
Map view	Map organisation

In relation to hypertext and hypermedia environments, the studies we report seem to suggest that to enhance wayfinding through these environments, the spatial relations of the Web should be structured so that it metaphorically corresponds to that of geographic space, or a process of spatialisation should be used to produce a space–time dimension that can be more easily navigated. In both cases, the rhizomic nature of the Web remains, but is interfaced by a form that appears more linear and which is cognitively accessible. Some of the spatialisations detailed in Chapter 6 were an explicit attempt to improve navigation through hypertext spaces. As yet, however, to our knowledge there have been no studies which have examined how effective different forms of spatialisation are at facilitating spatial understanding. As discussed in relation to geographic space, geographic visualisations do have a significant impact on spatial knowledge acquisition.

In this chapter, our discussion has primarily focused on visual virtual worlds and hypertext environments. There is also a need to chart how people learn to navigate through other virtual spaces, and what navigational aids facilitate such wayfinding practice. As with those spaces discussed, determining the processes used in cognising virtual space is made more difficult by the spatial nature of the space itself. As noted, most of the spaces of cyberspace do not have the formal geometrical properties of geographic space. Some spaces are discontinuous and fragmented, and lack concepts such as distance and direction. It is therefore possible that new methods of spatial learning, exploration and choice-making are developed, new processing heuristics employed, and new types of knowledge structure constructed in such scenarios.

10 Imaginative mappings of cyberspace

Central to our thesis so far has been the contention that we have entered into a condition of late-modernity, a condition that has arisen in part due to the transformative agency of ICTs and cyberspace. In combination, ICTs and cyberspace, we have argued, are changing socio-spatial relations and producing new modes of communication. In this chapter, we further this analysis through an examination of the imaginative geographies of cyberspace and the information society. In particular, we provide a detailed study of the science-fiction genre, cyberpunk, and other forms of cyberfiction, thereby examining the extent to which the modern is transforming into the late-modern, and exploring the spatialities of future geographies, both on- and offline.

We believe that such a reading is important for four reasons: (1) cyberpunk writings, in particular, have received widespread academic praise for their recognition and understanding of the socio-spatial processes underlying the late-modern condition now prevalent in Western societies, and their future visions of the new spatialities this condition will evoke (they have also, as we note, received criticism); (2) these writings provide an informed view of possible futures given present trends, futures that are imaginatively constructed and free of the constraints of academic prediction-making; (3) cyberfiction provides cognitive spaces, informed 'sites of contemplation', in which to examine present-day society and formulate critical, resistive practices (see Haraway 1991); (4) a number of recursive relationships exists between authors and readers, and there is clear evidence that the fiction provides an imaginal sphere in which to conceive technological development (Stone 1991; Tomas 1991). As such, some sections of society seek to make real socio-technical futures articulated in the narratives (and in some cases have succeeded – fiction *is* becoming reality[1]).

Our analysis is based on a detailed reading of thirty-four novels and four collections of short stories.[2] All but one of the novels are by North American writers[3] and all are by men bar two.[4] Initially, novels were identified by searching online book catalogues for fiction relating to cyberspace, virtual reality and cyberpunk. Other books were discovered whilst browsing bookstores, and a few were recommendations. In the first instance, our study was to be solely concerned with examining the writer's textual descriptions of virtual spaces for comparison to material in Chapters 5 to 8. Here, the explicit focus was to be on spatial relations, geometries and forms of virtual interaction. Before commencing the reading, the brief was expanded to include any (undefined) aspect that related to future geographies. As each book was read, passages containing 'geographically-related/spatial' descriptions and narratives were identified and notes were made.[5] In order to allow the data to 'speak for itself', so as to illustrate the visions of future geographies as expressed by the authors, the following account contains a number of passages from the stories.

It seems to us that there were three possible ways to structure our discussion and analysis of these texts. The first was to provide a systematic account of the observations of different authors. The second was to discuss the imaginative geographies of the information society, followed by a discussion of the imaginative geographies of cyberspace. The third approach was to divide the discussion into thematic sections. In our opinion, the first approach, whilst instructive, is too unstructured to provide the type of analysis required. The second approach reproduces the artificial spatial divide between geographic space and cyberspace by forcing a discussion that views each as an autonomous spatial arena. As we have argued throughout the book, these two spaces exist along a continuum. It therefore makes sense to structure our analysis using the third path. We use a dialectical approach to explore the interrelationship of cyberspace and geographic space, and how cyberspace is changing the nature of socio-spatial relations (these binary opposites for the most part match those outlined at the end of Chapter 1). We acknowledge that our analysis is partial, and that a more complete and thorough analysis of the texts is desirable. We start, however, by introducing the imaginative geographies of science fiction and the utility of studying cyberfiction.

Imaginative geographies of science fiction

In his book *Worlds Apart*, Malmgrem (1991) contends that fiction displaying scientific imagination transformed into the recognisable genre of science fiction (SF) with the publication of Mary Shelley's *Frankenstein* in 1818. This transformation occurred, he argues, because the age of Enlightenment provided a shift in systems of thought about how the world works. For Malmgren, SF as a genre is predicated on the assumptions of Enlightenment thought. Rational scientific practice, the industrial revolution, and accompanying technological and social change demonstrated how people, through the use of science, could advance society. Moreover, 'the possibility that the present had evolved from the past and that the future could be extrapolated from the present', opened up the future to narrative imagination (Malmgren 1991: 4). This imagination throughout the nineteenth and twentieth centuries was founded on a number of principles, namely, a use of scientific rationalism, humanism, linear time, and an understanding that the external world is both real and phenomenal. Writers thus sought to balance the fantastical with a scientific rationale that domesticated the implausibility of the narrative; estrangement is balanced by plausible scientific explanation. Malmgren (1991: 6) thus notes, 'SF rigorously and systematically "naturalises" or "domesticates" its displacements and discontinuities'. Estrangement is induced through the introduction of a totalising novum (novelty, innovation) in the form of extrapolation and speculation (Suvin 1979). In turn, SF uses its narrative to say something about the present condition, so that Bloch (cited in Suvin 1979: 54) states 'the real function of estrangement is – and must be – the provision of a shocking and distancing mirror above the all too familiar reality'. SF thus creates a cognitive space, an estrangement between real and fictional worlds, which the reader must negotiate (Malmgren 1991).

Both Malmgren (1991) and Suvin (1979) argue that central to the SF genre, and what makes it distinctive, are not the story lines, but rather their examination of worlds (this one or otherwise). In other words, it is the spatiality these novels evoke. Armitt (1996: 5) thus contends 'it is the spatial that determines the realm of textual dynamics'. As such, Armitt (1996) suggests that the space has become a central metaphor of examination in understanding SF and fantasy fiction. Here it is recognised that estrangement is bound within spatial metaphors such as being 'out-of-place'; space being invaded by strangers; the construction of new, unfamiliar spaces; territorial identity being disrupted, and is thus suitable to, as Armitt contends, psychoanalytical readings of boundary negotiations and other social, critical analyses. Given the centrality of space, it is perhaps not surprising that to date the imaginative geographies of science fiction have been little explored by geographers or other spatial theorists (although, see Kneale 1999; Warren *et al.* 1998).

In our analyses we concentrate on a reading of the spatialities of one particular genre of SF that we call 'cyberfiction'. Cyberfiction is any form of literary fiction set in the near-future within which cyberspace technologies, such as virtual reality, telemediation, computer intelligence, surveillance or person–machine relations such as cyborging, are a central part of the story. The novels which fall under this umbrella term are diverse but can generally be classified into two camps: cyberpunk and mainstream. Whilst we include analysis of both types of fiction, we concentrate on cyberpunk.

Of the four classes of science fiction identified by Malmgren (1991) – alien encounter, alternate society, gadget, and alternate world – cyberpunk, we feel, predominately espouses the second group; alternate society, an alternate spatiality. Cyberpunk's novum was an estranged socio-spatial order rooted in a dystopian framework:

> Cyberpunk is concerned with models of social order and disorder; narrative structures based on perception and spatial exploration; and . . . a mapping of compacted, decentered, highly complex urban spaces.
>
> (Bukatman 1993: 142)

Cyberpunk is predominately an extrapolative fiction (what if . . .), taking the present and projecting it forward, but it is also contains elements of speculation. However, whilst retaining classic SF processes of estrangement, such as technological innovation and defamiliarisation (making strange the familiar), cyberpunk broke away from modernist traditions of SF. Cyberpunk is more postmodern in its formulation (e.g., identity is fluid, boundaries are permeable) and to us represents a transformation as dramatic as Shelley's *Frankenstein*. This is because cyberpunk is a fictional genre that not only challenges the modernist style of science fiction but captures the essences of the contemporary postmodern condition. Whilst much of SF has played with such concepts as immortality, identity and reality, it is most often from a essentialist position; one that ultimately accepts as natural the distinction between us/them, life/death, real/imaginary, with humans retaining a privileged central position (Hollinger 1991). Traditional SF characterises future societies within modernist structures, failing to appreciate that the transformations in technology to make such societies possible will in themselves change society; that modernity will transform into a condition of postmodernity. Cyberpunk recognises this transformation, and explores the disruption of essentialist notions of knowledge. As such, cyberpunk helps to disrupt the modernist dualisms which are at the core of traditional SF: self–other; self–society; nature–technology; nature–civilisation; rational–irrational; order–chaos. It is these processes of defamiliarisation and disruption that provide the reader with a new view of the present. Bukatman (1993: 10) thus suggests that cyberpunk provides 'spaces of accommodation' where the shock of the new/future can be 'aestheticised and examined'. Cyberpunk is 'the apotheosis of postmodernism' (Csicsery-Ronay Jn 1991: 193), and in turn has been appropriated by a number of postmodernist theorists, as Armitt (1996: 9) states:

Postmodernism is strewn with the discourse of SF, for of course technological advances have ensured that we are, more than ever, at a moment in history when the fantastic surrounds us on all cultural levels. Now space is neither 'out there' nor 'the final frontier.' In its impatience it has come to greet us. So we live in a world of the 'hyperreal', citizens of a giant computer game.

Indeed, during the late 1980s and early 1990s, cyberpunk caught the attention of academics who hailed it as 'postmodern science fiction' (McCaffery 1991), offering privileged insights into contemporary culture (Jameson 1991). For them, cyberpunk recognised and explored the postmodern condition (the rise in multinationalism, the creation of hyperreal places, the merging of technology and nature, etc.), through a literary vehicle that is itself decidedly postmodern (narrative has aesthetic tendencies, thematic impulses, blends of narrative styles; see McCafferty 1991). Ross (1991: 147) states that: 'cyberpunk sketched out the contours of the new maps of power and wealth with which the information economy was colonising the global landscape'. Unlike other forms of modernist extrapolative fiction, which emphases the science in SF, the mechanics of technology, cyberpunk focuses on the everyday appropriation of technology; its use rather than design, the interface of technology and human subject (Bukatman 1993). It thus heralds a posthumanist fiction (Hollinger 1991), where technology is no longer the background to the narrative; instead, the narrative concerns the interconnections between human and the technological. Bukatman (1991: 22) contends that within cyberpunk 'a new [human] subject has emerged: one constituted by electronic technologies, but also by the machineries of the text'. Bukatman (ibid.: 31) renames cyberpunk 'terminal identity fictions', to articulate how it signifies the end of the subject and the construction of a new subjectivity constructed through electronic media; the creation of a 'cyborg discourse'.

Cyberpunk was a 1980s genre, like punk rock's initial blast against corporate, stylised and manufactured music of the 1970s, a cry from a 'generation without a future'. Cyberpunk was a reaction to formulaic, modernistic SF and its inability to recognise the transformation of Western societies into a new postmodern condition (McLafferty 1991b). As such, cyberpunk was, as with earlier SF genres, a product of socio-technical relations at a particular time and place (e.g., 1980s, North America) (Ross 1991) – 'a product of the Eighties milieu' (Sterling 1986); a genre that extended and built on the postmodern undertones of 'new wave' science fiction of the 1960 and 1970s (McLafferty 1991a). Sterling (1986: viii) contends that cyberpunk was 'rooted in the tradition of science fiction but, like punk with progressive rock, strip[ped] away "symphonic elegances"'. As with punk rock, however, cyberpunk mellowed and matured throughout the 1990s, and as with many radical movements its central protagonists, style and themes, have become part of mainstream SF. This has been aided by Hollywood's (often failed) attempts to bring the stories to the big screen.

Mainstream cyberfiction, in contrast, often lacks the dark, edgy style that characterises cyberpunk. Whereas all cyberpunk fits within the SF genre, as it contains at least one novum, the same cannot be said for all cyberfiction, which is often set in the present. Moreover, whereas cyberpunk is decidedly postmodern in its narrative and focus, mainstream cyberfiction is mostly modernistic in formulation. Here, we refer to the spate of 'cyberthrillers' and romantic 'You've Got Mail' novels. The mainstreaming of the technological elements of cyberpunk is in part due to a shift of ideas from the margins (SF) to the centre, but mainly because of the popularisation of cyberspace in present-day society. As detailed, the Internet is now used by over 195 million people, many of whom use it on a daily basis, and it has come under the gaze of the popular media (there is a whole swath of magazines and guides devoted to the Net), film-makers (e.g., films like *Lawnmower Man, Johnny Mnenomic, Hackers, The Net*) and academia. Moreover, personal computer ownership is increasing rapidly, as is the sophistication of software that provides three-dimensional gaming spaces which can be explored. Consequently, many fiction writers from outside the cyberpunk genre have turned their focus on cyberspatial technologies and their relationship to society. As with cyberpunk, some of these writers have projected forward in time this relationship, though often through a modernistic gaze. However, both sets of fiction are united by several commonalties including the adoption of David versus Goliath narratives, and shared dystopian visions of the future in relation to social relations, governance and political-economy. These visions often contrast strongly with the predictions of utopian, academic commentators such as Makridakis (1995), Martin (1978), Naisbett (1984), Negroponte (1995), Stonier (1983), and Toffler (1980), who all foresee a prosperous future centred around a thriving information economy.

It is our contention that it is particularly instructive to examine the writings of cyberfiction authors for a number of reasons, not least because they provide a cognitive space from which to examine current postmodern spatialities and reveal potential future spatialities; they provide dystopian fables of what society may become if it follows certain paths; they open up 'sites of contemplation'. It would be easy to dismiss such writing as nothing more than fantastical imagination, if it were not for the fact that most science-fiction writers (particularly writers of cyberfiction) seek to focus on the possible and probable rather than the fantastical. They tread a tightrope between scientific realism, taking current ideas and technology and projecting forward using the predictions of academics and scientists, and their imagination (Kneale 1999). This fiction, in turn, is read by a global audience, and the ideas contained within influence, to an unknown extent, the thoughts and motivations of its readers, including those who are funding, developing and deploying cyberspace technologies. Here, the cognitive spaces of cyberfiction are grounded and the defamiliarisation of the narrative is refamiliarised. Consequently, some interesting recursive relationships are developing between novelists, on one side, and academic scientists, professional engineers, computer programmers, the military, social scientists, politicians, musicians and 'lifestyle communities' on the other side, linking the present with imaginative futures.

An interesting example of this recursive relationship centres on the writer William Gibson, the inventor of the term cyberspace, who has both drawn from, and inspired, all five groups (academic scientists/professional engineers/computer programmers; social scientists; politicians; entertainers; lifestyle groups). Gibson, in his debut novel *Neuromancer* (1984), envisioned a networked, Cartesian, visual dataspace through which users navigated and interacted. With little experience of computers or computing, he came up with the idea whilst watching children play arcade games. Despite Gibson's lack of computing knowledge, within a few years computer scientists, drawing explicitly on his writing, had started to make Gibson's vision reality, first with the development of the World Wide Web and then with the development of virtual reality applications accessible across the Internet. For example, in 1988 John Walker launched the Autodesk (leading VR developers) 'Cyberpunk Initiative' (Chesher 1994). In a white paper titled 'Through the Looking Glass: Beyond User Interfaces', he invoked Gibson and proposed a project to produce a 'doorway into cyberspace' within sixteen months (Chesher 1994). Moreover, as noted in Chapter 6, cartographers of cyberspace are deploying Gibson's use of spatial metaphors to visualise informational spaces and make them navigable. Likewise, Neal Stephenson's novel *Snowcrash*, written in 1992, was an obvious inspiration in the development of a three-dimensional virtual world, accessible across the Internet by summer 1995. AlphaWorld, as discussed in Chapter 8, is life imitating art, in a very literal sense. Academics such as Tomas (1991) and Stone (1991) openly turn to Gibson to credit his foresight and acknowledge his influence in shaping the 'information society'. Whilst Gibson's novels certainly did not provide the technological blueprint (cyberpunk being more concerned with the everyday use of technology rather than technical details), they suggest that recent developments in both computing and society can be seen as an attempt to put Gibson's visions into practice. Indeed Stone (1991: 95) states that *Neuromancer*: 'provided . . . the *imaginal public sphere* and reconfigured discursive community that established the grounding for the possibility of a new kind of interaction' (our emphasis).

A similar recursive relationship exists between Gibson and some postmodern theorists. Gibson drew inspiration for his dystopian visions of urban and political-economy from such journals as *The Architectural Review* and Mike Davis' book *City of Quartz*, using Davis' descriptions of public space and surveillance in Los Angeles as the basis for his 1992 novel *Virtual Light*. Moreover, he draws on themes developed in postmodernist analysis of life at the end of the late twentieth century, exploring such discourses as globalisation, social fragmentation, urban reconfiguration, and changing social relations, to construct his vision of the future. In turn, Davis (1992), and other postmodern analysts such as Jameson (1991), have drawn on Gibson's writing to explain changes in today's society. Here, contemporary science fiction has been adopted as a resource for understanding contemporary social and cultural changes (Featherstone and Burrows 1995). Burrows (1997: 38, 45) goes so far as to argue:

> The themes and processes which a symptomatic reading of cyberpunk reveal are a good deal more insightful than those offered by what now passes for the theoretical and empirical mainstream. . . . I think that one gets a clearer analytical understanding of contemporary urban processes from a reading of Gibson or Stephenson than one does from a reading of Sassen or Castells.

Indeed, Gibson (1989) himself acknowledges that his work is social criticism and analysis of contemporary society: 'What's most important to me is that it's about the present. It's not really about an imagined future. It's a way of trying to come to terms with the awe and terror inspired in me by the world in which we live'.

Not that interpretations are always received as intended. Many utopian analysts and politicians have drawn on Gibson's writing in formulating their own visions of the future and to justify investment in information and communications technologies. This is paradoxical given that Gibson's work paints the future world as a dark, amoral, despotic, violent place ruled by large, all-powerful corporations. The irony of this utopian, technological re-interpretation is not lost on Gibson himself:

> I was delighted when scientists and corporate technicians started to read me, but I soon realized that all the critical pessimistic left-wing stuff just goes over their heads. The social and political naiveté of modern corporate boffins is frightening, they read me and just take bits, all the cute technology, and miss about fifteen levels of irony.
> (Gibson 1989)

In turn, science-fiction writers, including Gibson, have explored potential cyberspatial utopian communities (e.g., Hendrix 1997) and political futures, particularly around themes of surveillance and governance (which we explore below).

At a fourth level, science-fiction writers within the genre of cyberpunk have drawn recursively from punk and rock music to formulate characters, to structure the tempo and style of writing, and to create a general atmosphere of unrest within the narrative (Sterling 1986). Many cyberpunk writers have said that the music of such bands as The Velvet Underground and The Sex Pistols has fed directly into their writings. For example, Gibson names and bases characters within *Neuromancer* on Chrissy Hynde (Molly), lead singer with The Pretenders, and Robert Quine (hacker Bobby Quine), guitarist with Richard Hell and the Voidoids. His term for prostitutes, 'meat puppets', was named after the Arizona punk trio of the same name (Dery 1996). Gibson has detailed how the music that he listened to while writing the novel inspired its content and atmosphere. Other writers such as John Shiner, John Shirley and Bruce Sterling all testify to the influence of punk and rock music on their writing and some confess a desire to be musicians and pop stars (Shirley, for example, has been in several punk bands). Sterling draws direct parallels to the punk movement of the late 1970s, suggesting that cyberpunk writers wanted to achieve a 'garage-band aesthetic', to strip science fiction of its polished finish. He argues that cyberpunk drew the two overlapping worlds of pop culture and science-fiction literature together to form a new culture. Bands such as Nine Inch Nails and Front 242 have in turn been inspired by science-fiction writing and have created a form of post-punk, weaving the strains of punk with technological, artificial sounds. These bands have inspired real-world cyberpunks; providing the music for hackers to listen to whilst glued to the screen (Dery 1996). Indeed, for Sterling (1986: xi), 'cyberpunk is very much a pop phenomenon . . . the realm where the computer hacker and rocker overlap'.

The final recursive relationship exists between authors and communities organised around cyberspatial technologies. The history of cyberspace has strong roots in the links between Californian counter-culture and Silicon Valley. Many of the leading commentators and developers of personal computing, the Internet and virtual reality have been, and remain part of, counter-culture movements, forming a complex intertwining of technology and community. One consequence of this is that by the late 1980s the term cyberpunk had been uprooted from its strict definition of a science-fiction genre and re-appropriated by the mainstream to refer to a diverse set of cultural forms based around futuristic ideas of computing and communication. It also became associated with cybercafés, nightclubs, rave, ambient and industrial music, smart or designer drugs, and calls for cultural and political change (Kitchin 1998; and Chapter 2). As such, for certain sections of society, cyberspatial images of the future hold resonance and they structure their lifestyles into a particular subculture which aims to live out and bring about selected aspects of cyberspace's promise (Fitting 1991; Featherstone and Burrows 1995). In these projects, the visions and forecasts of writers like Gibson are central. Again, Gibson and his contemporaries draw inspiration from these movements, and their novels are replete with techno-subcultures, intertwining technology with spiritualism and defined lifestyle practices.

Mapping the present/mapping the future

In the light of these recursive relationships, the extent to which the narratives seem to reflect the socio-spatial processes operating within current Western societies (especially North America), and in particular, the influence of the ideas expressed on technological developments and social relations (that people are using their imaginal spheres to help guide techno-social futures), an examination of imaginative geographies, in our opinion, is an essential element in the mapping of cyberspace. Such a reading is not, however, unproblematic. The fiction is just that – fiction, describing geographies that have not as yet occurred and might never occur. Moreover, feminist and other analysts have critiqued the novels for their narrow plotlines, the reproduction of patriarchal relations and their failure to acknowledge race relations, environmentalism and other cultural forces and social movements (Clute and Nichols 1995; Roberts 1993; Ross 1991). As Ross (1991: 152) contends, counter-politics in cyberpunk literature generally consists of 'youthful male heroes with working-class chips on their shoulder and postmodern biochips in their brains'. The fiction, then, are situated within wider social relations (of authors and readers) that need to be acknowledged. As such, because they are fiction, and because they are situated visions of possible futures, they should be acknowledged as a set of partial and selective readings of the future – partial readings that none the less inform our understanding of our present late-modern condition, and potential future socio-spatial relations.

Real/virtual

As discussed in Chapter 1, a number of academic commentators contend that the distinction between real and virtual is destabilising: memories from media and actual experiences are conflated, places are becoming simulacra, hyperreal, and transformed through a process of real virtuality. As such, one of the foundational epistemologies of modernist thought, the separation of real and virtual, is being systematically undermined. This blurring of real and virtual and its consequences on how people live their lives is a common theme explored within many cyberfiction novels. In general, narratives focus on two different sorts of blurring: the conflation of real (offline) and virtual (image/online); the virtualisation of real space.

In the first instance, characters conflate virtual and real spatialities, failing to recognise the differences between the two: 'She wondered for a moment whether Stephen understood the difference between net and real life' (Williams 1996: 33).

In Stephen's case, he is a young boy brought up in a world where a ubiquitous Internet is interfaced by an immersive VR, where the images are largely indistinguishable from viewing real space. Moreover, the use of VR bodysuits and force feedback devices provide an embodied space in which actions in VR have partial, material consequences in real space (your real body controls the actions of your avatar and in turn responds to forces placed on the avatar in cyberspace). Visually, then, the virtual and real are indistinguishable, and as a consequence the mind is tricked into accepting the virtual as real, as the central character in Foy's *The Shift* describes:

> the clarity and depth and immediacy of the Virtix are so fantastic that my body believes, in some dimension both shallower and deeper than logic, it is actually standing in [a VR simulation of] New York in 1850.
>
> (Foy 1996: 38)

Indeed, some novelists predict that the visual sophistication of future VR environments will be so great that it will be impossible to tell them apart from real spaces, as Foy (1998: 24) details:

> And I was in a clearing in the mountains of Idaho. There's no other way to describe it. I *was* there . . . every damn detail of the scene in front of me was three-dimensional

and perfect. . . . The sense of depth, of objects being separated by a distance you could touch and walk, was so great that I was forced at once to treat the world, in all the brain areas that mattered, as a place of actual geodesics.

This conflation of real and virtual is so persuasive that the central character in *The Shift*, Alex Munn, keeps trying to invoke the rules of his virtual life in an effort to switch off or cope with situations in real life:

> The fingers of my right hand twitch, wanting to tap at whatever keyboard controls this game. My mind insists loudly that this is fiction. Something as ludicrous as Riker's [prison] simply cannot happen to a well-paid middle-class professional like me . . . 'Rewind,' my mind mumbles urgently, 'hit default, punch the bail-out macros!' But there are no keyboards and this is not a game any longer.
>
> (Foy 1996: 217)

In other cases, characters forsake real space so that the virtual realm becomes their primary domain. In these cases, characters structure their lives around their virtual encounters, problematising the question as to what reality they occupy. As Gibson (1996: 89) describes: 'A multi-user domain. It is his obsession. Like a drug. He has a room here. He seldom leaves it. All his waking hours he is in Walled City. His dreams, too, I think.' And Williams (1996: 39) writes: 'A decade or two ago you used to hear about a netboy or netgirl dying every couple of weeks – too long under simulation, forgot they needed real food and water.'

In the second theme, authors explore the extent to which real space is increasingly virtualised through the use of cyberspace. Here it is recognised that data and actions in cyberspace have consequences in real life; the realm of cyberspace is not seen as a separate, discrete space, but as an extension of 'real space'. As such, interactions within cyberspace, whether it involves altering records, conducting business or monitoring the city, have consequences on people in real space. The distinction is further blurred through the use of telepresence-operated robots where a machine in the 'real world' is directed by operators working in a virtual environment. For example: 'The surgeon on the right . . . is located at Johns Hopkins Medical Center in Baltimore. The subject . . . he's at Cedar Sinai Hospital in Los Angeles. The purpose of this telerobotic surgery is . . .' (Harry 1996: 2–3).

Here, virtual and real become entwined as the relationship between computer, city and people deepens.

The consequence of this conflation of the real and virtual is that in many narratives the virtual becomes a hyperreality, 'more real than reality' (Anderson and Beason 1996: 9), to the extent that 'half the population has given up on objective reality' (Sterling 1996: 339). The destabilisation of categories of real and virtual allows authors to question the importance and validity of the distinction, and the nature of reality itself. For example, in Hendrix's *Lightpaths*, Jhana one of the central characters, ponders the concept of reality after experiencing a seamless VR:

> The illusion of the virtual reality about her was so flawless that it made her question whether any reality she had ever known was truly real – or if the reality she had taken for granted her whole life was also only virtual.
>
> (Hendrix 1997: 314)

Calder (1996) and Besher (1998) take these thoughts to their logical conclusion, questioning the metaphysics of being:

> Everything you see around you may be no more that an enormous string of 0s and 1s. It follows, therefore, that we ourselves may be a program – cosmic software powerful enough to create the illusion of reality – that is being run on a gigantically powerful computer.
>
> (Calder 1996: 257–8)

> Who would have thought that we are part of a virtual world that is being generated elsewhere by some other species? We are their VR.
>
> (Besher 1998: i)

Fabi (1998: 507) thus asks 'what is "real" anyway? Maybe reality is what you experience.' Harry (1996: 478) similarly asks, 'What is it that sets those memories [real/virtual] apart? What is it that differs – qualitatively – between the memories you have in real life and those you have of cyberspace?' Here, the whole premise of objective, neutral and detached modernist thought is undermined through the suggestion that reality as an objective construct is unobtainable, it is merely the product of experience; 'theater of the mind' (Harry 1996: 131). The experience of the virtual being indistinguishable from the real thus carries the suggestion that any environment is potentially unreal in the traditional sense, thus creating 'spaces of anxiety' where nothing can be accepted as truth. In Platt's *The Silicon Man*, this anxiety underwrites the main character's disillusionment and unease as the certainty of rational thought is dissolved by life in the machine. His wife tries to rationalise this anxiety by asking, 'if you can't tell the difference – what does it matter?' (Platt 1991: 322), though it clearly does, as the basis of Bayley's identity is fundamentally altered.

The destabilisation of real and virtual, and the creation of simulacra, is explored in most detail in Calder's *Cythera*. The story charts the quest of a young man and his simulacrum lover, Dahlia Chan, to find Cythera where real (flesh) and virtual (image) become one:

> 'Cythera, of course. We are trying to find Earth3.' . . . for wasn't Cythera supposed to be a synthesis of Earths 1 and 2, a place where the real and the artificial became indistinguishable, where the body and spirit became as one?
>
> (Calder 1998: 15)

Dahlia is a pastiche, a ghost, a virtual construct, created out of thousands of digitised photographs, films, critic reviews and fan-mail of a real child star; a virtual simulcrum projected into real space:

> my ghost's burgeoning incorporeality (she, one of the fibersphere's damned; a copy divorced from its original; fame evolved into a separate, alien form of life; a new morphology congealed out of the mediascape, the hyperuniverse that interpenetrates our own). . . . She was an aftertrace, a myth without a medium, a representation that had been copied so many times that it had become disassociated from its original. . . . Had she really ever had an original, or had she always been so: a copy of a copy? A mirror within a mirror within a mirror?
>
> (Calder 1998: 5, 25)

Dahlia Chan, then, is a construct composed of meta-data so rich that its sum can create a sentient creature in the form of a downloaded computer projection; a product of the fibresphere (cyberspace) so real that emotional attachments to her mirror those between people. The main character has such emotional attachments, seeking a place where they can be together without stigma, whilst trying to keep her image in the real world alive through regeneration:

> I could almost hear The Wound calling to her, that locus in space–time where information became live, where the fibersphere bled into our own world; the site of sites that was everywhere and nowhere and which – her simulacrum disintegrating – was consuming her, deporting her piecemeal back to the collective images from which she had sprung.
>
> (Calder 1998: 8)

In *Cythera*, virtual constructs are not contained within virtual space, as with VR. Instead they join hyperreal places, virtualising real space. Here, making the distinction between real and virtual becomes more taxing as there is no real space to retreat to; it is not simply a conflation of virtual (online) and real (offline). Every space within which we reside becomes an indeterminate blend of real and virtual. Modernist notions of neutral, objective, measurable space thus dissolves as the systems of knowledge which support such notions collapse. Here, space, in human geographic terms, is meaningful only as spatiality: produced, contested and ephemeral.

Public/private spaces

It has been widely contended that a dominant spatial process of the late twentieth century has been the erosion of public space in Western societies. For example, shopping is moving from the public street to privately regulated malls, and shops remaining on the public street are increasingly subject to the gaze of surveillance. Moreover, city spaces are closing themselves off from the public sphere retreating into gated communities. Similarly, it is reported that the public spaces of democracy are being lost to corporate power, so that modes of regulation and governance transfer from the public sphere to the private. These trends are explored in detail in cyberfiction, in particular, the spatiality of society under different modes of surveillance and governance, both on- and offline. In this section we explore these spatialities.

As detailed in Chapter 2, there has been a large amount of recent writing by academic commentators on the rise of a 'surveillance society'. Analysts report that it is increasingly difficult to take part in everyday life without leaving a digital trace that can be monitored. Writers of cyberfiction have projected concerns of the consequences of surveillance into the future. They predict that technological advances will make the world a giant panopticon, so that all behaviours both off- and online can be collected, stored and cross-referenced. Here, public space has all but vanished, constantly monitored and policed by state and private concerns to maintain hegemony. For example, Harry (1996) details a geographic space where surveillance is total: '"So there are cameras constantly watching you when you're in public?" she asked.

"And infrared, thermal, low-light, microphones, ground-motion detectors, pressure sensors, feedback from things like light switches"' (Harry 1996: 83).

Here it is impossible to undertake any activity outside and inside the home without that activity being externally monitored by fixed surveillance systems. In Stephenson's *Snow Crash* (1992), the surveillance is mobile and collected by people for profit:

> Gargoyles represent the embarrassing side of the Central Intelligence Corporation. Instead of using laptops, they wear their computers on their bodies, broken up into separate modules that hang on the waist, on the back, on the headset. They serve as human surveillance devices, recording everything that happens around them.
>
> (Stephenson 1992: 124)

All the data collected by gargoyles and others is stored in a massive central database, accessible to those online and who can pay. Each time a piece of data is used its collector receives a royalty payment. Information is capital, and is sustained through libertarian spying: 'a true Information Society is a society made of informers' (Sterling 1996: 92). Indeed, Stephenson's novel, *Snow Crash*, is about libertarianism run riot, with the collapse of nation-states into corporate franchises with defined territories which co-exist amongst many others. A burbclave is thus: '[a] city-state with its own constitution, a border, laws, cops, everything' (1992: 6).

In Stephenson's second novel, *Diamond Age*, a host of clave types exist: '"are you a member of any signatory tribe, phyle, registered Diaspora, franchise-organized quasi-national entity, sovereign polity, or any other form of dynamic security collective claiming status under the CEP [Common Economic Protocol]?"' (1996: 33).

Here, notions of community and society are tied to franchises rather than places, and individual/franchise prosperity prioritised – survival of the fittest. This exploration of libertarianism, particular corporate libertarianism where governance and power is controlled by multinational regimes, is common to much cyberfiction. Indeed, whilst many of the books examine the negative consequences of centralised, totalitarian power, it is power wielded through authoritarian capitalism rather than government structures (as in Orwell's *1984*). As the main character in Cythera states: '"I'm not sure if I have a culture anymore. I'm not sure if anyone does. Authoritarian capitalism: it's conquered the world"' (Calder 1998: 45).

This rise in power of multinationals in today's society is widely documented (e.g., Harvey 1989; Castells 1996). Indeed, there is little doubt that there has been a major restructuring of economies at both local and global levels, due to processes such as the globalisation of trade and labour, back-officing, and teleworking, with multinational companies increasingly dominating markets and influencing local, regional and national development within nation-states (Daniels 1995; and see Chapter 2). In cyberfiction, this rise in power is taken to its logical conclusion. As Sterling (1988: 179) describes in *Islands in the Net*: 'Modern governments are weak. We [multinationals] have made them weak. Why pretend otherwise? . . . They need us worse than we need them.' And in *Holy Fire* (1996) he details a society governed by a medical-industrial complex, where status and privileges are determined by body health which is strictly surveyed. Cyberspace is seen as essential to the demise of nation-states as it undermines the territoriality on which they are based: cyberspace knows no borders. As Sterling (1994: 174) describes in *Heavy Weather*: 'You and me both know the border doesn't mean anything anymore. There are no more borders. Just free and open markets!'

Similarly, Stephenson writes (1996: 273) in the *Diamond Age*: 'That's one reason the nation-states collapsed – as soon as the media grid was up and running, financial transactions could no longer be monitored by governments, and the tax collection systems got fubared.'

The result is a new political-economic order, with a re-territorialised political geography:

> There'd been countries as big as anything: Canada, USSR, Brazil. Now there were lots of little ones where those had been. Skinner said America had gone that route without admitting it. Even California had all been one big state, once.
>
> (Gibson 1992: 72)

It also results in a new spatiality and systems of justice:

> Peacock Bank supports a global network of clean, safe, and commodious workhouses, so if any unforeseen circumstances should befall you during our relationship . . . you can rely on being housed close to home while you and the bank resolve any difficulties.
>
> (Stephenson 1996: 11)

This political-economic order is regulated and controlled through systems of power in which surveillance and defensible space are central.

Surveillance in future societies takes many forms. As described above these include an array of fixed-motion monitors and Stephenson's gargoyles. They also include mobile tracking devices:

> three hundred and fifty tag mites remained in his flesh and were later extracted during the course of our examination. As usual the tag mites were equipped with inertial navigation systems that recorded all of the suspect's subsequent movements.
>
> (Stephenson 1996: 100)

Other systems include enhanced personal identification such as DNA-encoded passports (in Gibson 1996) and bar-coded visas:

> Y.T. has a visa to everywhere. It's right there on her chest, a little bar code. A laser scans it as she careens toward the entrance and the immigration gate swings open for her.
>
> (Stephenson 1992: 32)

There are also identification numbers (note the depersonalisation of numbering): '"sixteen and SINless", meant she hadn't been assigned a SIN when she was born, a Single Identification Number, so she'd grown up on the outside of most official systems' (Gibson 1988: 64).

Surveillance can also take the form of productivity tracking of work practices by the constant monitoring of activities:

> The central computer notices just about everything. Keeps track of every key you hit on the keyboard, all day long, what time you hit it, down to the microsecond, whether it was the right key or the wrong key, how many mistakes you make and when you make them.
>
> (Stephenson 1992: 282)

> If there was no paper, they had a record of every call, every image called up, every keystroke.
>
> (Gibson 1996: 29)

And finally, longer-term monitoring such as the cross-referencing of databases to provide a character profile:

> They made a business of abstracting, condensing, indexing, and verifying . . . marriage certificates – divorces – charge cards, names, addresses, phones . . . newspapers, scanned over twenty, thirty years, by computers, for every single mention of your name . . . I've seen their dossier on you. On Laura Webster. All kinds of photos, tapes, hundreds of thousands of words. . . . It's really weird . . . I know you so well, I feel like I'm inside your head, in a way.
>
> (Sterling 1988: 43, 104)

These technologies are used in conjunction with one another to provide layers of surveillance covering all spatial scales and arenas, effectively destroying the notion of public space and individual privacy. All people are monitored, but access and social position determine the degree of concern over the personal infringement created by surveillance. For example, surveillance is 'sold' on the notion that it protects – as long as you are not doing anything illegal you have nothing to fear. Such notions thus lead to the creation of defensible spaces.

> The mall a wall-mall built in a new style, with three floors underground and five over, incorporating not only shops but clinics, an arboretum, a motel, a bus station; surrounded by razor wire, perimeter lights and interlocking-arc security cameras.
>
> (Foy 1997: 275)

Defensible spaces are a common feature of cyberfiction. This is because the new political-economic order of libertarian capitalism creates a system of haves and have-nots, the powerful and the powerless; a dual society. As Gibson (1992: 123) notes: 'There's only two kinds of people. People can afford hotels like that, they're one kind. We're the other. Used to be, like, a middle class, people in between. But not anymore.'

Only those with the right privileges, such as wealth, can visit the mall described above by Foy (1997). The undesirables are disenfranchised, kept outside by the security arrangements. Thus Foy (1997: 212) continues:

> No UCC-card for these people, with the bar code to hold your bank and Visa account and office key code; no credit or reference for these men and women to bootstrap themselves into sudden productivity. The forms changed but the substance remained the same. Far from heralding a new world, the coming of the millennium had shipped large portions of the American population straight back to the 1800s.

Similarly, online spaces are monitored and regulated by private concerns, protected by firewalls that only permit access to those who have the correct password. Again, these narratives draw on trends in today's society. Commentators such as Castells (1996) have argued that Western society is becoming increasingly separated into the haves and the have-nots. Thus employment is either well paid, stable, rewarding and full-time, or else it is part-time, casual, menial and poorly paid, with middle-class jobs and status being eliminated through corporate restructuring (see Chapter 2). Consequently, spatiality becomes increasingly polarised, with wealth concentrated within certain locations which maintain and increase their status through defensible means. In cyberfiction, then, the urban landscape of Western society mirrors this dual economy, with cities divided into rich and poor areas; gleaming mirrored spires and gated suburban housing standing in contrast to shanty towns (see below).

Hard/soft city

As present day analysts, such as Graham and Marvin (1996), note, restructuring of the economic landscape and the increasing centrality of ICTs to city functioning is leading to a change in urban landscapes as we witness the playing out of corporate trade-offs between urban fixity and electronic mobility. As described in detail in Chapter 2, cities are torn by tensions of centralisation and decentralisation and are becoming virtualised, composed of and controlled by distributed networks of computers.

Many cyberfiction writers examine these processes of urban-regional restructuring, charting the shift from modern to late-modern cities – from hard to soft cities – extrapolating trends to provide visions of future urban form. For example, a number of writers explore the tensions developing between the decentralisation and centralisation of urban space. In Stephenson's novels, *Snow Crash* and *The Diamond Age*, the processes of decentralisation, fuelled by a collapse in place-based politics, win out to produce a sprawling, centreless urban landscape composed of small claves, where 'old cities were doomed, except possibly as theme parks' (Stephenson 1996: 71). However, in most other narratives, such as those by Gibson, urban space becomes a large, decentralised sprawl with pockets of highly centralised and dense city spaces:

> Home was BAMA, the Sprawl, the Boston-Atlanta Metropolitan Axis. Program a map to display frequency data exchange, every thousand megabytes a single pixel on a very large screen. Manhattan and Atlanta burn solid white. Then they start to pulse, the rate of traffic threatening to overload your simulation. Your map is about to go nova. Cool it down. Up your scale. Each pixel a million megabytes. At a hundred million megabytes per second, you begin to make out certain blocks in midtown Manhattan, outlines of hundred-year-old parks ringing the old core of Atlanta.
>
> (Gibson 1984: 57)

In the urban cores, space is at a premium, the cityscape is corporate, highly centralised and extremely dense both structurally and in terms of population. The value of space forces development both upwards and underground, to produce a vertical spectrum of stylised, mirrored, postmodern architecture – a riot of glass and steel. Here, Besher in *RIM* (1994: 211, 213) describes Tokyo:

> Sure enough, immense mounds dotted the landscape as far as the eye could see. Gobi guessed these were underground cities.
>
> The freeway suddenly dipped. To Gobi's surprise, they were now traveling though the guts of one these mound cities. The elevated maglev freeway had suddenly become a transparent artery.
>
> They flew through a tube at a height about 30 stories above base level. All along both sides of the tube were rows of internal high rises. These high rises were spread-eagled over a series of parks and urban work-play centers. . . .

He caught his breath. They had finally arrived in downtown Neo-Tokyo, the circuit-board heart of the rim.

Gobi saw wave after wave of towers.

Some of them were 500 stories tall, soaring to a point almost above the earth's atmosphere.

He saw the famous Aeropolis sky-rise, much larger than life but no different than the postcard image that was famous all over the world. Like a skeletal Mt. Fuji constructed of living tubes, it was a man-made volcano that pulsed and breathed in an awesome symmetry of life and death. Half-a-million people lived on its top floors, and commuted from one vector to another.

Similarly, Sterling (1988: 215) writes of a futuristic Singapore:

It was like downtown Houston. But more like Houston than even Houston had ever had the nerve to become. It was an anthill, a brutal assault against any sane sense of scale. Nightmarishly vast spires whose bulging foundations covered whole city blocks. Their upper reaches were pocked like waffle irons with triangular bracing. Buttresses, glass-covered superhighways, soared half a mile above sea level.

Story after story rose silent and dreamlike, buildings so unspeakably huge that they lost all sense of weight; they hung above the earth like Euclidean thunderheads, their summits lost in sheets of steel-gray rain.

These buildings are more than mere glass and steel, however. They are virtualised through the incorporation of computer networks which render them 'smart'; they are 'buildings with advanced infrastructure, buildings with the late twenty-first century embedded in their diamond bones and fiber optic ligaments' (Sterling 1996: 139). As Fabi (1996: 187) details:

This is a 'smart building' . . . 'Totally state-of-the-art; we just built it last year. Carry those cards with you, and a central computer knows where you are at all times. It'll open doors for you, turn on lights, adjust the climate control, everything. Your guest cards are all preprogrammed to average settings, but you can adjust the settings for things like temperature, illumination level, even what kind of Muzak plays when you're on the elevator.

In this quotation, the features are used to enhance habitation but, as noted above, virtualisation such as this can also serve purposes of surveillance and control.

In these narratives the centre is usually the home to 'the haves'; those who have wealth and power. The edges of these concentrations and 'the sprawl' are predominately composed of the disenfranchised; those who are on the outside of the information economy (although part of the sprawl is also the defensive, suburban homes of the super-rich). This dichotomy between corporate centres of wealth and power, and struggling hinterlands mirrors current regional developments which sees cities such as London, New York and Tokyo continue to grow in political and financial power, as companies decentralise predominately their lower level services, typically those requiring less employee skill. This is accompanied by gentrification of city locales as people who have become rich in the information economy move back into inner-city spaces. The product of the current postmodern condition for cyberfiction writers is thus the formation of a dual spatiality: the gleaming, mirrored landscapes of corporate affluence and the abject poverty of slums and homelessness:

> Orlando scrunched down in his seat so he could see the hammock city. He had long been fascinated by the multi-level shantytowns, sometimes called 'honeycombs' by their residents – or 'rats' nests' by the kind of people who lived in Crown Heights. . . . Long ago, he had discovered, during the first great housing crisis at the beginning of the century, squatters had begun to build shantytowns beneath the elevated freeways, freeform agglomerations of cardboard crates, aluminum siding, and plastic sheets. As the ground beneath the concrete chutes filled up with an ever-thickening tide of the dispossessed, later arrivals began to move upward into the vaulting itself, bolting cargo nets, canvas tarpaulins, and military surplus parachutes onto the pillars and undersides of the freeway. Rope walkways soon linked the makeshift dwellings, and ladders linked the shantytown below with one growing above. Resident craftsmen and amateur engineers added intermediary levels, until a marrow of shabby multilevel housing ran beneath nearly every freeway and aqueduct.
>
> (Williams 1996: 510)

This dual economy is a central theme of Gibson's work, whose vision of the future is shaped in part by the analyses of Mike Davis and his view of Los Angeles as a 'laboratory of the future'. Davis' (1990) influential book *City of Quartz* exposes the dual economy of LA: on one side is the corporate, predominately white, middle-class working in mirrored-offices and living in defensible spaces in suburbia protected by wealth and law; on the other side is the underclass, predominately Black and Latino, first-, second-generation immigrants working in menial and casual jobs in the service and manufacturing sectors, whose adolescents and young adults roam in menacing gangs. This summary description denies the complex spatiality and socio-spatial processes that (re)produce LA's geographies, revealed in Davis' analysis, but it is a stereotypical picture of any American city in 1990s often portrayed on television or in Hollywood movies. Gibson extrapolates these themes so that in *Neuromancer*, Case resides in Chiba City, a seedy, low-rent, criminalised Toyko edge-city; in *Count Zero*, Bobby lives in The Projects, which are run-down, forgotten and disenfranchised, large-scale public housing, home to the underclass and gang culture; in *Virtual Light* home for two of the main characters is the top of a reclaimed Golden Gate bridge, now the residence of part of the city's homeless:

> The integrity of its span was rigorous as the modern program itself, yet around this had grown another reality, intent on its own agenda. This had occurred piecemeal, to no set plan, employing every imaginable technique and material. The result was something amorphous, starkly organic. At night, illuminated by Christmas bulbs, by recycled neon, by torchlight, it possessed a queer medieval energy. By day, seen from a distance, it reminded him of the ruin of England's Brighton Pier, as though viewed through some cracked kaleidoscope of vernacular style.
>
> (Gibson 1992: 58–9)

It is in *Virtual Light*, the book most influenced by Davis, where the urbanism accompanying the rise in a dual economy is most fully explored. In this future, any piece of land or structure in the city not being used by the rich is reappropriated for use by the poor: 'past the haunted island, the wingless carcass of a 747 housed the kitchens of nine Thai restaurants' (Gibson 1992: 59). However, whilst the inner cities are overcrowded and filled with contests over space, places away from the centre have become financially unviable and form new, twenty-first century ghost towns: 'Where he was had the feel of one of those fallen-in edge-cities, the kind of place that went down when the Euro-money imploded' (Gibson 1992: 245).

These new forms of spatiality, as with Stephenson's vision, lack forms of public space, and are situated in modes of governance and territoriality underscribed by libertarian capitalism. In contrast to Stephenson, the spatial logic of libertarian capitalism is seen to be a process of centralisation and decentralisation at both local and global scales, rather than a simple fracturing and fractaling of political geography. By whichever process, however, it is clear to cyberfiction writers that the basis of the modern city is changing rapidly through large-scale restructuring, that the city is transforming in all aspects (form, function, process) from modern to postmodern.

Some of the more radical views of future urbanism concern planning and development in the light of new modes of governance and building technology. For example, in *Virtual Light*, the story centres around a corporation's search for a missing database that contains the plans to redevelop San Francisco from the ground up using nanotechnology. This technology was used in the reconstruction of Tokyo after an earthquake, as described in *Idoru*:

> They're going to rebuild San Francisco. From the ground up, basically. Like they're doing to Tokyo. They'll start by layering a grid of seventeen complexes into the existing infrastructure. Eighty-story office/residential, retail/residence in the base. Completely self-sufficient. Variable-pitch parabolic reflectors, steam-generators.
>
> New buildings, man; they'll eat their own sewage.
>
> Who'll eat the sewage?
>
> The buildings. They're going to grow them, Rydell. Like they're doing now in Tokyo. Like the maglev tunnel.
>
> (Gibson 1992: 230)

> Sure, but they did it all so fast, mostly with that nanotech, that just grows? Eddie got in there before the dust had settled. Told me you could see those towers growing, at night. Rooms up top like a honeycomb, and walls just sealing themselves over, one after another. Said it was just like watching a candle melt, but in reverse. That's too scary. Doesn't make a sound. Machines too small.
>
> (Gibson 1996: 46)

Here the city is literally virtualised, created from the programmed building of nanobots; grown like a virtual culture. Similarly, in *The Diamond Age*, Stephenson details a process whereby new geographic space can be created to order by a multinational conglomerate:

> Apthorp was not a formal organization that could be looked up in a phone book; . . . it referred to a strategic alliance of several immense companies, including Machine Phase Systems Limited and Imperial Tectonics Limited. . . . MPS made consumer goods and ITL made real estate. . . . Imperial Tectonics had geotects, and geotects could make sure that every new piece of land possessed the charms of Frisco, the strategic location of Manhattan, the feng-shui of Hong Kong, the dreary but obligatory Lebensraum of L.A. It was no longer necessary to send out dirty yokels in

coonskin caps to chart the wilderness, kill the abos, and clear-cut the groves; now all you needed was a hot young geotect, a start matter compiler, and a jumbo source.

(Stephenson 1996: 19)

Whilst mere speculation, as opposed to the usual extrapolation, these radical visions give interesting accounts of how future urbanism may develop given industrial advances. In all cases, however, they are decidedly postmodern in formulation.

Place/placeless and space/spaceless

Thus far we have concentrated on the effects of cyberspace on geographic space. In this section, however, we turn our attention to the spatialities of cyberspace itself and their implications for how we conceptualise space and place. Cyberfiction writers, drawing on Gibson's portrayal in *Neuromancer*, often describe cyberspace to be placeless and spaceless, a 'non-place' or 'non-space'. For example, in *Neuromancer*, Gibson famously describes cyberspace as a visual, Cartesian non-space; the appearance of space but not space – a visual metaphor abstracted from real space but lacking in tangible substance:

A graphic representation of data abstracted from the banks of every computer in the human system. Unthinkable complexity. Lines of light ranged in the nonspace of the mind, clusters and constellations of data. Like city lights receding . . . He punched himself through and found an infinite blue space ranged with color-coded spheres strung on a tight grid of pale blue neon. In the nonspace of the matrix, the interior of a given data construct possessed unlimited subjective dimensions.

(Gibson 1984: 67, 81)

As such, Gibson (1988: 55) describes that 'There's no there, there. They taught that to children, explaining cyberspace'. In Gibson's Sprawl trilogy,[6] in cyberspace there is no 'space', no places and thus no spatiality. Cyberspace is merely a distributed, corporate information system that uses a spatial metaphor as its interface and which is accessed individually. The spatial metaphor is used solely to aid navigation and provide a tangible interface to data constructs too difficult to comprehend otherwise (as detailed in Chapter 6). As Gibson writes in *Mona Lisa Overdrive*:

all the data in the world stacked up like one big neon city, so you could cruise around and have a kind of grip on it, visually anyway, because if you didn't it was too complicated, trying to find your way to a particular piece of data you needed . . . the neon gridlines of cyberspace, ranged with bright shapes, both simple and complex, that represented vast accumulations of stored data.

(Gibson 1988: 22, 254)

Gibson's descriptions are mirrored by other authors such as Foy (1997: 498) in *Contraband*:

A web of patterns appeared on three screens. The patterns shifted, flowed, rearranged themselves like colored glass in a kaleidoscope. The lines were green and black, as on the ECM's scan, but much finer and denser than the ECM showed. Most of the servers were unlabeled. Wildnet. Overhead, the lights and shine of the Web bubbled like a heavy cover of altocirrus with every molecule of vapor lit from inside.

It is only in Gibson's later books, *Virtual Light* and *Idoru*, that individuals meet and interact in cyberspace, using the medium as an alternative social space. However, it is this use of cyberspace as a social space that problematises the notion of it as somewhere that is placeless and spaceless. As soon as people use the non-space of cyberspace to interact it gains a spatiality – people produce visual non-spaces to facilitate certain kinds of inter-action, and, in turn, people's online behaviour is mediated by the spatial arena in which they are situated. Cyberspace thus gains spatiality and recognisable places, such as 'The Marketplace of Ideas' (Bethke 1995), 'Virtuopolis' (Besher 1994) and 'Metaverse' (Stephenson 1992) which have characteristics similar to real-world places. Indeed, a com-mon visual metaphor to aid both navigation and user-to-user interaction is the use of virtual cityscapes. Besher thus describes 'Virtuopolis' as a full online city offering all the business functionality of any offline city, but one that can be accessed instantaneously from anywhere in the world:

> Satori City, better known as 'Virtuopolis', or 'Virtualopolis,' was developed in the year 2017 by the Satori Group as the world's first on-line VR city. . . . Virtuopolis offers fully-equipped virtual reality office buildings – VR-rises – with discrete cells for short- or long-term lease; interactive convention centers; pay-per-use R and D labs in all the major nano-industries, from desk cold fusion plants to bio-origami pulp processing facilities, to name a few applications.
>
> (Besher 1994: 53)

The 'Marketplace of Ideas' (Bethke 1995) and 'Lambda Mall' (Williams 1996), similarly utilise a spatial mall metaphor to provide a structure to online shopping:

> The Marketplace of Ideas is big. Way big. . . . You can see it virtual miles away: huge discount price structures sprawling out across the datascape, soaring vertical market-ing schemes reaching up to disappear in the haze of high-fashion advertising, and everywhere banners, billboards, and spam.
>
> (Bethke 1995: 89)

> Lambda Mall, the main tradeground of the entire net, surrounded them completely. The mall was a nation-sized warren of simulated shopping districts, a continent of information with no shore.
>
> (Williams 1996: 66)

As detailed in Chapters 1 and 3, these spaces are entirely defined and generated in the computer, and as such, whilst their geography can resemble and function as if a geographically-rooted city, they can utilise new space–time geometries. For example, Egan (1994: 84) details a virtual world where primary stakeholders can alter their virtual geography to suit their needs:

> He'd constructed an auxiliary geography – or architecture – for his private version of Frankfurt; an alternative topology for the city, in which all the buildings he moved between were treated as being stacked on top of the other, allowing a single elevator shaft to link them all. His house 'in the suburbs' began sixteen stories 'above' his city office; in between were board rooms, restaurants, galleries and museums.

Fabi comments that cyberspace's spatiality and the spatial geometries that underlie it represent a 'world where you can defy all the normal boundaries of ordinary existence: travel faster than the speed of light, journey backward or forward in time, bring fairy-tale creatures to life' (Fabi 1996: 507). As such, whilst spatial geometries and forms are often contiguous to real-world geometries and forms they can also differ to form new space–time configurations. In general, these space–time alterations, such as teleporting, are initially spatially confusing but quickly assimilated into normal practice. For example, a common feature of William's 'Lambda Mall' is a small shop frontage, due to the expense of having a 'shop front' on a mall regularly accessed by over 10 million people, that when entered opens into a huge shopfloor:

> Frontage space on the mall is expensive so the exterior displays tend to be small, but the commercial node itself isn't behind it as it would be in a real market. We've just moved into another location on the information network.
>
> (Williams 1996: 71)

In the case of 'Heaven' (Bethke 1995) and 'Treehouse' (Williams 1996), the deviation from spatial convention is more radical and occurs in spaces created and occupied by hackers, rather than the everyday user. These are spaces that are meant to be understood and used only by those 'in the know', people with the skill to interpret, use and survive in them; secret hideaways from the Net police and the masses of ordinary users. Here, Bethke (1995: 84) describes 'Heaven':

> (Some advice here. Don't try to visualise this. The basic geometry of Heaven was designed by the legendary 'cowboy bret' Bollix, and it is *not* Euclidean. That wall at the back that seems to recede into an infinite black void really does, and the pools of lambent light over the tables are just that: pools of monodirectional light, with no source. There are places in Heaven where gravity is purely local; invisible private rooms you can get into only by starting in *exactly* the right spot and then walking exactly the right sequence of steps or turns; there's even a phased-space room, where who you meet and what you see depends on which door you came in. And be careful where you step in the Jobs Memorial Lounge: some of those black floor tiles are actually virtual teleport pads that will deposit you in some really *embarrassing* places in the InfoMall).

And Williams describes the spatial geometry of 'Treehouse' as 'higgle-piggle' (1996: 528):

> There was no up or down – that was the first and most disorientating thing. The virtual structures of TreeHouse connected with each other at every conceivable angle. Neither was there a horizon. The ragged mesh of building like shapes stretched away in all directions . . . Unlike the commercial spaces of the net, which care-fully enforced certain real-world rules such as horizon and perspective, TreeHouse seemed to have turned its collective back on petty Newtonian conventions.
>
> (Williams 1996: 496, 522)

In the case of Fabi (1996), the radical departure from spatial convention is a central feature of the quest the protagonists are set. Their aim is to complete a series of tasks in a variety of MUDs (see Chapter 8) in order to prevent a worm from infecting and destroy-ing the global Internet. In this case, textual MUDs are translated into visual, three-dimensional worlds using a piece of revolutionary code. In this context, the space of the textual MUDs become latter-day labyrinths and mazes in which their Arthurian-style quest takes place, using the imaginary world of the computer to make real the fantastic. Here, space is deliberately used in a confusing manner:

> Going inside was a profoundly unsettling experience. The space inside the spire was much wider than would seem possible, its dimensions having been viewed from the outside, but that was not what was most disorientating. Staircases rose in various areas, some hugging the inner walls of the tower, others rising in the center of the floor as if spurning the support of the walls. But the orientation of the stairs vis-a-vis the floor did not remain constant; they joined up with other flights at a variety of odd angles, so that above their heads some stairs seemed to be turned sideways, and others were actually upside down.
>
> (Fabi 1996: 354)

As detailed, these fantastical geometries are few and far between, with the city metaphor predominating, reinforced by the use of traditional space–time 'rules'. This is probably most clearly articulated in the 'Metaverse' of Stephenson's *Snow Crash*. The Metaverse is an online globe, circumscribed by an arterial routeway, accessed by avatars in ways that mirror real-world interaction. So, for example, in the Metaverse 'an avatar can't exist in two places at once' (1992: 103), 'can't be any taller than you are' (p. 41), and must travel from one place to another in a manner consistent with real-world travel (i.e., without teleporting, hence the routeway). As such,

> You can't just materialize anywhere in the Metaverse. It would break the metaphor. Materializing out of nowhere (or vanishing back into Reality) is considered to be a private function best done in the confines of your own House.
>
> (Stephenson 1992: 36)

Stephenson's vision is particularly interesting because, as detailed in Chapter 8, it has been realised to a large degree through the creation of AlphaWorld. Here, Internet users can interact with a visual, three-dimensional, online world that largely mimics the real world, but which has its own unique geography and spatiality.

It is this spatiality which further challenges modernist systems of thought. Within such systems of thought, media such as literature, television and cyberspace are merely representative and imaginative – they portray a spatiality that reflects real-world spatiality or exists only as a fantastical construct, one which can only be experienced through imagination. Cyberspace, however, presents the opportunity to engage with and construct a new spatiality, one in part divorced from real space. Here the division between real and imaginary, real and media, becomes blurred. As such, the spatiality of cyberspace raises various questions concerning the relationship between place, identity and com-munity. As discussed in Chapters 2 and 3, cyberspace undermines the connection of place to identity and community by providing a space of identification free of location, and communities based on affinities rather than on shared geographical space. This unrooting of place and identity and the disintegration of geographic communities, in the logical conclusion of Relph's (1976) sense of placelessness, is a central theme of much cyberfiction.

As such, the narratives often centre on the search for a place of belonging, both on- and offline.

Nature/technology

Several key tenets of modernist thinking are the separation of the body from the material world, of body from mind, of life from death, of nature from culture. Analysts and cyberfiction writers however, contend that cyberspace, as in the cases noted above, disrupts these distinctions by questioning and challenging the logic of their separation through an examination of processes such as cyborging and genetic engineering and the technological promise of cyberspace (see Chapter 1). Such an examination reveals that these dualistic distinctions are not 'natural' or 'God-given' but are constructed through a particular world view. Indeed, as discussed in Chapter 1, it has recently been argued that technological developments render the distinction between natural and technological obsolete. Technology, in the guise of tools, instead of being mere aids to people in daily living, now modify (e.g., genetic engineering) and replace (e.g., artificial limbs) the natural (Haraway 1991; see also Chapter 1). People, it is contended, are increasingly becoming cyborgs; the product of nature *and* technology. Here, nature and technology collide and merge into one. An historical analysis reveals that both technology and nature have always been social constructs, with technology being more than a mere neutral tool, but it is the range and extent of present-day developments which makes the essentialist distinction unsustainable. This unsustainability is a central narrative of a number of works of cyberfiction, where two discourses predominate: technology to improve/supplement/replace the body (natural/technological), and technology to escape the body (mind/body, life/death).

In the first discourse, cyberfiction writers explore the possibilities and implications of cyborging, what Maddox (1991) calls 'post-Darwinian conceptions of life' (324) or 'posthumanity' (p. 327). Here, human evolution as a 'natural' process is replaced by a technological process that rapidly accelerates species development. This process of technically-mediated evolution consists of two different forms: genetic engineering/cloning and technological implants. In Sterling's Shaper/Mechanist stories, written in the 1980s (see Sterling 1990), humans who use genetic engineering (shapers) to evolve contest future society with humans who use technical implants (mechanists). More commonly, however, these two forms of technological enhancement are often combined to produce human bodies uniquely reconfigured with technological enhancements, as in Gibson's cyberspace trilogy. In this future, DNA is recoded, flesh is grown, and body parts are replaced, both legitimately and in back-street parlours, as the following quotations from Gibson's work shows:

> His primary hedge against aging was a yearly pilgrimage to Tokyo, where genetic surgeons re-set the code to his DNA, a procedure unavailable in Chiba.
>
> (Gibson 1984: 20)

> he stared through the glass at a flat lozenge of vatgrown flesh.
>
> (Gibson 1984: 23)

> The blue eyes were inhumanly perfect optical instruments, grown in vats in Japan. She was both actress and camera.
>
> (Gibson 1986: 151)

He realized that the glasses were surgically inset, sealing her sockets. . . . She held out her hands, palms up, the white fingers slightly spread, and with a barely audible click, ten double-edged, four centimeter scalpel blades slid from their housings beneath the burgundy nails. . . . 'I can see in the dark, Case. Microchannel image-amps in my glasses. . . . Got a readout chipped into my optic nerve.'

(Gibson 1984: 36, 37, 44)

Turner . . . tapped the socket behind his ear. 'It's a fully-integrated system. They'll sell you the interface software and I'll jack straight in.' The microsoft Conroy had sent filled his head with its own universe of constantly shifting factors; air-speed, altitude, attitude angle of attack, g-forces, headings. The plane's weapon delivery information was a constant subliminal litany of target designators, bomb fall lines, search circles, range and release cues, weapons counts.

(Gibson 1986: 127)

For Gibson, and writers like him, the progression from human to cyborg is accepted as inevitable, as Rucker in his seminal cyborg novel, *Software*, details: 'One could legitimately regard the sequence human – bopper [robot] – meatbop [cyborg] as a curious but inevitable zigzag in evolution's mighty steam' (Rucker 1982: 71).

Rucker continues:

'It's not so unreasonable,' Cobb protested. 'It's a natural next evolutionary step. Imagine people that carry mega-byte computing systems in their heads, people that communicate directly brain-to-brain, people who live for centuries and change bodies like suits of clothes!'

(Ibid.: 73, 138)

This progression is not uncritically accepted, however. For example, in *Holy Fire* Sterling explores a society controlled by a medico-industrial complex which determines who can receive life-lengthening medical treatments. In this case, one of the characters concludes that the price of an extended, sickness-free life is too high, given the lifestyle constraints imposed by the complex – posthumanity, as outlined in Gibson's novels, comes with conditions, namely, libertarian capitalism dominated by multinationals.

The other discourse explored is not of technological enhancements to the body, but of transcendence – abandoning the body to achieve immortality in the machine; technology will make humankind's ultimate dream of immortality realisable. Here, the blurring between nature and technology is complete as they become indistinguishable. In this context, cyberspace is a technology of promise, a utopian landscape, as detailed by Platt (1991: 302):

'Some day,' he said, 'our minds will make the final transitions – from organic entities that evolved to ensure the survival of our physical bodies, to electronic entities of pure intellect. The man/machine distinction will break down completely. There'll be no further need to satisfy the old animal desires for food, shelter, and sex.'

Unlike the process of cyborging, which is already a reality, disembodiment into the machine exists as nothing more than an experiential state of those using VR. In these cases, however, the mind seemingly becomes freed of the body as it interacts with the space within the machine. It is thus not uncommon for cyberspace to be described as a disembodied space, a space occupied by the mind not the body: 'jacked into a custom cyberspace deck that projected his disembodied consciousness into the matrix' (Gibson 1984: 11). Here, the mind/body dualism becomes meaningless because one half of the equation, the body, is missing. Whereas in modern society all spaces were embodied (bodies occupied spaces), in the age of cyberspace analysts contend, this is no longer the case (although, see Chapter 1). Uploading consciousness into the machine is the imaginative extension and promise of this illusionary experience.

Cyberfiction writers explore a number of scenarios in which a person's consciousness and memories are uploaded into a machine, in order to contemplate the essence of what it means to be human. Inevitably, the body is deemed to be expendable, it is the mind that defines humanity: 'The soul is the software, you know. The software is what counts, the habits and the memories. The brain and the body are just meat, seeds for the organ-tanks' (Rucker 1982: 66).

In *Software*, Rucker (1982) examines the capture and storage of a human mind for uploading into a robot, and in *The Silicon Man* (Platt 1991), *Permutation City* (Egan 1994) and *Lightpaths* (Hendrix 1997), the authors consider uploading into a stand-alone computer and then a distributed network. In all these cases, the mind becomes analogous to computer software, literally becoming composed of zeros and ones, mimicking the computational and connectionist models of psychology (Kitchin and Blades, forthcoming):

> Cobb was definitely out of the picture, just a frozen S-cube sitting on a shelf in the Nest's personality storage vaults. HUMAN SOFTWARE-CONSTRUCT 225-70-2156: COBB ANDERSON. An unread book, a Platonic form, a terabyte of zeroes and ones.
> (Rucker 1988: 67)

Rather than simply abandoning one world for another, however, the intention of the use of telepresence robots in all three novels is to allow the uploaded consciousness to interact with space outside the computer. 'telepresence robots will let Copies interact with the physical world as fully as if they were human. Civilization wouldn't have deserted reality – just transcended biology' (Egan 1994: 34). 'There's vehicles like that all over the place, now. You can rent one, pipe your mind into it, and go wherever you want if you still need to interact with the real world (Platt 1991: 322).

In essence, the computer construct becomes the third site of posthumanity (along with genetic engineering and technological implants).

Summary

Our analysis of cyberfiction provides conclusions that in many interesting ways mirror those drawn from earlier chapters. Cyberspace, in the view of these authors, has a rich spatiality and is a transformative agent of late-modernity. As such, these fictions suggest that modernistic structures and modernist thought are being radically reconfigured as we move towards a new socio-spatial nexus. Cyberspace is a particularly powerful transformative agent because it disrupts a number of systems of modernist thought. As detailed, cyberspace blurs the boundary between real/virtual, public/private, place/placelessness, natural/technological, mind/body, life/death, nature/culture, often on a number of levels. This is not to say that this disruption is 'good'; as we have outlined, the consequences of a late-modern society often intensifies unequal development and injustice. Instead, our analysis highlights how the fundamental tenets of Western society are changing at the end of the twentieth century, allowing us to think about the consequences and possibilities of these changes. Indeed, some of the most influential academic analysis of recent years has used these 'spaces of disruption' to explore critical, political theory (notably Donna Haraway's [1991] examination of the possibilities of the merging of natural and technological in the development of her cyborg politics). We believe that cyberfiction is important because it opens up these spaces to provide sites for wider contemplation; they are partial and selective, dystopian fables of the future, extrapolated from the present.

At present, these 'sites of contemplation' are remarkably consistent in their general vision of future spatiality and their agreement as to the socio-spatial processes that under-lie this spatiality. Whilst the exact form of any future socio-spatial nexus varies, these authors collectively envisage a future dominated by libertarian capitalism, where global wealth and power are the preserve of multinationals and nation-states are weak or gone; where a dual economy flourishes and is enforced through corporate modes of governance and surveillance; where society is increasingly urbanised within fragmented, divided, simulacra cities; where the body is enhanced through the use of genetic engineering and technical implants. The wider challenge for present-day critical theorists and on-the-ground activists is to construct a set of oppositional politics that prevents this dystopian vision becoming reality; that engages with the erosion of public space and contests the new spatialities of surveillance, governance, and poverty being constructed; that identifies and promotes the utopian possibilities of late-modernity (as discussed by Haraway [1991] and Plant [1996]).

11 Future mappings of cyberspace

> Early in the next millennium . . . the digital planet will look and feel like the head of a pin.
> (Negroponte 1995: 6)

There is a tendency as a new technology develops to concentrate on technical details and the promises such technology offers (Cheesman *et al.* forthcoming). It is only at a later date that we usually examine critically the implications of technological development. In this book we have sought to provide such a critical examination from a geographic perspective, prompted in part by the populist rhetoric that cyberspace spells the 'death of geography'. Our aim has been twofold: (1) to describe and explain the importance of space and spatiality in seeking to understand cyberspace and the infrastructure that supports it; (2) to actually chart some of these spatialities and geometries and provide an overview of the research conducted to date. The scope of the topics discussed, we believe, illustrates the utility and importance of adopting a geographic approach which provides a unique and powerful lens for analysing techno-social changes in society.

Mapping cyberspace, as we have demonstrated, is a multifaceted project; one that consists of elements that are philosophical, theoretical, empirical and practical. It is a project that is only partially captured by our overarching cover questions:

- What does cyberspace look like?
- How is cyberspace changing social relations?
- Will cyberspace make geography obsolete?

Indeed, it is broad in conception, concerned with identifying and analysing the spatialities and spatial forms, structures and geometries of online spaces, how cyberspace is altering social, political and economic relations in geographic space, and the interrelationships between online and offline socio-spatial processes and forms. It is also a project that is in its early stages. As such, although we have attempted to be as comprehensive as possible, our discussion in previous chapters should be viewed as initial, partial and selective.[1] Indeed, as one of us has noted elsewhere (Kitchin 1998b), the geographies of cyberspace remain largely uncharted and a whole series of research agendas and questions need to be addressed.

Agendas for future studies

Philosophical, theoretical, empirical

Whilst there has been significant progress in theorising the utility and effects of cyberspace, much analysis and commentary remains utopian and underlain by a naive form of technological determinism. In nearly all cases, the focus of attention has been solely on cultural, social, political and economic relations. The role of space is little considered, taken as a natural or given element, rather than a facet that is itself constructed and one that provides the fundamental context within which such relations occur. In an attempt to counter the use of atheoretical approaches and the general neglect of the role of space, our discussion has been grounded in a theoretical framework that mixes elements of postmodern thought with social constructivism and political economy, and explicitly recognises that an understanding of socio-spatial relations is central to comprehending cyberspace and its effects. This approach, we contend, has utility because it acknowledges the inseparability of technology and society, as well recognising that cyberspace usage (on- *and* offline) and development is socially constructed at the local scale through the interactions of individuals and institutions and mediated within a globalised political economy (ownership, regulation, power geometries). These socio-spatial relations and processes are not universal and, as such, the adoption of a postmodernist perspective seeks to acknowledge the differences that exist between people and places, and provide an approach centred on readings and interpretations of cyberspace and its relationship to geographic space.

Here, space is recognised as being (re)produced through social relations and in turn shaping those relations: space is constructed by people; space situates the lives of people. We contend that space, both geographic and cyber, is a relational concept. A major aspect of the project of mapping cyberspace is to examine further the relational nature of space and its translation into spatiality. This venture, however, as Bingham (1999) argues, should not treat cyberspace as a sublime ('unknowable') space or a paraspace ('other' space), but rather it should seek comparisons between how space is produced in geographic space and cyberspace, and their interrelationships; how each is shaped and embodied by the other.[2] The adoption of a critical perspective is important in this regard, as interrogating the implicit and explicit meanings and consequences of cyberspace development and usage provides a basis from which to resist and challenge hegemony. Using this approach, we feel that the series of wider theoretical issues relating to the impact of cyberspace detailed in Chapter 1 (each sub-section within the section 'Why Cyberspace Matters') should form the basis for sustained empirical research.

One particularly important aspect of this part of the project is to chart the historical geographies of both the information society and cyberspace. As we detailed in Chapter 1, the development of cyberspace is one component in a long evolution of communication media. The Internet itself is the product of a geopolitical crisis in the form of the Cold War. As a consequence, cyberspace has a historical geography that is particularly American in nature, but one that is in need of further examination, especially through the form of detailed case studies (such as Stein's 1999 analysis of the early development of the telephone network in London). Moreover, the geographic diffusion of the Internet and the development of other networks, especially those outside of an American context, needs more analysis. Further, the historical geographies of the socio-spatial relations of cyberspace itself are largely unknown. Research to date has tended to prioritise the charting of technical developments rather than socio-spatial relations. One fruitful area of future research, therefore, is to determine the sociological and spatial evolution of various kinds of online social spaces.

In addition, there is a need to examine how individual and collectives outside of Western academia, corporate media and institutions view and understand ICTs and cyberspace. Research should consist of three interrelated strands. First, as examined in Chapter 10, there is a need to explore how artists, writers and film-makers conceptualise cyberspace and its implications. Second, there is a need to conduct empirical research on how people within society understand, utilise and reappropriate ICTs and cyberspace. To date, sociologically informed research has tended to examine the 'world in the wires', the people online and the social relations they form. Very little research has examined how online social relations impact on social relations in geographic space, or how the general population views developments, or the technophobia that exists among different populations (although, see Holloway *et al.* forthcoming, for an example with a geographic perspective). Third, there is a need to understand and contrast Western and non-Western conceptions of ICTs and cyberspace. Our analysis has been framed by our own Western-situated knowledge, shaped through our education and our reading of English-language texts. Our discussion, then, provides a selective analysis, one that might benefit significantly from an understanding of non-English language and non-Western knowledge.

One aspect that needs further consideration, but which we only discussed in relation to mapping ICTs or spatialising cyberspace, is empirical research methodologies. Whether knowledge-production is sought through inductively building, or deductively testing theories, both strategies rely on sound data generation or high-quality secondary data. At present, little is known about the validity and reliability of traditional research methods when applied in virtual space, especially considering the claims about cyberspace's ability to 'hide the truth' and allow people to play with their identity and personae. A key priority, therefore, is to determine the external and internal validity of methods used to generate data online.[3] As detailed in Chapter 4, the same issues apply to data gathered by large agencies or self-presented by network owners or ISPs. As we noted, data quality is generally poor and availability is restricted, with much of the available data created or collected by those with vested interests in the expansion and increased usage of cyberspace. The scale of these data is often limited to country aggregations and it is difficult to get greater geographic resolution. This leads to potential ecological fallacies by masking local experiences, variations and inequalities. This situation is compounded by the fact that ICTs and cyberspace lack both central planning and a controlling authority that monitors and gathers statistics on their operation and use, and forms of data standardisation.

General philosophical, theoretical and empirical research questions

- How should cyberspace be theorised?
- How should space be conceptualised in relation to cyberspace?
- To what degree is cyberspace transforming conventional spatial relations, and what are the consequences of such transformations?
- To what extent is cyberspace rendering geographic space placeless, and how does this affect the relationships between place, identity and community?

- To what degree has the 'mode of capital' been replaced by a 'mode of information', that is, to what degree has Western society changed from an industrial economy to an post-industrial (information) economy?
- How is public space eroded by ICTs, and how does such erosion vary across different locales, and different forms of public space?
- How do ICTs alter the nature of mass communication, what new modes of communication do they imbue, and what implications does this have on social relations?
- How is the fabric of reality warped and reconfigured by ICTs?
- To what degree are ICTs leading to a reconfiguration of how we understand the relationship between nature and technology?
- Is cyberspace a 'paraspace', a 'sublime' space, or an 'embodied' space?
- What are the historical geographies of ICT development?
- What are the historical geographies of online spatiality?
- To what extent is cyberspace altering our geographical imagination?
- How do geographical imaginations feed into the development and design of ICTs and virtual spaces, and vice versa?
- How valid are different research methodologies in the context of online research?
- How can data quality be improved?

Geographies of the information society

It is now clear that cyberspace is a key transformative agent, changing the way we live at the start of a new millennium. As we demonstrated in Chapter 2, the extent of these changes is immense, affecting many aspects of daily life, regardless of whether an individual wants them to or not. A key aspect of mapping cyberspace then is to further chart the nature and extent of these changes, particularly through case study analysis that provides empirical evidence of the effects of ICTs in relation to different aspects of society, in differing locales. Analysis and empirical research needs to concentrate on three main areas: culture and society; politics and polity; and economic and urban development.

As documented in Chapter 2, cyberspace is transforming cultural and social relations. The key questions are to what extent and in what ways? At present, the evidence suggests that ICTs are accelerating and deepening the processes of global culturalisation with a move towards a homogenised society based on Western (that is, American) values and cultures of consumption. However, as Morley and Robins (1995) note, ICTs do not render the local insignificant. As such, the connections and interplay between global and local processes, and the interrelations of social, economic, political concerns across spatial scales, need to be deconstructed, teasing apart the complex relations and contradictions of capitalism, hyperreality, simulacra, institutions, individuals and re-localism. Furthermore, the specific effects of ICTs and cyberspace on social relations, and in particular, on identity and community, need to be examined. Much of the writing around these issues is hypothetical, anecdotal and utopian in nature. Whilst much of it seems intuitive, empirical evidence is needed before more definitive conclusions can be drawn. Indeed, those studies which followed the hype of the early 1990s have revealed a picture far more complex than the simple placelessness thesis initially offered by commentators such as Rheingold (1994).

Whilst cyberspace undoubtedly provides spaces in which to explore identity and a space in which to construct an alternative community, analysts are beginning to recognise that cyberspace is a supplementary space, not a replacement. A key area for research, then, is to examine the ways in which people use cyberspace *and* how this connects with other aspects of their lives. To date, this is an aspect that is poorly understood, with most studies directed at understanding social relations online. In addition, there is a need to further scrutinise the relationship between place and community, including how places and offline communities are utilising cyberspace to create a stronger communal spirit and ties, reconnecting community members. In a sense, rather than merely ask the question, 'to what extent is geographic space made placeless by cyberspace?', we also need to ask, 'to what extent is geographic space placed through cyberspace?' Again, the contradictory forces of 'placelessness' and 'placing' need to be deconstructed through case studies.

Specific questions in relation to geographic space, culture and society

- How do the interrelations between global and social processes feed into social relations at different spatial scales?
- To what extent does cyberspace render place meaningless?
- To what degree is geographic space being 'placed' by cyberspace, and what is its role in community development and place promotion?
- How is cyberspace being used to reconnect community members in geographic space?
- To what degree does spatial and social behaviour in virtual spaces 'spill over' and affect behaviour in the real world, and in what ways?

Similarly, the interrelations between cyberspace and politics and polity need further empirical examination. As we detailed, cyberspace disrupts the geometries of power in geographic space in a number of important ways. For example, cyberspace potentially subverts traditional, placed-based political and legal systems by undermining the territorial units on which they are based, and by providing new media of communication, lobbying and voting; it opens up new modes of surveillance, broadening and deepening the means by which people can be monitored; and it reproduces socio-spatial exclusion through its regulation. As yet, however, there is little empirical evidence to support the rhetoric and it is difficult to determine the extent to which cyberspace is changing political structures and power relations.

Specific questions in relation to geographic space, politics and polity

- How is cyberspace affecting current place-based political systems?
- To what extent is the concept of the nation-state (and other territorial units) under threat because cyberspace knows no borders?
- How effective are PENs (Public Electronic Networks) and other online political fora, and how do they feed into placed-based political systems?
- How are institutions using ICTs to manage geographic space, and in what ways are they being resisted?

- How is cyberspace utilised by counter-hegemonic forces, and how successful a media are they for mobilising political and social movements?
- Does cyberspace form the effective basis for a super-panoptican and Orwellian 'surveillance society'?
- What are the ethical implications of geo-demographic industries?
- What are the socio-spatial implications of a 'surveillance society'?
- Is cyberspace providing an effective space in which to challenge patriarchal relations and other power relations (e.g., racism, homophobia)?
- To what extent is cyberspace a new space of deviancy and crime?
- In what ways, and to what extent, do online deviant spaces intersect with geographic space, and how can they be effectively policed?
- What are the material consequences of crimes in cyberspace?
- Will cyberspace be a reality for people with low incomes and those living in non-Western societies, and what are the consequences of their exclusion?

In comparison to the issues discussed above, there has been a wealth of studies that have examined the economic consequences of ICTs and their effect on urban-regional restructuring. As described in Chapter 2, this research has focused on a number of topics including the globalisation of trade, office automation and back-offices, teleworking, organisation and employment restructuring, urban restructuring, the development of soft cities, and the rise of a dual economy society. Although all these issues have attracted empirical research, each one is in need of further examination to provide a wider insight into the developments that are taking place. In general terms, there is a need to shift from analysis centred on the scrutiny of secondary sources of data (especially that produced by the pundits from ICT industries themselves) to the administering of primary data generation, and in particular adopting research strategies that move beyond surveys dominated by closed-questions. Again, detailed case studies that utilise a range of research methodologies including qualitative interviews with key actors and end-users would help to shed light on the socio-spatial processes ICTs are helping to instigate, and the social relations that shape ICT adoption and appropriation.

Specific questions in relation to geographic space, economies and urban development

- What is the nature, extent and influence of the 'space of flows', and what will be the new 'spatial fix' of capitalism?
- To what extent are ICTs contributing to processes of globalisation, and to what degree and how are these being tempered by processes of re-localism?
- How are ICTs affecting local, regional, national and international labour markets, and what are the long-term local employment patterns resulting from ICT usage?
- How are ICTs affecting organisational structures, and what are the long-term consequences of this restructuring?
- What are the social and economic implications of office automation, back-officing and telework?

- To what extent are ICTs leading to the dematerialisation and dissolution of the city, and what are the processes underlying the trade-off between urban fixity and electronic mobility?
- To what extent are city structure and ICTs joined in a recursive relationship (the computer is the city; the city is the computer)?
- How will ICTs affect spatial behaviour and the social aspects of the city such as leisure and entertainment pursuits – to what extent are there changes in shopping patterns (home buying), travel patterns (home working), entertainment (online video), and education (home studying)?
- How do ICTs help reproduce and deepen dual economy societies?
- What are the consequences of urban-regional restructuring on other aspects of urban economies such as the housing market and retail?

One aspect of the project of mapping cyberspace, which we only discussed in brief in Chapter 4, is the extent to which geographic space is represented and presented online, with wider questions relating to verisimilitude and mimesis. Traditional forms of mapping geographic space are now available online, in forms that can be interrogated and questioned. When one site, MapQuest, produces 1.5 billion maps a year, this clearly has implications for map companies, especially when this data can be supplemented with other data such as route guides, VR-style fly-throughs and real-time weather and traffic reports. As yet, however, these implications have not been thoroughly researched. Cyberspace also offers new modes of representing and modelling geographic space, for example, through the development of interactive three-dimensional models or the creation of sound maps for visually impaired users. This can provide a relatively sophisticated, distributed simulation media in which users can explore possible scenarios. In cases where sophisticated VR systems are used, these can employ visualisations that are becoming barely visually indistinguishable from geographic space. Access to these visualisations is widening as they can now be produced on affordable, standard PC equipment, rather than million-dollar supercomputers. Again, whilst research is concentrating on technical questions relating to modes of representation, little consideration has yet to be directed at wider questions concerning the use of this technology.

Specific questions in relation to geography online

- How can cyberspace technologies be exploited to enhance and improve the representation and modelling of geographic data?
- What are the implications of geographic space being presented and modelled online to traditional forms of geographic media and the corporations that produce such media?

Geographies of cyberspace

In recent years there has been significant attention directed at charting the social relations of cyberspace; what life online is like. Less attention has been paid to the role of space in relation to these relations; of the construction and role of spatiality. As we argued in

Chapter 3, we believe that the spatiality of cyberspace is an important element in its use and appropriation. Therefore, a key aspect in the project of mapping cyberspace is to examine its spatiality with particular reference to identity and community. To map out the ways in which geographic metaphors are used implicitly and explicitly to create a 'sense of place' and foster social interaction, to chart the socio-spatial processes which operate in different spaces, and to assess the degree to which cyberspace is dislocated and disembodied – to consider the ways in which social relations online are grounded in, or free of, the practices, relations and memories of geographic space.

This mapping must be undertaken in relation to the full range of different cyberspaces (e.g., email, mailing lists, MUDs, etc.) and across a range of different types of community. For example, we know little about how socio-spatial relations vary across communities centred on different interests, or which draw their members from different parts of society or nations. For example, do similar socio-spatial relations exist in communities founded around sex as around pet care, or in communities predominately composed of Japanese users as opposed to American users? This process of mapping has to be systematic and rigorous and extend well beyond anecdotal accounts that presently dominate the liter-ature. It is particularly important to try and study individuals in both online and offline contexts so that a full picture of online identity 'play' can be examined – studies which concentrate solely on online identity and social relations, where the subjects are known only by their words, lack context and reveal only a partial and selective picture of how cyberspace is being used by individuals who ultimately are sat at computer terminals in geographic space. In a sense, there is a need to move from conducting virtual fieldwork from the comfort of the office to actually engaging with people in the field.

In addition, there is a need to further deconstruct the complex power geometries that have started to emerge online. Cyberspace is clearly not an egalitarian space. It is a space regulated through a set of informal customary laws, market-led governance, social prac-tices (e.g., trolling) and the expression of cultural ideologies (e.g., patriarchy). The use of these forms of regulation and their development over time needs to be more fully researched. Moreover, how these forms of institutional and social regulation intertwine, the effect of their interplay on online socio-spatial relations, and how these regulations relate to, and feed off, corporate and individual practices and strategies in geographic space also need to be documented. This should be achieved through a systematic study and comparison of how different ISPs regulate their services, how governments are seeking to legislate for cyberspace, and how different communal online spaces are governed and regulated and the strategies of resistance adopted by users.

Specific questions in relation to cyberspace, culture and society

- How are geographic metaphors used online, and what metaphors are employed in particular virtual spaces?
- How is space implicated in the construction and development of social relations online?
- To what extent does cyberspace provide a disembodied and dislocated space in which users can 'play' with their identity?
- To what degree is cyberspace fostering the development of placeless communities?
- How do the identities adopted online and communities to which individuals be-long compare to the offline lives of users?

- How do socio-spatial relations vary across different virtual spaces, communities founded on different interests, or temporally across users online at different times?
- How are social relations online shaped by geographies of domination and resistance?
- What particular strategies are employed in the social regulation of virtual spaces and communities, and what tactics are employed in resistance?
- How are different virtual spaces/communities regulated?
- To what extent are virtual spaces regulated by forces that operate in geographic space, such as network owners?
- How are people disciplined, how effective is this disciplining, and how does it differ between companies and spaces?
- How are governments and other institutions seeking to legislate cyberspace?

The bulk of our analysis in this book has concentrated on how to conceptualise the spatial forms, structures, and geometries of cyberspace, and how the use of cyberspace can be augmented by the use of spatialisations. Whilst we have endeavoured to provide a wide ranging and comprehensive discussion, it should be clear that this project is in its infancy. There is still much work needed to be done in relation to conceptualising space in the context of cyberspace, the mapping of ICTs and the spatialisation of cyberspace. Indeed, quantitative measures of the extent and usage of ICTs/cyberspace, using the lens of geography at various scales, is an area ripe for study.

In relation to the mapping of ICTs there is still much work to be done on improving the nature, form and sophistication of visualisation techniques. In particular, research needs to concentrate on the development of interactive and dynamic maps, fully utilising the system itself as a means of data presentation and data collection. The latter issue of real-time data gathering and analysis is particularly important in order to chart and understand the daily fluxes in cyberspace usage and infrastructure.

As we noted in the final section of Chapter 5, the spatio-temporal aspects of cyberspace are poorly understood. As a consequence, the time geographies of cyberspace, for different locales (e.g., position on the network, geographic location of host computer), domains (e.g., email, bulletin boards, MUDs), and temporal sequences (e.g., minutes, hours, days, weeks, months) need to be addressed. Moreover, at present, research has concentrated on the technical implementation of visualisation, not on what these visualisations tell us about cyberspace. The 'views' these maps of ICTs reveal also need to be documented – detailing how a system is configured, who controls the system, how it is developing, who has access to it, where it is being accessed from, how much traffic flows through each system, how representative they are, what is excluded and how systems are linked – to provide a comprehensive picture of the material power geometries and demographics of cyberspace.

Similarly, spatialisations of cyberspace are in their infancy and considerable technical research is needed in relation to the implementation of different forms of spatialisation. This needs to be accompanied by considered evaluation and usability studies to ensure that the adopted techniques communicate in the manner intended. At present, many of the visually most interesting maps have limited usability, with many being dead-end prototypes and graduate experiments.

In relation to information spaces, the sophistication and practical usability of mapping needs to be improved, again with a move to interactive and dynamic displays, particularly those employing three-dimensional metaphors, and for temporal aspects to be incorporated. In particular, there is a need to learn from advances in computer games, where significant developments have been made in creating navigable, interactive landscapes. Many of the examples we present in this book, while interesting, have limited practical application. Despite advances in spatialisation, we still navigate and interact in cyberspace using the desktop metaphor, static webpages, and typing text. Whilst three-dimensional spaces seem the logical way forward, more consideration needs to be given to the effects of interacting with three-dimensional spaces on a two-dimensional screen, using solely a mouse and keyboard. Further, visual and cartographic tools need to be more fully integrated into the browser itself. One aspect that has received relatively little attention is the creation of spatialisations to aid decision-making of the 'finds' generated from search engines, seeking ways to replace long, text listings of matches.

The scope of the information spaces spatialised needs to be widened to incorporate all forms of data. In relation to asynchronous and synchronous media there needs to be a wider application of spatial metaphors to aid interaction and understanding. This is particularly the case with asynchronous media because, as yet, there have been few attempts to spatialise domains such as email and bulletin boards. Again, the sophistication of those spatialisations so far attempted needs to be improved, with increased functionality, modes of display, and wider application beyond test data.

As detailed in Chapters 6 to 8, the relative crudity of present-day spatialisations is to be expected due to their experimental nature. The next stage in their development is the testing of their utility and formalising of their use. As such, as with maps of ICTs, research centred on these spatialisations needs to move beyond technical development and implementation to be actually used as tools to analyse the various forms of cyberspace they spatialise. Furthermore, the implications of these spatialisations on the social interactions within these spaces needs to be documented. For example, how does the use of Chat Circles alter the nature of social interaction within chat rooms?

The use and development of these maps and spatialisations cannot be accepted uncritically. As detailed in Chapter 4, a key aspect largely absent from current work is a critical analysis of the maps and spatialisations produced. Consequently, there is a need for a systematic analysis of each form of visualisation in relation to the five questions outlined in Chapter 4: How 'accurate' is the map? Is the map interpretable? What does the map *not* tell us? Why was the map drawn? Is the map ethical?[4]

As we note above, a fundamental requirement for studies of cyberspace is an improvement in the quality, supply and scope of data. Maps and spatialisations are only as 'accurate' as the data used in their construction. One approach that would benefit data quality and availability is the identification and implementation of data standards. In addition, given the growth and flux of cyberspace, more 'accurate' and comprehensive supplies of dynamic data are required. This will be added by advances in metadata, enabling data structures and mappings that are separate from the actual content. Further, it has been argued that data on, and our knowledge of, the Net is getting worse as it grows faster than it can be monitored (for example, even the largest search engines are failing to keep pace; see Lawrence and Giles 1998). Consequently, a key area for research is to determine ways to dynamically 'mine' cyberspace for data concerning infrastructure characteristics, user demographics and traffic details. This data mining needs to be temporally and spatially referenced to maximise utility.

In addition, the effectiveness of each map and spatialisation as a means of communication and navigation needs to be assessed. At present, most of these visualisations are designed in ways that seem intuitive to the designers. Most of the maps that are produced are not designed by trained cartographers, and there is a danger of 'reinventing the wheel', when in fact many specific design issues could be countered by drawing on generations of cartographic experience. Moreover, most visualisations are extremely complex and the cognitive ability of users to understand them is questionable, especially given that an increasing percentage of users of cyberspace are 'naive' and not technically or computer literate. Consequently, as we discussed in detail in Chapter 9, their utility needs to be formally assessed using traditional cognitive mapping techniques, and the findings from these studies used to make necessary improvements in spatial legibility. Similarly, the spatial legibility of different virtual environments and the navigation strategies used within them need to be assessed. At present only a few studies have examined how users navigate and find objects and information in cyberspace. These studies have concentrated on visual, virtual environments and hypermedia (where the spatialisation is explicit), and studies need to be extended to other virtual realms such as textual MUDs and bulletin boards. Similarly, the effects of different modes of interacting with a representation such as navigation strategies (e.g., teleporting) and spatial views (e.g., restricted visibility of only 60m in AlphaWorld) on users' spatial conceptions and their ability to keep track of their location needs to be assessed.

Maps and spatialisations also need to be interrogated as to the messages they convey. Both types of visualisation generalise and classify the data in order to present a particular picture. As cartographers are well aware, different methods of generalisation and classification can reveal different aspects of the data. As we discussed in Chapter 5, different strategies of aggregation can lead to alternative conclusions, and in some cases may be inappropriate (such as plotting data into territorial units defined by national borders). In the case of ICTs, determining the degree of verisimilitude can be achieved by comparing the data included to that excluded (as determined by comparing the map to geographic space). In relation to spatialisations this is more problematic and can only be achieved by using different methods to plot the same data. At present, systematic comparisons of different forms of spatialisation have not been undertaken, so that for a user unfamiliar with the data or the form of spatialisation, it is impossible to know what is 'missing' from a spatialisation.

As is implied from the above discussion, maps and spatialisations are not neutral artefacts; they are creations, constructed by people, and algorithms written by people. As such, maps and spatialisations are imbued with the values and judgements of the individuals who constructed them. A vital part of the project of mapping cyberspace, then, is to deconstruct the rhetorical power, the 'second text', of each visualisation – to examine why the map was constructed, who funded it and the message it is seeking to communicate and the motivation behind this message. Whilst this can be achieved to a certain extent by examining the map in relation to knowledge of the producers' background (e.g., based on company material), a more productive approach will be case studies involving primary data in the form of interviews with the map creator and its sponsors. In this way, the explicit, as opposed to implied, 'second text' might be revealed.

The final question that needs to be considered relates to ethics. Maps and spatialisations are constructed from data that often refers to individuals, institutions and corporations. The process of spatialisation might reveal information about those users that some of them might not want to be revealed. For example, in the course of researching this book we produced a number of plates relating to bulletin boards and email which revealed individuals' email addresses and showed messages they had written. Whilst the group is open, and anyone could have read that message, it is generally accepted by the poster that his/her message is being read by those interested in the topic. Our reprinting opens up the message and the user to wider surveillance and scrutiny. Spatialisations of virtual spaces might shift semi-private spaces to public spaces, and open up a community to a wider audience. As we noted in Chapter 4, it is important to consider the ways in which spatialisations might change the nature of community – the extent to which they are 'responsible artefacts'. Clearly, spatialisations such as Chat Circles and Loom have the potential to alter the nature of social interaction, and provide a means by which conversations usually consider ephemeral can be examined in detail. Therefore, users' behaviours may be changed in complex ways. Here, users' opinions of these technical developments needs to be researched as part of their overall assessment of utility. In other words, technical developments should not be assessed purely on technical grounds, but should also consider the social implications of their implementation.

Given these concerns and the current limitations with the maps and spatialisations so far produced, we have been asked at conferences whether there is any point in seeking to develop them. In other words, are current maps and spatialisations merely 'eye-candy'? That is, nice to look at but totally impracticable, and unreliable and invalid due to problems with data. Accompanying questions usually include: Will the maps ever be based on accurate data? Will spatialisations ever become the norm in searching and navigating the Net? Are there fundamental design and data issues that will mean we never use spatialisations as an interface to cyberspace? Clearly, only time and research will answer these questions, but one thing is certain, they will continue to be developed. We would certainly accept that many are 'eye-candy', merely prototypes and experiments. Most people still surf with a conventional browser, search with keywords and directories, chat with scrolling text and email in text.

As we argued in Chapter 4, we believe that, despite the problems of data, maps of ICTs reveal important information concerning the development of infrastructure and its users. As such, whilst many are currently flawed to varying degrees, they still provide valuable insights into how social and spatial relations are being transformed. Moreover, they will continue to be developed for marketing purposes, showing the extent and range of network services. Similarly, spatialisations, although having limited use at present, provide a useful experimental basis from which practical methods of spatialisations with widespread application will emerge. Therefore, it useful to consider this period as analogous to the mapping of the New World in the Middle Ages, where the charts produced displayed considerable error but provided the basis for the sophisticated mapping that followed. Spatialisations are presently crude for navigation, but are still powerful visual frames and their potential is that they may change how cyberspace is viewed, works and is navigated through. At a fundamental level, so many researchers and corporations would not be investing so much time and money into their development if they did not believe that they have a future practical use and financial reward. Indeed, the financial reward for the group who develop the 'killer' means of navigating, searching and researching cyberspace will be immense. In the same manner, the development of the graphical Web browser in the early 1990s set the current interface paradigm and made many people rich. On this basis alone, we expect that many more spatialisations will be produced and that their use will become more widespread, in turn changing the nature of cyberspace and how people use it.

Specific questions in relation to cyberspace, spatial forms, structures and geometries

- How can the nature, form and sophistication of visualisation techniques be improved?
- How can data generation be improved so that it becomes more comprehensive, dynamic, and is spatially and temporally referenced?
- What are the time geographies of cyberspace across different locales and domains?
- How can maps and spatialisations be used to understand the different geographies of cyberspace – what do they tell us about the spatial nature and spatiality of cyberspace?
- How do spatialisations affect interactions with a virtual space, and how does this differ across spaces?
- How 'accurate' are maps of ICTs and spatialisations of cyberspace?
- How 'interpretable' (spatially legible) is each map or spatialisation?
- What does a map or spatialisation *not* tell us?
- Why was a map drawn and what is its 'second text'?
- How ethical is the map in its construction and the data it portrays?
- Does a map or spatialisation fulfil its designated utility (e.g., improving navigation or understanding)?
- How can the spatial legibility of maps and spatialisations be improved?
- How do different modes of interacting with a space affect user's conceptions of that space and their ability to interact with it and other users within it?
- Are the maps and spatialisations of cyberspace just 'eye-candy'?

Final words

In answer to one of our initial questions in the Introduction, it is clear that space and geography continue to matter, and in some senses they have taken on more importance. Cyberspace is inherently spatial. Its adoption and utility are founded on its spatial qualities: either because it produces radical space–time compression, or because it provides new spaces in which social relations can develop and thrive. Moreover, its form and modes of operation are largely built on foundations that often depend on spatial metaphors and spatialisations. This given, it is perhaps a little surprising that the spatialities and spatial qualities of cyberspace are relatively ignored. We hope that this book and the questions we have outlined in this chapter will stimulate cross-disciplinary, theoretically informed, empirical research that rectifies this situation.

Notes

1 Introducing cyberspace

1 Some of the material in this chapter is condensed and updated from previous work (Kitchin 1998).
2 Although the cost of access, particularly telephone charges, varies greatly from country to country (see OECD 1999; Petrazzini and Kibati 1999).
3 http://www.inktomi.com/new/press/billion.html
4 http://www.activeworlds.com/
5 For a full account, see Kitchin (1998).
6 It is a popular misconception that the linking of computers was to ensure continuation of the network after a nuclear attack. The system did, however, use the idea formulated to overcome this eventually, as proposed by Paul Baran (see Hafner and Lyons 1996).
7 WELL stands for Whole Earth 'Lectronic Link. Found on the Web at, http://www.well.com/
8 See the MUD Connector for information on active MUDs, http://www.mudconnect.com/
9 Rheingold traces virtual reality's roots back further to the visual, immersive qualities of Cinerama, three-dimensional films and Sensorama. These technologies were the first in which you not only saw the images but where you experienced them, they projected the illusion of reality. As broadcast media, however, they did not allow interactivity.
10 This discussion takes a form similar to that outlined in Mitchell (1995: Chapter 3) and is an extension of earlier analyses undertaken by ourselves and colleagues (see Cheesman et al. forthcoming).
11 One of our reviewers suggested that this use of binaries was 'having our cake and eating it' – although we agree, we could find no other obvious way to achieve our ends.
12 In Chapter 4 we provide a similar examination of issues relevant to understanding the cartographies and spatialisations of cyberspace.
13 Some analysts prefer the term postmodernity. However, the term we use 'late-modern', as whilst there are significant and radical transformations, it is a condition that it is still underlain by the mode of capital/production.
14 Some spaces such as university and NGO sites and individual homepages remain relatively public.
15 For an interesting fictional account, see Julian Barnes' *England, England* (Picador, London, 1998).
16 This is contrary to what one of us has previously argued, where it was contended that online identities were 'shallow and distract from life in the real world' (Kitchin 1998: 82).
17 Paraspace means 'other space' – a sublime space that has forms and practices alien to that in geographic space.
18 Note that the map is an operative semiotic device that is the territory for people who have never been there (Francis Harvey, personal communication with the authors).

2 Geographies of the information society

1 Some parts of this chapter, notably in the sections 'Urban, regional and global restructuring' and 'Geographies of power', have been redrafted and updated from Kitchin (1998).
2 See Total New York (http://www.totalny.com/) and Citysearch San Francisco (http://www.citysearch.com/).

3 The political power of these groups should not be underestimated, with even the smallest of groups effecting change through their online presence. For example, the Geography and Disability mailing list has united a number of geographers interested in disability, and email exchanges and the cross-fertilisation of ideas has led to a number of changes within the discipline, notably the formation of a Disability Speciality group which has successfully campaigned for accessible conference venues.

4 Goudy (in Warf forthcoming) reports that in 1994, 4 million Americans were under 'correctional supervision' at home, monitored by such transponders.

5 Mitterer and O'Neill (1992) report that in 1990 the US government maintained 910 computer systems detailing records of citizens, 292 of which did not meet the Privacy Act regulations; Roszak (1994) estimates that these systems hold between 2 and 4 billion overlapping files on individual citizens.

6 Supermarkets are beginning to build consumer profiles which are then used to target junk mail.

7 Kirsh *et al.* (1996) report that marketing information is a $183 billion industry in the US alone.

8 A major consumer geo-demographic company.

9 We acknowledge that this is a partial review, dictated by space constraints; for a full review, see Castells 1996, Graham and Marvin 1996, Kitchin 1998.

10 As of 1999 (IBM).

11 Figures are numbers of permanent and contract staff for 1997. Other large manufacturing and assembly firms include Gateway 2000 (1,400, with 800 in teleservices), 3COM (325), Fujitsu (200), Kao Infosystems (370), Ericsson Systems (260), Stratus Computer (212). Many of these have subsequently expanded.

12 These include, in 1997, Microsoft (1,000 staff members), Lotus (470), Digital (430), Kindle Banking Systems (355), EDS (315), Oracle (310), Ericsson Software (300). More recent additions include Novell and Netscape.

3 Geographies of cyberspace

1 This is an application of Goffman's thesis presented in *The Presentation of Self in Everyday Life* (New York, Doubleday, 1959).

2 Spamming involves sending the same message to many lists. This practice is discouraged as traffic is needlessly multiplied, occupying valuable bandwidth.

3 A mail bomb consists of posting a massive file to the offender's account which then overloads the mailer's ability to process it. Mass flaming involves sending many messages to an account so that it becomes full and inoperable. The most famous case of mass flames concerns the law firm Canter and Siegel. The firm sent out an advertisement to a number of news list groups, and in return received over 30,000 messages, crashing their machine and causing them to be removed by their Internet supplier.

4 Some individuals have taken it on themselves to act as Net police. One famous example is Cancelmoose who deleted unwanted spam from net lists. Cancelmoose always justifies his deletes and also provides information to system operators to reinstall the offending messages if they want. Cancelpoodle, however, has taken it on him/herself to delete all messages from the Church of Scientology (Maltz 1996).

5 For example, Shapiro (1995) reports that the company Prodigy actively restricts dissenting speech. In 1990, it raised its charges for users posting large amounts of email. Some of the more prolific posters protested that they were being penalised for speaking frequently and sent email to Prodigy's online advertisers threatening a boycott. Prodigy, however, intercepted, read and destroyed the messages and dismissed dissenting members from the service. The members had no legal recourse and no means to picket Prodigy's services. In a *New York Times* opinion piece, Prodigy was adamant that it would continue to restrict speech as it saw fit.

6 It should be noted that some women help to replicate patriarchal relations and gender stereotypes, adopting Barbie-like or movie star descriptions to match the posturing by males (O'Brien 1999).

7 A rhizome is a tangled root system that develops horizontally, and in a non-hierarchical fashion.

4 Introducing the cartographies of cyberspace

1 Geodetically (latitudinally and longitudinally); planimetrically (distances and bearings between places); and topographically (quantity and quality of information about places in the map).

2 In cases where there are no spatial attributes, we use the terms 'map' and 'mapping' metaphorically.

3 It should be noted that in cases where users can create their own knowledge through the application of algorithms to their own data, the process of mapping contains an element of empowerment. Whilst constrained by the data and algorithm, the monopoly power of the professional cartographer is severely undermined. That said, the maps produced still have a second text in need of consideration.

4 http://www.mapquest.com/; http://www.mapblast.com

5 http://terraserver.microsoft.com/

6 http://www.truflite.com/ (other companies include WoolleySoft and Rapid Imaging Software: http://www.woolleysoft.co.uk/software.html; http://www.landform.com/).

7 http://www.casa.ucl.ac.uk/vuis/

8 http://www.ge.ucl.ac.uk/vucl

9 http://www.vtourist.com/

10 http://www.scit.wlv.ac.uk/ukinfo/uk.map.html (created by Peter Burden at the University of Wolverhampton).

11 http://www.irational.org/cctv/

12 http://www.channel.org.uk/metropolis/frameset.html

13 http://www.ems.psu.edu/VRStuff/QTVR.html

5 Mapping information and communication technologies

1 Hillis argues that this invisibility has been a key reason why ICTs have received scant attention in human geography, as the focus has been on visible phenomena (Hillis 1998). He contends that 'For a discipline firmly rooted in an empirical and visually dependent understanding of the facts, too often, if it can't be seen "it's not geography"' (Hillis 1998: 544).

2 The first four maps in Figure 1.2 (December 1969, June 1970, December 1970 and September 1971) are redrawn from Salus (1995); the remaining two maps (August 1972 and June 1975) are original maps scanned from CCR (1990).

3 A couple of useful meta-lists on the Web that index the maps of the main US backbones are Russ Haynal's ISP page (http://navigators.com/isp.html) and Randy Benn's page (http://www.clark.net/pub/rbenn/isp.html).

4 http://www.iGuide.co.il/maps.htm

5 UUNET is one of the largest Internet backbone operators. Its latest map of UK infrastructure can be found on its website at http://www.uk.uu.net/network/ukmap

6 From ftp://ftp.cs.wisc.edu/connectivity_table/

7 Produced in 1997 by Chapman Bounford and Nico Macdonald for World Link. It is a large poster-size world map (measuring 32″ × 54″). The subtitle on the map states that it was produced 'on the occasion of the 1997 Annual Meeting of the World Economic Forum in Davos, Switzerland'. The map was sponsored by Hewlett Packard and Novell, two major multinational IT companies.

8 http://www.mids.org/mapsale/world/index.html

9 MIDS (Matrix Information and Directory Services) is a research consultancy based in Austin, Texas, run by John S. Quarterman, Internet cartographer extraordinary. He has been involved in computer networking and the Internet from the late 1970s, having begun work with Bolt Beranek and Newman (BBN) on ARPANET. Since the mid-1980s, he has been tracking the geography of wide-area computer networks (Quarterman and Hoskins 1986) and his 1990 book *The Matrix* is now an important historical record of the development of computer networks in the late 1980s. MIDS gathers, analyses and presents a wide range of data on the infrastructure and demographics of the global Internet. Of particular interest is the fact that MIDS makes considerable use of geographic maps to present its results (for early examples, see Quarterman *et al.* 1993, 1994). In terms of static maps of Internet infrastructure, MIDS produces a whole series of maps in its Matrix Maps Quarterly publication for different geographic regions and scales, and also employs a number of distinct cartographic styles including shaded area maps, surfaces and bar graphs. In recent years, MIDS has also been active in developing more interactive maps of the Internet, using dynamically collected data.

10 BITEARN was a group of computers on various regional BITNET networks, which was a low-cost computer network used by academia and research, that used slow leased line (BITNET

stands for 'Because It's Time Network'). FidoNet was a very low-cost network built by home computer enthusiasts using DOS-based PCs and dial-up connections. UUCP (Unix-to-Unix Copy Protocol) was a simple network between UNIX computers, mostly via dial-up connections. All three networking systems have been superseded by the Internet; in fact, MIDS icon maps since 1997 only show Internet hosts.

11 There are a number of common domain names that structure their use. Some of the most common are: .com (commercial), .edu (education), .gov (US government), .net (network providers), .org (non-profit organisations).

12 Each country has a unique two-letter country identifier designated by the International Standards Organisation in its ISO3166 standard (e.g., fr (France), de (Germany), au (Australia)). For comprehensive information on all the different agencies around the world that administer different county code domains, see the AllWhoIs website at http://www.allwhois.com/

13 The precise figure is 4,294,967,296. The current version of IP addresses is called IPv4. It is in the process of being replaced by the next generation of addresses called IPv6 (Hoffman and Claffy 1997; Semeria 1997; Boroumand 1998). This space, defined by an 128-bit number, is a truly astronomic number, giving 340,282,366,920,938,463,463,374,607,431,768,211,456 possible unique locations. By way of analogy to grasp the sheer size of this, if the whole of the Earth were made of fine beach sand, each grain could be allocated around 386,000 unique addresses.

14 A notable exception is the OECD report which has some details on IP address allocations by country (OECD 1998). There is also some work by researchers in San Diego, who have visualised the usage of IP address space, but not onto a real-world geographical framework. Instead, they use a regular two-dimensional grid that better suits their purpose (Braun 1997; McCreary and Claffy 1998).

15 In an ideal situation it would be possible to map the actual locations in which IP addresses are actually used, but this is impossible with publicly available data.

16 RIPE is responsible for the technical co-ordination of the Internet in Europe and surrounding countries, this includes managing the allocation of IP address space. More details on RIPE's work are available from its website (http://www.ripe.net/).

17 A good estimate is provided by NetNames Ltd (http://www.domainstats.com/). Its global domain registrations statistics show that 15.7 million domain names were registered around the world as of February 2000, the vast majority in the generic top-level domains, particularly .com (9.5 million).

18 Given the current technical definition of domain names used by InterNIC, they can be 22 characters long and consist of 37 legal characters (26 letters, 0–9 and the minus sign) which allows for 37^{22} permutations. This is a very large number indeed! Of course, many of these would not be meaningful or memorable to humans, but still there is plenty of unique locations in name space (Holtzman 1997).

19 The maps are available from their website (http://www.internet.org/), as well as other summary statistics such as domain names per city and per country.

20 This is part of his Ph.D. research into the Internet industry (Zook 1998, forthcoming).

21 Differential access to cyberspace is discussed in detail in Chapters 1 and 2.

22 http://www.geom.umn.edu/~worfolk/SaVi/

23 The animation itself can be seen at http://www.mappingcyberspace.com/

24 http://www.teledesic.com/

25 http://www.caida.org/Tools/Mapnet/

26 Cooperative Association for Internet Data Analysis (CAIDA) is a collaborative research effort by academic, government and commercial communities to 'promote greater co-operation in the engineering and maintenance of a robust, scalable global Internet infrastructure' (Monk and Claffy 1997). The CAIDA initiative is focused on measuring and visualising the infrastructure and traffic flows of the global Internet, presenting many of the results via interactive Web-based mapping and graphing tools (http://www.caida.org/).

27 Unfortunately, we can only reproduce a single static screen shot in this book of one particular view of the map; to see the full potential of the Cesnet VRML, download the map from http://www.ces.net/cesnet/cesnet.wrl

28 http://www.cesnet.cz/

29 See Reid (1997, Chapter 4) for a readable history of the birth of VRML; and Pesce (1995) provides full technical details on the format.

30 For a short, non-technical explanation of the MBone, see Steinberg (1996), while Halabi (1997) provides full technical details.

31 VRML models can be downloaded from the 'Planet Multicast' website at http://oceana.nlanr.net/PlanetMulticast/

32 It should also be noted that there have been a number of interesting attempts to map the topological structure of the MBone network infrastructure using non-geographic approaches which we discuss in Chapter 6.

33 The original traceroute tool was created by Van Jacobson, a computer scientist working at Lawrence Berkeley national laboratory in the US in 1988.

34 http://www.neotrace.com/

35 http://www.traceroute.org

36 http://www.visualroute.com/

37 http://visualroute.datametrics.com

38 GeoBoy began life as a project developed by computer science students at the Curtin University of Technology, in Perth, Australia. The original utility was called GeoTraceMan (Brown 1994; Pleitner and Brown 1995).

39 http://www.ndg.com.au

40 For a fuller discussion of Usenet, see Chapter 7.

41 A range of the maps are available, via FTP, in postscript format from ftp://gatekeeper.dec.com/pub/maps/. Also, a number of Reid's Netmap maps have been reproduced in Quarterman (1990: 236, 351–53, 359), and Salus (1995: 138, 149–51).

42 TeleGeography is a research consultancy based in Washington, DC, started by Gregory C. Staple, which specialises in measuring and analysing the geography of international telecommunications (see http://www.telegeography.com/). In its reports and wall posters, it employs charts and geographic maps in a range of styles to represent the cyberplace of infrastructure and traffic flows of cyberspace, including cartograms and arc-node maps.

43 The full animation can be accessed from the MappingCyberspace.com website.

44 In addition to geographic visualisation of network traffic, the research team at Bell Labs are interested in many other aspects of information visualisation, including application in data mining of large databases and software source code (Ball and Eick 1996). The commercial potential of their visualisation research and applications is being exploited through a spin-off company called Visual Insights (http://www.visualinsights.com/).

45 Plate 2B is, in fact, a single frame from an animation of two years' worth of Internet traffic.

46 See chapter one in Reid (1997) for the history of Mosaic Web browser developed at NCSA.

47 http://cello.cs.uiuc.edu/cgi-bin/slamm/ip2ll/

48 A CAVE is an expensive, state-of-the-art virtual reality environment which uses a room-sized cube of back-projected screens to display the scene. Users can walk around in the CAVE and the scene updates in response to their movements (Cruz-Neira *et al.* 1993).

49 For example, WebTrends, http://www.webtrends.com/

50 The application is named after a set of magic stones in Tolkien's *The Lord of the Rings* which, 'connected to each other forming a web. Anyone looking into the stone could see what was going on by the other stones' (Papadakakis *et al.* 1998).

51 http://sappho.ics.forth.gr:9000/

52 http://www.mids.org/weather/

53 http://www.sensorium.org/webhopper/

54 WebHopper was one of a number of experimental projects developed by a Japanese collective of Web designers and artists called Sensorium (http://www.sensorium.org/). The Sensorium group originally formed in 1995 to produce online exhibits for the Japanese pavilion at the Internet World Fair in 1996, and their guiding concept was 'to expand the possibilities of the Internet as a tool to sense the world' (Sensorium 1997). They treat the Internet as a living system and, in their art, seek to perceive the 'Webness' of it (Takemura *et al.* 1998).

6 Spatialising cyberspace

1 These four pioneering companies are spins-off from cutting-edge research at Bell Labs-Lucent Technologies, MIT Media Lab, Xerox PARC, and Pacific Northwest National Laboratories.

For further information, see their websites at, http://www.visualinsights.com/, http://www.perspecta.com/, http://www.inxight.com/ and http://www.cartia.com/

2 Then Director of IPTO in ARPA and responsible for the development of ARPANET. http://www.ziplink.net/~lroberts/index.html

3 This map is from Salus (1995), originally provided by Alex McKenzie.

4 The map is from Heart *et al.* (1978) and was scanned by Larry Press.

5 http://www.oss.buffalo.edu/Info/Maps/MapsBySubnet.html

6 http://www.iij.ad.jp/network/backbone-e.html

7 http://www.aih.com/

8 See http://www.cs.berkeley.edu/~elan/mbone.html, for a large version of the map.

9 These can be looked up in an index to find their actual address and domain name.

10 Bill Cheswick is a senior researcher at Bell Labs-Lucent Technologies and Hal Burch is a computer science graduate student at Carnegie Mellon University. See their *Internet Mapping Project* homepage, Bell Labs. http://www.cs.bell-labs.com/~ches/map/index.html

11 Their spatialisations provide some of the most comprehensive topological views of Internet infrastructure. Indeed, their maps are being commercially sold as glossy wall-posters by Peacock Maps, with the billing 'The Whole Internet'. See http://www.peacockmaps.com/

12 http://www.caida.org/Tools/Skitter/

13 http://www.caida.org/Tools/Plankton/

14 From the movie *Stars Wars* (1977).

15 See http://moat.nlanr.net/, technical details and the full animation are available at http://moat.nlanr.net/ASx/

16 Note that here we refer to directories, not search engines.

17 According to Media Metrix's website popularity measurements, Yahoo! and its associated online properties received 38 million unique visitors in July 1999. See http://www.mediametrix.com/TopRankings/TopRankings.html

18 http://ai2.bpa.Arizona.edu/ent/

19 http://www.newsmaps.com/

20 ThemeScape technologies are the outcome of several years of research at the Pacific Northwest National Laboratories, in Washington State, US, to develop visual analysis tools for US intelligence agencies, enabling analysts to cope with growing 'information overload' from the welter of textual information generated today. From the beginning, they used spatialisation techniques and their first prototype, using a landscape metaphor, was called SPIRE (Spatial Paradigm for Information Retrieval) (Wise *et al.* 1995). How the ThemeScape system turns documents like webpages into browseable maps is not made explicit as the techniques are commercial secrets.

21 http://www.cartia.com/

22 http://lislin.gws.uky.edu/Sitemap/

23 http://heiwww.unige.ch/girardin/cgv/

24 http://websom.hut.fi/websom/

25 http://www.smartmoney.com/marketmap/

26 http://ucsu.colorado.edu/~fabrikas/research/spatialization.html

27 http://www.hilton.com/sitemap/map.html

28 The map was in use on the Yell website (http://www.yell.co.uk/) in 1997 but, unfortunately, it is no longer available.

29 See http://www.inxight.com/Products/Web/SLS.html. The Site Lens mapping system is the commercialisation of research into information visualisation by scientists at Xerox PARC (Lamping and Rao 1995)

30 http://www.ibm.com/java/mapuccino/

31 http://www.merc-int.com/products/astrasitemanager/

32 http://www.clearweb.com/, http://www.microsoft.com/siteserver/site/, http://world.isg.de/World/2_Internet/Visual_Web/index.html, http://www.incontext.com/WAinfo.html

33 http://footprints.media.mit.edu/

34 http://www.dcs.gla.ac.uk/~matthew/

35 http://www.dynamicdiagrams.com/

36 http://www.hyperwave.com/

37 See his homepage at http://www.textuality.com/

38 These questions are still relevant to academic researchers and commercial pundits today (see, for example, Woodruff *et al.* 1996; Lawrence and Giles 1998 and 1999; Huberman *et al.* 1998; Albert *et al.* 1999; Clever 1999).

39 Comprising a mere 11 million pages from about 90,000 sites, compared to 180 million plus in 1999 (Lawrence and Giles 1999).

40 In contrast, the quick demise of the Mosaic browser has probably drastically reduced the visibility of UIUC.

41 Ziggurats are ancient stepped pyramidal temples.

42 http://www.asymptote.net/

43 Semantic Constellation can be accessed from Chen's homepage at, http://www.brunel.ac.uk/~cssrccc2/

44 Early examples of 'surf maps' used two-dimensional graphs, for example Peter Dömel's Webmap system from 1994 and Eric Ayers and John Stasko's MosaicG system in 1995. More recent, Romain Zeiliger, at the Centre National de la Recherché Scientifique, France, has developed a browser extension called Nestor Navigator (http://www.gate.cnrs.fr/~zeiliger/nestor.htm), which provides navigational support through dynamic concept maps and a personal graphical overview (Eklund *et al.* 1999). In terms of surf maps that employ three-dimensional spatialisations, there was notable work from Sougata Mukherjea and his Navigation View Builder system (Mukherjea and Foley 1995), although WebPath presents a much more striking visual appearance. Perhaps the ultimate surf map was the WWW3D prototype developed by Dave Snowdon and colleagues which did away with the conventional browser page view completely and visualised both the webpages and their local structure in a multiuser, immersive VR spatialisation (Snowdon *et al.* 1996).

45 http://www.sgi.com/software/sitemgr.html

46 Apple ended research into HotSauce and stopped supporting it in 1997. However, the plug-in and documentation are still available from the Lightbulb Factory, see http://www.xspace.net/hotsauce/

47 MCF has evolved in the Resource Description Framework (RDF), a proposed standard for the Web's underlying metadata infrastructure that would support advanced cataloguing and searching tools. See http://www.w3c.org/RDF/ for more information on RDF.

48 http://www.perspecta.com/

7 Mapping asynchronous media

1 http://www.icq.com/

2 Directories of available public lists are available at, http://www.reference.com, http://www.listz.com and, http://www.mailbase.ac.uk/

3 http://www.wired.com/news/

4 The name is based on the leading list software called LISTSERV, by L-Soft International, http://www.lsoft.com/listserv.stm

5 See Randall (1996) and Rheingold (1993) for historical overviews of Usenet, while Quarterman (1990) provides useful technical detail, including some facts and figures on Usenet growth through the 1980s.

6 These groups contain soft and hardcore pornographic images.

7 For example, alt.2eggs.sausage.beans.tomatoes.2toast.largetea.cheerslove

8 These technical protocols are defined in Internet RFC documents. Protocol for exchanging news is called NNTP (Network News Transfer Protocol) and is defined in RFC 977 (ftp://ftp.ripe.net/rfc/rfc977.txt) from 1986. The standard for the format of articles is specified in RFC 1036 (ftp://ftp.ripe.net/rfc/rfc1036.txt) from 1987.

9 WinVN is maintained by Sam Rushing and Jim Dumoulin, of the NASA/Kennedy Space Center, and is available from http://www.ksc.nasa.gov/software/winvn/winvn.html

10 Note that portions of Usenet have been archived since around 1995 by the DejaNews service, http://www.deja.com/, which provide powerful tools to search the archival resource.

11 The personal details of the poster have been blanked out.

12 Perhaps the most noteworthy example is the *Time* magazine 'Cyberporn' issue, from July 1995, with the sensational tagline 'Exclusive: a new study shows how pervasive and wild it really is. Can we protect our kids – and free speech?' See Randall 1998, chapter 16, for an interesting discussion of this article and the reaction to it by the Internet community.

13 http://netscan.research.microsoft.com/
14 See http://www.media.mit.edu/~kkarahal/loom/

8 Mapping synchronous social spaces

1 There are a large number of number of IRC chat clients available for all major platforms. For technical details on IRC clients and connecting, see the excellent online help at IRChelp.org, http://www.irchelp.org/. The IRC protocol is defined in RFC1459 (ftp://ftp.ripe.net/rfc/rfc1459.txt).
2 Available from http://www.mirc.co.uk/
3 It is possible to log the conversation of the channel locally from the client. However, channels are not archived on mass at the server end.
4 http://chatcircles.media.mit.edu/
5 See the MUD FAQ for more details: http://muds.okstate.edu/~jds/mudfaqs.html (note that specifics vary from MUD to MUD).
6 The MUDline by Burka (1995) provides a detailed time-line of MUD history up to 1995: http://www.apocalypse.org/pub/u/lpb/muddex/mudline.html
7 Interestingly, the founders are still actively involved in MUD development with their company MUSE (http://www.mud.co.uk/).
8 The article has been widely reprinted, see for example Stefik (1996). It is also available online at, http://www.levity.com/julian/bungle.html
9 A list of active MUDs is available at the MUD Connector (http://www.mudconnect.com).
10 Dr Who's famous Tardis (Time And Relative Dimension In Space) was the size of a police telephone box on the outside but contained a much greater space within. For more information see http://www.tardis.ed.ac.uk/~abr/drwho/tardis/
11 Dibbell says, 'there was the Earth itself, spinning quietly on the axes of a globe in the entrance hall, medium-sized to all outward appearances but of planetary magnitude once you stepped into its atmosphere – and growing more capacious all the time as newly arriving MOOers added fond simulations of their hometowns, home states, and home countries to the globe's open database' (Dibbell 1999: 49).
12 http://discworld.imaginary.com:5678/
13 From the story 'of exactitude in science', in Jorge Luis Borges (1983), *A Universal History of Infamy*.
14 Rooms which are only linked to the main MUD structure in one direction. One can enter the room, but then has no means of exiting (except by direction teleportation to another location if allowed or by quitting and restarting). Some of these topological 'black-holes' are designed as deliberate pranks to trap the unwary player, a nice example from BayMOO is Alcatraz, a series of rooms that can be entered, but with no means of escape (Anders 1998).
15 Source: http://www.dragonmud.org/dragonmud/aboutdragonmud.html
16 http://discworld.imaginary.com:5678/atlas/atlas.htm
17 These are conceptually similar to the topological graphs discussed in Chapter 6.
18 http://baymoo.sfsu.edu:4242
19 http://www.zuggsoft.com/zmud/zmudinfo.htm
20 The first virtual world system was called Habitat and it fused together a graphical interface with avatars and a game-style MUD environment (Morningstar and Farmer 1991). It was a pioneering effort, developed with the limited home computers of the late 1980s (the Commodore 64).
21 AlphaWorld is owned and managed by Activeworlds.com, Inc., a small firm based in Newburyport, MA, in the US (http://www.activeworlds.com/).
22 V-Chat: http://www.microsoft.com/windows/ie/chat/?/ie/chat/vchatmain.htm; InterSpace, http://www.is.ntts.com/ispace.html; Worlds Chat, http://www.worlds.net/; WorldsAway, http://www.worldsaway.com/; The Palace, http://www.thepalace.com/; Deuxième Monde, http://www.2nd-world.fr/homevisiteurs.htm; Blaxxun CyberGate, http://www.blaxxun.com/; Onlive Traveller, http://www.onlive.com/. Damer (1997) provides a populist travel guide-style overview of a number of the virtual worlds.
23 A term first used in the pioneering Habitat system of the late 1980s (Morningstar and Farmer 1991), and popularised by Neal Stephenson's 1992 science-fiction novel *Snow Crash*.
24 Unlike other virtual world systems, such as the Palace, participants in AlphaWorld cannot create their own unique avatar (Suler 1997). Instead, users can select an avatar from a range of approximately thirty models, all of which are humanoid in form.

25 For example, there is the AlphaWorld Historical Society, with a museum (http://www.awcommunity.org/awhs/). There is also a 'national' newspaper called the *New World Times* (or *NWT* for short), run by AlphaWorld citizens. The first edition of the *NWT* was published only a couple of months after AlphaWorld opened, with regular editions appearing since then. An archive of all the editions is available on the Web providing valuable historical documents (NWT 1998).

26 Although the software only runs on Windows95/NT and a citizen registration fee is needed to gain access to all facilities.

27 AlphaWorld first came online on 28 June 1995.

28 http://awmap.vevo.com/densmap.html

29 http://awmap.vevo.com/

30 105.4 North, 187.8 East.

31 http://www.peacekeeper.net/

32 http://www.casa.ucl.ac.uk/30days/

10 Imaginative mappings of cyberspace

1 Neal Stephenson's novel *Snowcrash*, written in 1992, partly inspired the development of a virtual, visual world, accessible across the Internet by the summer of 1995. AlphaWorld, an online, visual virtual world where people can interact using avatars, is life imitating art, in a very literal sense (see chapter 8, Dodge 1998)

2 The following novels were read for this chapter: Anderson, K.J. and Beason, D. (1996) *Virtual Destruction*. New York: Ace Books; Bear, G. (1985) *Blood Music*. London: Vista; Besher, A. (1994) *RIM*. London: Orbit; Besher, A. (1998) *MIR*. London: Orbit; Bethke, B. (1995) *Headcrash*. London: Orbit; Cadigan, P. (1998) *Tea From an Empty Cup*. London: HarperCollins; Calder, R. (1998) *Cythera*. London: Orbit; Egan, G. (1994) *Permutation City*. London: Millennium; Fabi, M. (1998) *Wyrm*. New York: Batham Books; Foy, G. (1996) *The Shift*. New York: Batham Books; Foy, G. (1997) *Contraband*. New York: Batham Books; Gibson, W. (1984) *Neuromancer*. London: HarperCollins; Gibson, W. (1986) *Count Zero*. London: HarperCollins; Gibson, W. (1988) *Mona Lisa Overdrive*. London: HarperCollins; Gibson, W. (1992) *Virtual Light*. London: Penguin; Gibson, W. (1996) *Idoru*. London: Penguin; Harrison, H. and Minsky, M. (1992) *The Turing Option*. London: Roc; Harry, E.L. (1996) *Society of the Mind*. London: Coronet Books; Hendrix, H.V. (1997) *Lightpaths*. New York: Ace Books; McLaren, J. (1997) *Press Send*. London: Simon and Schuster; Piercy, M. (1993) *He, She and It*. Fawcett Crest; Platt, C. (1991) *The Silicon Man*. San Francisco: Wired Books; Rucker, R. (1982) *Software*. New York: Avon Books; Rucker, R. (1988) *Wetware*. New York: Avon Books; Rucker, R. (1997) *Freeware*. New York: Avon Books; Stephenson, N. (1992) *Snow Crash*. New York: Bantam Books; Stephenson, N. (1995) *The Diamond Age*. New York: Bantam Books; Sterling, B. (1988) *Islands in the Net*. New York: Ace Books; Sterling, B. (1994) *Heavy Weather*. New York: Bantam Books; Sterling, B. (1996) *Holy Fire*. New York: Bantam Books; Williams, T. (1996) *Otherland*. London: Orbit.

   Short collections:
   Gibson, W. (1994) *Burning Chrome*. London: HarperCollins; Greenburg, M.H. and Segriff, L. (eds) (1996) *Future Net*. New York: DAW Books; Sterling, B. (ed.) (1986) *Mirrorshades: The Cyberpunk Anthology*. London: HarperCollins; Sterling, B. (1993) *Globalhead*. London: Millennium.

3 Richard Calder.

4 Pat Cadigan and Marge Piercy.

5 On completion of reading all the books, the identified passages were transcribed and imported into NUD*IST 4.0 (a qualitative data analysis package). NUD*IST allows qualitative data to be easily managed, cross-referenced, and analysed using simple Boolean operations to identify patterns. Using NUD*IST, the data were analysed using the prescription detailed in Dey (1993) and Kitchin and Tate (1999). This prescription is structured and rigorous, consisting of three primary stages – description, classification, connection, – that are operationalised through a sequence of standardised tasks. First, each discrete passage was annotated, detailing potential category allocation. Next, the data were sorted into categories of related material. Where relevant, data were assigned to more than one category. To aid the process of connection, the data categories were then split (divided into new discrete categories) or spliced (merged to form new, more generalised categories) to create new sorted categories of related data. Where appropriate,

links between sorted categories of data were then examined using the Boolean operations within NUD*IST. Finally, the interpretations drawn from the data within each sorted category were corroborated in relation to evidence within other sorted categories. This process of analysis means that it was possible to explore the richness of the passages in a constructive and rigorous manner, allowing the identification of common trends within the novels. It is appreciated that to some the process adopted will seem excessively mechanical, but given the sheer volume of data it was felt that this approach was the most appropriate. The resultant database consisted of hundreds of cross-referenced passages.

6   *Neuromancer, Count Zero, Mona Lisa Overdrive.*

11 Future mappings of cyberspace

1   Our analysis is centred on a Western conception of both space and society, and our discussion draws predominately from English-language texts. Moreover, we acknowledge that our discussion is relatively weak on the contributions of architecture and information design. Our analysis, then, is partial and selective. One future part of the project of mapping cyberspace is to consider non-Western views and analysis of cyberspace.

2   As argued in Chapter 1, the tendency to treat cyberspace as a sublime or paraspace is exacerbated by the noted 'invisibility' of cyberspace infrastructure in the landscape, particularly compared to earlier technological developments such as the railway or motor car.

3   Internal validity concerns whether the results from a study can be interpreted in different ways; can different conclusions be drawn from the same results? External (ecological) validity concerns whether the results from one study can be applied to populations beyond the study sample.

4   Here, map is synonymous with spatialisations.

Bibliography

Abbate, J. (1993) *An Archaeology of the ARPANET*. Unpublished paper, University of Pennsylvania. http://www.wam.umd.edu/~abbate/papers/Arch.html

Abbate, J. (1999) *Inventing the Internet*. MIT Press, Cambridge, Mass.

Adam, A. and Green, E. (1998) 'Gender, Agency, Location and the New Information Society'. In Loader, B. (ed.) *The Cyberspace Divide*. Routledge, London, pp. 83–97.

Adams, P.C. (1992) 'Television as Gathering Place'. *Annals of the Association of American Geographers* 82, 1: 117–135.

Adams, P. (1995) 'A Reconsideration of Personal Boundaries in Space–Time'. *Annals of the Association of American Geographers* 85, 2: 267–285.

Adams, P. (1996) 'Protest and the Scale Politics of Telecommunications'. *Political Geography* 15, 5: 419–441.

Adams, P. (1997) 'Computer Networks and Virtual Place Metaphors'. *Geographical Review* 87, 2: 3, 5.

Adams, P. (1998) 'Network Topologies and Virtual Place'. *Annals of the Association of American Geographers* 88, 1: 88–106.

Adams, P.C. (forthcoming) 'Application of a CAD-based Accessibility Model'. In Janelle, D. and Hodge D. (eds) *Information, Place and Cyberspace: Issues in Accessibility*. Elsevier, Amsterdam.

Albert, R., Jeong, H., and Barabási, A-L. (1999) 'Diameter of the World Wide Web'. *Nature* 401: 130–131.

Allen, G. (1985) 'Strengthening Weak Links in the Study of the Development of Macrospatial Cognition'. In Cohen, R. (ed.), *The Development of Spatial Cognition*. Erlbaum Lawerence, Hillsdale, New Jersey, pp. 301–321.

Allen, G.L. (1988) 'The Acquisition of Spatial Knowledge Under Conditions of Temprospatial Discontinuity'. *Psychological Research* 50: 183–190.

Allen, G. and Kirasic, K. (1985) 'Effects of the Cognitive Organization of Route Knowledge on Judgments of Macrospatial Distance'. *Memory and Cognition* 13: 218–227.

Allen, G., Siegel, A.W., and Rosinski, R.R. (1978) 'The Role of Perceptual Context in Structuring Spatial Knowledge'. *Journal of Experimental Psychology: Human Learning and Memory* 4: 617–630.

Alles, P., Esparza, A., and Lucas, S. (1994) 'Telecommunications and the Large City–Small City Divide: Evidence from Indiana cities'. *Professional Geographer* 46: 307–316.

Amir, E. (1993) *Carta: A Network Topology Presentation Tool*. CS268 Project Paper, 26 May 1993, Computer Science Division, University of California Berkeley. http://www.cs.berkeley.edu/~elan/mbone/carta.ps.gz

Anders, P. (1996) 'Envisioning Cyberspace: The Design of Online Communities'. Paper presented at 5Cyberconf, 1996, Madrid. http://www.telefonica.es/fat/eanders.html

Anders, P. (1998) *Envisioning Cyberspace: Designing 3D Electronic Space*. McGraw-Hill, New York.

Anderson, B. (1983) *Imagined Communities*. Verso, Oxford.

Anderson, K.J. and Beason, D. (1996) *Virtual Destruction*. Ace Books, New York.

Anderson, R.H., Bikson, T.K., Law, S.A., and Mitchell, B.M. (1995) *Universal Access to E-mail: Feasibility and Societal Implications*. RAND, Santa Monica, CA.

Andreessen, M. (1999) 'Innovators of the Net: Ramanathan V. Guha and RDF'. *Netscape TechVision column*, 8 January 1999. http://www31.netscape.com/columns/techvision/innovators_rg.html

Andrews, K. (1995) 'Visualising Cyberspace: Information Visualisation in the Harmony Internet Browser'. *Proceedings of Info-Vis '95, IEEE Symposium on Information Visualization*. IEEE Computer Society Press, Los Alamitos, CA., pp. 97–104. ftp://ftp.iicm.edu/pub/papers/ivis95.pdf

Andrews, K., Pichler, M., and Wolf, P. (1996) Towards Rich Information Landscapes for Visualising Structured Web Space. *Proceedings of Info-Vis '96, IEEE Symposium on Information Visualization*. IEEE Computer Society Press, Los Alamitos, CA., pp. 62–63. ftp://ftp.iicm.edu/pub/papers/ivis96.pdf

Anooshian, L.J. and Seibert, P.S. (1996) 'Diversity within Spatial Cognition: Memory Processes Underlying Place Recognition'. *Applied Cognitive Psychology*, 10: 281–290.

Aoki, K. (1994) 'Virtual Communities in Japan'. Paper presented at the Pacific Telecommunications Council Conference.

Appleyard, D. (1970) 'Styles and Methods of Structuring a City'. *Environment and Behavior*, 2: 100–117.

Argyle, K. and Shields, R. (1996) 'Is There a Body in the Net?' In Shields, R. (ed.) *Cultures of Internet: Virtual Spaces, Real Histories and Living Bodies*. Sage, London, pp. 58–69.

Armitt, L. (1996) *Theorising the Fantastic*. Arnold, London.

Aronowitz, S. (1994) 'Technology and the Future of Work'. In Bender, G. and Druckery, T. (eds) *Culture on the Brink: Ideologies of Technology*. Bay Press, Seattle, WA, pp. 15–29.

Aycock, A. (1995) 'Technologies of the Self: Foucault and Internet Discourse'. *Journal of Computer Mediated Communications* 1, 2. http://www.ascusc.org/jcmc/vol1/issue2/aycock.htm

Ayers, E.Z. and Stasko, J.T. (1995) 'Using Graphic History in Browsing the World Wide Web'. *Proceedings of the Fourth International World Wide Web Conference*, December 1995, Boston. http://www.w3.org/Conferences/WWW4/Papers2/270/

Bachiochi, D., Berstene, M., Chouinard, E., Conlan, N., Danchak, M., Furey, T., Neligon, C. and Way, D. (1997) 'Usability Studies and Designing Navigational Aids for the World Wide Web'. *Proceedings of the Sixth International World Wide Web Conference*, 7–11 April 1997, Santa Clara, CA., US. http://decweb.ethz.ch/WWW6/Technical/Paper180/Paper180.html

Badgett, T. and Sandler, C. (1993) *Internet: From Mystery to Mastery*. MIS Press, NewYork.

Ball, T.A. and Eick, S.G. (1996) 'Software Visualization in the Large'. *IEEE Computer*, April 1996, 29, 4: 33–43.

Barkow, T. (1996) 'The Domain Name System: Let Your Resolver do the Walking'. *Wired*, September 1996: 105.

Barratt, N. (1996) *State of the Cybernation*. Kogan Paul, London.

Bartle, R. (1990) *Early MUD History*. 15 November 1990. http://www.mud.co.uk/richard/mudhist.htm

Batty, M. (1990) 'Invisible Cities'. *Environment and Planning B: Planning and Design*, 17: 127–130.

Batty, M. (1991) 'Urban Information Networks: The Evolution and Planning of Computer-Communications Infrastructure'. In Brotchie, J., Batty, M., Hall, P., and Newton, P. (eds) *Cities of the Twenty-first Century*. Longman, London, pp. 139–158.

Batty, M. (1995) 'The Computable City'. In Wyatt, R. and Hossein, H. (eds) *Proceedings: Fourth International Conference on Computers in Urban Planning and Urban Management*, University of Melbourne, Australia, 11–14 July 1995, pp. 1–18.

Batty, M. (1997) 'Virtual Geography'. *Futures*, 29, 4/5: 337–352.

Batty, M. and Barr, B. (1994) 'The Electronic Frontier: Exploring and Mapping Cyberspace'. *Futures*, 26, 7: 699–712.

Batty, M., Dodge, M., Doyle, S., and Smith, A. (1998) 'Modelling Virtual Environments'. In Longley, P.A., Brooks, S.M., McDonnell, R., and Macmillan, B. (eds) *Geocomputation: A Primer*. Wiley, Chichester, UK, pp. 139–161.

Baudrillard, J. (1983) *Simulacra and Simulations* (trans. Foss, P., Patton, P., and Beitchman). Semiotext(e), New York.

Baym, N.K. (1995) 'The Emergence of Community in Computer-Mediated Communication'. In Jones, S.G. (ed.) *CyberSociety: Computer-Mediated Communication and Community*. Sage Publications, Thousand Oaks, CA, pp. 138–163.

Baym, N.K. (1997) 'Interpreting Soap Operas and Creating Community: Inside an Electronic Fan Culture'. In Kiesler, S. (ed.) *Culture of the Internet*. Lawrence Erlbaum Associates, New Jersey, USA, pp. 103–120.

Bechar-Israeli, H. (1995) 'From <Bonehead> to <cLoNehEAd>: Nicknames, Play and Identity on Internet Relay Chat'. *Journal of Computer Mediated Communications* 1, 2. http://www.ascusc.org/jcmc/vol1/issue2/bechar.html

Becker, R.A., Eick, S.G., and Wilks, A.R. (1991) 'Basics of Network Visualization'. *IEEE Computer Graphics and Applications* 11, 3: 12–14.

Becker, R.A., Eick, S.G., and Wilks, A.R. (1995) 'Visualizing Network Data'. *IEEE Transactions on Visualization and Computer Graphics* 1, 1: 16–28.

Benedikt, M. (1991a) 'Introduction'. In Benedikt, M. (ed.) *Cyberspace: First Steps*. MIT Press, Cambridge, Mass., pp. 1–26.

Benedikt, M. (1991b) 'Cyberspace: Some Proposals'. In Benedikt, M. (ed.) *Cyberspace: First Steps*. MIT Press, Cambridge, Mass., pp. 119–224.

Benford, S., Snowdon, D., Greenhalgh, C., Ingram, R., Knox, I., and Brown, C. (1995) 'VR-VIBE: A Virtual Environment for Co-operative Information Retrieval'. *Proceedings of Eurographics '95*, 30 August–1 September 1995, Maastricht, The Netherlands, pp. 349–360.

Benford, S., Snowdon, D., Brown, C., Reynard, G., and Ingram, R. (1997) 'Visualising and Populating the Web: Collaborative Virtual Environments for Browsing, Searching and Inhabiting Webspace'. *Proceedings of the Eighth Joint European Networking Conference (JENC8)*, 12–15 May 1997, Edinburgh. http://www.rare.nl/conf/jenc8/papers/123.ps

Bennahum, D.S. (1998a) *Extra Life: Coming of Age in Cyberspace*. Basic Books, New York.

Bennahum, D.S. (1998b) 'The Hot New Medium is . . . Email'. *Wired* 6.04, April 1998.

Berman, M. (1992) 'Why Modernism Still Matters'. In Lash, S. and Friedman, J. (eds) *Modernity and Identity*. Blackwell, Oxford, pp. 33–58.

Berners-Lee, T. (1999) *Weaving the Web: The Past, Present and Future of the World Wide Web by its Inventor*. Orion Business.

Berstein, M. (1991) 'The Navigation Problem Reconsidered'. In Berk, E. and Devlin, J. (eds) *Hypertext / Hypermedia Handbook*. McGraw-Hill, New York, pp. 285–297.

Besher, A. (1994) *RIM*. Orbit, London.

Besher, A. (1998) *MIR*. Orbit, London.

Bethke, B. (1995) *Headcrash*. Orbit, London.

Bilstad, B.T. (1996) 'Obscenity and Indecency on the Usenet: The Legal and Political Future of Alt.Sex.Stories'. *Journal of Computer Mediated Communication* 2, 2. http://www.ascusc.org/jcmc/vol2/issue2/bilstad.html

Bingham, N. (1999) 'Unthinkable Complexity? Cyberspace Otherwise'. In Crang, M., Crang, P., and May, J. (eds) *Virtual Geographies*. Routledge, London, pp. 244–260.

Blakemore, M. and Harley, J.B. (1980) 'Concepts in the History of Cartography A Review and Perspective'. *Cartographica* 17: 14.

Boardwatch, (1999) *Directory of Internet Service Providers*. Boardwatch Magazine. Golden, Colorado. http://boardwatch.internet.com/isp/index.html

Boroumand, S. (1998) 'Demand > Supply: Why We Need IPv6'. *InterNIC News*, February 1998.

Bowman, C.M., Danzig, P.B., Manber, U., and Schwartz, M.F. (1994) 'Scalable Internet Resource Discovery: Research Problems and Approaches'. *Communications of the ACM* 37, 8: 98–107.

Brail, S. (1996) 'The Price of Admission: Harrassment and Free Speech in the Wild, Wild West'. In Cherney, L. and Reise, E.R. (eds) *Wired Women: Gender and New Realities in Cyberspace*. Seal Press, Seattle, pp. 141–157.

Brake, D. (1997) 'Lost in Cyberspace'. *New Scientist* 154, 12: 12–13.

Brants, K., Huizenga, M., and van Meerten, R. (1996) 'The New Canals of Amsterdam: An Exercise in Local Democracy'. *Media, Culture and Society* 18: 233–247.

Braun, H.W. (1997) 'BGP-System Usage of 32 bit Internet Address Space'. *Measurement and Operations Analysis Team, NLANR*, 15 November 1997. http://moat.nlanr.net/IPaddrocc/

Bray, T. (1996) 'Measuring the Web'. *Fifth International Conference World Wide Web*, 6–10 May 1996, Paris. http://www5conf.inria.fr/fich_html/papers/P9/Overview.html

Breathnach, P. (1998) 'Exploring the "Celtic Tiger" Phenomenon: Causes and Consequences of Ireland's Economic Miracle'. *European Urban and Regional Studies* 5, 4: 305–316.

Breathnach, P. (forthcoming a) 'Globalisation, Information Technology, and the Emergence of "Niche" Transnational Cities: The Growth of the Call Centre Sector in Dublin'. *Geoforum*.

Breathnach, P. (forthcoming b) 'Gender Segmentation and the International Division of Labour in Office Employment: The Growth of the Teleservices Sector in Ireland'. *Journal of Gender Studies*.

Breeze, M. (1997) 'Quake-ing in my Boots: <Examining> Clan Community Construction in an Online Gamer Population'. *Cybersociology Magazine* 2, November 1997. http://members.aol.com/Cybersoc/is2breeze.html

Brewster, S. and Blades, M. (1989) 'Which Way to Go? Children's Ability to Give Directions in the Environment and From Maps'. *Environmental Education and Information* 8: 141–156.

Bromberg, H. (1996) 'Are MUDs Communities? Identity, Belonging and Consciousness in Virtual Worlds'. In Shields, R. (ed.) *Cultures of Internet: Virtual Spaces, Real Histories and Living Bodies*. Sage, London, pp. 143–152.

Brown, D. (1994) *Geographical Trace Route V1.0*. Technical Report, Department of Computer Science, Curtin University of Technology, Perth, Australia, February 1994.

Bruckman, A. (1993) 'Gender Swapping on the Internet'. Paper presented at *INET '93 Conference*, August 1993, San Francisco. ftp://ftp.cc.gatech.edu/pub/people/asb/papers/gender-swapping.txt

Bruckman, A. (1996) 'Finding One's Own Space in Cyberspace'. *MIT Technology Review* 96, 1: 48–54. http://www.techreview.com/articles/jan96/Bruckman.html

Bruckman, A. and Resnick, M. (1995) 'The MediaMOO Project: Constructionism and Professional Community'. *Convergence* Spring 1995, 1, 1: 94–109. http://www.cc.gatech.edu/fac/asb/papers/convergence.html

Brunn, S.D. (1998) 'GEOGED as a Virtual Workshop'. Paper presented at the *National Council for Geographic Education, 83 Annual Meeting*, 11–14 October 1998, Indianapolis, USA.

Bukatman, S. (1993) *Terminal Identity: The Virtual Subject in Postmodern Science Fiction*. Duke University Press, London.

Burch, H. and Cheswick, B. (1999) 'Mapping the Internet'. *Computer*, April 1999: 97–98, 102. http://computer.org/computer/IntWatch0499.html

Burka, L.P. (1995) *The MUDline*, maintained by Lauren P. Burka. http://www.apocalypse.org/pub/u/lpb/muddex/mudline.html

Burkhart, G.E. and Goodman, S.E. (1998) 'The Internet Gains Acceptance in the Persian Gulf'. *Communications of the ACM*, March 1998, 41, 3: 19–25.

Burkhart, G.E., Goodman, S.E., Mehta, A., and Press, L. (1998) 'The Internet in India: Better Times Ahead?' *Communication of the ACM*, November 1998, 41, 11: 21–26.

Burrough, P.A. (1986) *Principles of Geographical Information Systems for Land Resources Assessment*. Clarendon Press, Oxford.

Burrough, P.A. and McDonnell, R. (1998) *Principles of Geographical Information Systems*. Oxford University Press, Oxford.

Burrows, R. (1997) 'Virtual Culture, Urban Social Polarisation and Social Science Fiction'. In Loader, B. (ed.) *The Governance of Cyberspace*. Routledge, London, pp. 38–45.

Butler, D.L., Acquino, A.L., Hissong, A.A., and Scott, P.A. (1993) 'Wayfinding by Newcomers in a Complex Building'. *Human Factors* 35: 159–173.

Cadigan, P. (1998) *Tea From an Empty Cup*. HarperCollins, London.

Cairncross, F. (1997) *The Death of Distance: How the Communications Revolution Will Change Our Lives*. Harvard Business School Press, Boston.

Calcutt, A. (1995) 'Computer Porn Panic: Fear and Control in Cyberspace'. *Futures* 27: 749–762.

Calder, R. (1998) *Cythera*. Orbit, London.

Canter, D., Rivers, R., and Storrs, G. (1985) 'Charaterizing User Navigation Through Complex Data Structures'. *Behaviour and Information Technology* 4: 93–102.

Card, S.K., Mackinlay, J.D., and Shneiderman, B. (1999) *Readings in Information Visualization: Using Vision to Think*. Morgan Kaufmann Publishers, Inc., San Francisco.

Carroll, J. (1997) '(D)RIVEN'. *Wired* 5.09, September 1997: 120ff.

Castells, M. (1988) *The Informational City: Information Technology, Economic Restructuring and the Urban-Regional Process*. Blackwell, Oxford.

Castells, M. (1996) *The Rise of the Network Society*. Blackwell, Oxford.

Castells, M. and Hall, P. (1995) *Technopoles: Mines and Foundaries of the Information Economy*. Routledge, London.

CCR (1990) 'Selected ARPANET Maps'. *Computer Communication Review*, October 1990, 20: 81–110.

Cerf, V. (1995) 'Foreword'. In Salus, P. (ed.) *Casting the Net: From ARPANET to Internet and Beyond. . . .* Addison Wesley, Reading, Mass.

Chalmers, M., Rodden, K., and Brodbeck, D. (1998) 'The Order of Things: Activity-Centered Information Access'. *Proceedings of the Seventh International World Wide Web Conference*, April 1998, Brisbane, Australia, pp. 359–367. http://www.dcs.gla.ac.uk/~matthew/papers/WWW7/www98.html

Charitos, D. (1997) 'Designing Space in Virtual Environments for Aiding Wayfinding Behaviour'. *Proceedings of UK VRSIG 97*. http://www.brunel.ac.uk/~empgrrb/VRSIG97/proceed/028/028.html

Cheesman, J., Dodge, M., Harvey, F., Jacobson, D., and Kitchin, R. (forthcoming) '"Other" Worlds: Augmented, Comprehensible, Non-Material Spaces'. In Fisher, P. and Unwin, D. (eds) *Virtual Reality in Geography*. Taylor and Francis, London.

Chen, C. (1997) 'Structuring and Visualising the WWW with Generalised Similarity Analysis'. *Proceedings of the Eighth ACM Conference on Hypertext (Hypertext '97)*, June 1997, Southampton, UK, pp. 177–186. http://www.brunel.ac.uk/~cssrccc2/papers/ht97.pdf

Chen, C. (1999) *Information Visualisation and Virtual Environments*. Springer Verlag, Berlin.

Chen, H., Houston, A.L., Sewell, R.R., and Schatz, B.R. (1998) 'Internet Browsing and Searching: User Evaluations of Category Map and Concept Space Techniques'. *Journal of the American Society for Information Science* 49, 7: 582–603.

Chen, H., Schuffels, C., and Orwig, R. (1996) 'Internet Categorization and Search: A Machine Learning Approach'. *Journal of Visual Communications and Image Representation* 7, 1: 88–102.

Cherny, L. (1995) *The MUD Register: Conversational Modes of Action in a Text-Based Virtual Reality*. Unpublished doctoral thesis, Stanford University, December 1995. http://portal.research.bell-labs.com/orgs/ssr/people/cherny/diss-overview.html

Cherny, L. and Weise, R. (eds) (1996) *Wired Women: Gender and New Realities in Cyberspace*. Seal Press, Seattle.

Chesher, C. (1994) 'Colonizing Virtual Reality: Construction of the Discourse of Virtual Reality, 1984–1992'. *Cultronix* 1(1). http://eserver.org/cultronix/chesher/

Cheswick, B. and Burch, H. (1999) *Internet Mapping Project*. Bell Labs, New Jersey, USA. http://www.cs.bell-labs.com/~ches/map/index.html

Choo, C.W. (1995) 'National Computer Policy Management in Singapore: Planning an Intelligent Island'. *Proceedings of the Fifty-eight Annual Meeting of the American Society for Information Science*, 9–12 October, Chicago, USA. Vol. 32: 152–156.

Churchill, E.F., Snowdon, D., Benford, S., and Dhanda, P. (1997) 'Using VR-VIBE: Browsing and Searching for Documents in Three-Dimensional-Space'. Paper presented at *HCI International '97: Seventh International Conference on Human-Computer Interaction*, 24–29 August 1997, San Francisco, USA.

Claffy, K. and Huffaker, B. (n.d.) 'Macroscopic Internet Visualization and Measurement'. Unpublished Article by *Cooperative Association for Internet Data Analysis (CAIDA)*, San Diego, USA. http://www.caida.org/Tools/Mapnet/summary.html

Claffy, K., Monk, T.E., and McRobb, D. (1999) 'Internet Tomography'. *Nature*, 7 January 1999. http://helix.nature.com/webmatters/tomog/tomog.html

Clarke, D. and Doel, M. (1999) 'Virtual Worlds: Simulation, Suppletion, S(ed)uction and Simulacra'. In Crang, M., Crang, P., and May, J. (eds) *Virtual Geographies*. Routledge, London, pp. 261–283.

Clement, P.C. (1998) *The State of the Net*. McGraw-Hill, New York.

Clerc, S. (1996) 'Estrogen Brigades and "Big Tits" Threads: Media Fandom Online and Off'. In Cherney, L. and Reise, E.R. (eds) *Wired Women: Gender and New Realities in Cyberspace*. Seal Press, Seattle, pp. 73–97.

Clever (1999) 'Hypersearching the Web'. *Scientific American*, June 1999: 44–52.
http://www.scientificamerican.com/1999/0699issue/0699raghavan.html

Clodius, J. (1994) *Concepts of Space and Place in a Virtual Community*.
http://www.dragonmud.org/people/jen/space.html

Clute, J. and Nicholls, P. (1993) *The Encyclopedia of Science Fiction*. St Martin's Press, New York.

Cockburn, A. and Jones, S. (1996) 'Which Way Now? Analysing and Easing Inadequacies in WWW Navigation'. *International Journal of Human-Computer Studies* 45: 105–109.

Cockburn, A. and Greenberg, S. (1999) *Beyond the 'Back' Button: Issues of Page Representation and Organisation in Graphical Web Navigation Tools*. Research Report 99-640-03, Department of Computer Science, University of Calgary, Calgary, Canada.
http://www.cpsc.ucalgary.ca/grouplab/papers/1999/99-WebView/graWeb.pdf

Contact Consortium (1998) *Sherwood Forest Community Project*.
http://www.ccon.org/events/sherwood.html

Cornell, E., Heth, D., and Alberts, D.M. (1994) 'Place Recognition and Wayfinding by Children and Adults'. *Memory and Cognition* 22: 633–643.

Cornell, E.H. and Hay, D.H. (1984) 'Children's Acquisition of a Route Via Different Media'. *Environment and Behavior* 16: 627–641.

Correll, S. (1995) 'The Ethnography of an Electronic Bar: The Lesbian Café'. *Journal of Contemporary Ethnography* 24: 270–298.

Cosgrove, D. (1994) 'Contested Global Visions: One-World, Whole-Earth, and the Apollo Space Photographs'. *Annals of the Association of American Geographer* 84, 2: 270–294.

Couclelis, H. (1996) 'The Death of Distance'. *Environment and Planning B: Planning and Design* 23: 387–389.

Couclelis, H. (1998) 'Worlds of Information: The Geographic Metaphor in the Visualisation of Complex Information'. *Cartography and Geographic Information Systems* 25: 209–220.

Couclelis, H., Golledge, R.G., Gale, N., and Tobler, W. (1987) 'Exploring the Anchor-Point Hypothesis of Spatial Cognition'. *Journal of Environmental Psychology* 7: 99–122.

Cox, D. and Patterson, R. (1992) *Visualization Study of the NSFNET*. Presentation by National Center for Supercomputing Applications, July 1992.
http://www.ncsa.uiuc.edu/SCMS/DigLib/text/technology/Visualization-Study-NSFNET-Cox.html

Cox, K.C. and Eick, S.G. (1995) '3D Display of Internet Traffic', In Gershon, N. and Eick, S.G. (eds), *Proceedings of the 1995 IEEE Symposium on Information Visualization*, 30 October 1995, pp. 129–131 (IEEE Computer Society Press).

Cox, K.C., Eick, S.G., and He, T. (1996) '3D Geographic Network Display'. ACM *Sigmod Record*, December 1996, 25, 4: 50–54.

Crampton, J. (1999) 'Maps as texts, maps as visualisations'. Department of Geography and Earth Science, George Mason University. http://geog.gmu.edu/gess/people/jwc.html

CRG (1997) *Improving the Legibility of Abstract Spaces*. Communications Research Group, University of Nottingham, 19 November 1997.
http://www.crg.cs.nott.ac.uk/research/technologies/visualisation/leads/

Critical Art Ensemble (1995) *Utopian Promises – Net Realities*.
http://www.well.com/user/hlr/texts/utopiancrit.html.

Cronin, M.J. (1994) *Doing Business on the Internet*. Von Nostrand Reinhold, New York.

Cruz-Neira, C., Sandin, D.J., and DeFanti, T. (1993) 'Surround-Screen Projection-Based Virtual Reality: The Design and Implementation of the CAVE'. *Computer Graphics* 27: 135–142.

Csicsery-Ronay Jnr., I. (1991) 'Editorial Introduction: Postmodernism's SF/SF's Postmodernism'. *Science Fiction Studies* 18: 305–308.

Curry, M. (1995) 'On Space and Spatial Practice in Contemporary Geography'. In Earle, C., Mathewson, K., and Kenzer, M. (eds) *Concepts in Human Geography*. Rowman and Littlefield Publishers, Lanham, pp. 3–32.

Curry, M. (1998) *Digital Places*. Routledge, London.

Curtis, P. (1996) 'Mudding: Social Phenomena in Text-based Virtual Realities'. In Stefik, M. (ed.) *Internet Dreams: Archetypes, Myths, and Metaphors*. MIT Press, Cambridge, Mass., pp. 265–291.

CyberAtlas (1998) http://www.cyberatlas.com/

Damer, B. (1996) 'Inhabited Virtual Worlds'. ACM *Interactions*, September–October 1996: 27.

Damer, B. (1997) *Avatars! Exploring and Building Virtual Worlds on the Internet*. Peachpit Press, Berkeley, CA.

Danet, B., Ruedenberg, L., and Rosenbaum-Tamari, Y. (1998) 'Hmmm . . . , Where's that Smoke Coming From? Writing, Play and Performance on Internet Relay Chat'. In Sudweeks, F., McLaughlin, M., and Rafaeli, S. (eds) *Network and Netplay: Virtual Groups on the Internet*. MIT Press, Cambridge, Mass., pp. 41–76.

Daniels, P. (1995) 'Services in a Shrinking World'. *Geography* 80: 97–110.

Danowitz, A.K., Nassef, Y., and Goodman, S.E. (1995) 'Cyberspace Across the Sahara: Computing in North Africa'. *Communications of the ACM*, December 1995, 38, 12: 23–28.

Darken, R.P. and Sibert, J.L. (1993) 'A Toolset for Navigation in Virtual Environments'. *Proceedings of ACM User Interface Software and Technology*, 1993, pp. 157–165.

Darken, R.P. and Sibert, J.L. (1995) 'Navigating Large Virtual Spaces'. *International Journal of Human-Computer Interaction*.
    http://www.seas.gwu.edu/faculty/sibert/darken/publications/Navigating_IJHCI95/navigating.html

Darken, R.P., and Sibert, J.L. (1996a) 'Wayfinding Strategies and Behaviors in Large Virtual Worlds'. *Proceedings of ACM SIGCHI 96*.
    http://www.seas.gwu.edu/faculty/sibert/darken/publications/Strategies_CHI96/strategies.html

Darken, R.P., and Sibert, J.L. (1996b) 'Navigating Large Virtual Spaces'. *International Journal of Human-Computer Interaction* 8, 1: 49–71

Davis, J. (1993) 'Cyberspace and Social Struggle'. *Computer Underground Digest* 28, 5 November 1993.

Davis, M. (1990) *City of Quartz*. Verso, London.

Dear, M. and Wolch, J. (1987) *Landscapes of Despair: From Deinstitutionalisation to Homelessness*. Polity Press, Oxford.

December, J. (1995) 'A Cybermap Gazetteer: Maps of the Online World for Browsing and Business'. In Staple, G.C. (ed.) *TeleGeography 1995*. TeleGeography, Inc, Washington, DC, pp. 74–82.

Denis, M. and Zimmer, M. (1992) 'Analog Properties of Cognitive Maps Constructed from Verbal Descriptions'. *Psychological Research* 54: 286–298.

Dery, M. (1996) *Escape Velocity: Cyberculture at the End of the Century*. Hodder and Stoughton, London.

Devlin, A. (1976) 'The "Small Town" Cognitive Map: Adjusting to a New Environment'. In Moore, G.T. and Golledge, R.G. (eds) *Environmental Knowing*. Dowden, Hutchinson and Ross, Stroudsberg, PA., pp. 58–66.

Dey, I. (1993) *Qualitative Data Analysis: A User Friendly Guide for Social Scientists*. Routledge, London.

Diamond, D. (1998) 'Whose Internet is It, Anyway?' *Wired*, April 1998: 172ff.

Dibbell, J. (1996) 'A Rape in Cyberspace: How an Evil Clown, a Haitian Trikster Spirit, Two Wizards, and a Cast of Dozens Turned a Database into a Society'. In Stefik, M. (ed.) *Internet Dreams: Archetypes, Myths, and Metaphors*. MIT Press, Cambridge, Mass., pp. 293–313.

Dibbell, J. (1999) *My Tiny Life: Crime and Passion in a Virtual World*. Fourth Estate, London.

Dieberger, A. (1996) 'Browsing the WWW by Interacting with a Textual Virtual Environment: A Framework for Experimenting with Navigational Metaphors'. *Proceedings of ACM Hypertext '96*, Washington, DC, March 1996, pp. 170–179.

Dieberger, A. (1997) 'A City Metaphor to Support Navigation in Complex Information Spaces'. http://www.sis.pitt.edu/~cosit97/abstracts.html

Dodge, M. (1997) 'A Cybermap Atlas: Envisioning the Internet'. In Staple, G.C. (ed.) *TeleGeography 97/98: Global Communications Traffic Statistics and Commentary*. TeleGeography Inc., Washington, DC., pp. 63–68.

Dodge, M. (1998) *Geography of Internet Address Space*. Centre for Advanced Spatial Analysis, University College London. http://www.geog.ucl.ac.uk/casa/martin/internetspace/

Dodge, M. and Kitchin, R. (2000) 'Exposing the "Second Text" in Maps of the Network Society'. *Journal of Computer Mediated Communication* in press.

Dömel, P. (1994) 'Webmap – A Graphical Hypertext Navigation Tool'. *Proceedings of the Second International World Wide Web Conference*, September 1994, Chicago, USA, pp. 85–97. http://www.ncsa.uiuc.edu/SDG/IT94/Proceedings/Searching/doemel/www-fall94.html

Donath, J.S. (1995) 'Visual Who'. *Proceedings of ACM Multimedia '95*, 5–9 November 1995, San Francisco, USA. http://judith.www.media.mit.edu/Judith/VisualWho/VisualWho.html

Donath, J.S. (1997) 'Inhabiting the Virtual City: The Design of Social Environments for Electronic Communities'. Unpublished Ph.D. thesis, MIT, February 1997. http://judith.www.media.mit.edu/Thesis/

Donath, J.S. (1999) 'Identity and Deception in the Virtual Community'. In Kollock, P. and Smith, M. (eds) *Communities in Cyberspace*. Routledge, London, pp. 29–59.

Donath, J., Karahalios, K., and Viégas, F. (1999) 'Visualizing Conversation'. Paper presented at the *Hawaii International Conference on System Science*, HICSS–32, 5–8 January 1999, Hawaii. http://persona.www.media.mit.edu/papers/VisualizeConv.pdf

Downs, R.M. and Stea, D. (1973) 'Theory'. In Downs, R.M. and Stea, D. (eds) *Image and Environment*. Aldine, Chicago, pp. 1–7.

Ducatel, K. and Halfpenny, P. (1993) 'Telematics for the Community? An Electronic Village Hall for East Manchester'. *Environment and Planning C: Government and Policy* 11: 367–379.

Durand, D. and Kahn, P. (1998) 'MAPA: A System for Inducing and Visualizing Hierarchy in Websites'. *Ninth ACM Conference on Hypertext and Hypermedia (HT '98)*, June 1998, Pittsburgh, USA. http://www.dynamicdiagrams.com/pdf/papers/mapaht98.pdf

Durham, T. (1999) 'Month of Building Dangerously'. *The Times Higher Education Supplement*, 15 January 1999: 12.

Dyson, E. (1999) 'Bricklayers and Emailers'. *Guardian*, Online section, 1 July 1999: 11.

Edwards, D.W. and Hardman, L. (1989) *Lost in Hyperspace: Cognitive Mapping and Navigation in a Hypertext Environment*. Intellect Books, Oxford.

Egan, G. (1994) *Permutation City*. Millennium, London.

Eick, S.G. (1996) 'Aspects of Network Visualization'. *IEEE Computer Graphics and Applications*, March 1996, 16, 2: 69–72.

Eklund, J., Sawers, J., and Zeiliger, R. (1999) 'NESTOR Navigator: A Tool for the Collaborative Construction of Knowledge Through Constructive Navigation'. In Debreceny, R. and Ellis, A. (eds) *Proceedings of Ausweb99, The Fifth Australian World Wide Web Conference*. Southern Cross University Press, Lismore, pp. 396–408. http://ausweb.scu.edu.au/aw99/papers/eklund2/

Elkin-Koren, N. (1996) 'Public/Private and Copyright Reform in Cyberspace'. *Journal of Computer Mediated Communication* 2, 2. http://www.ascusc.org/jcmc/vol2/issue2/elkin.html

Emberley, P. (1988) 'Technology, Values and Nihilism'. *Science, Technology and Politics* 3: 41–58.

Escobar, A. (1994) 'Welcome to Cyberia: Notes on the Anthropology of Cyberculture'. *Current Anthropology* 35: 211–231.

Evans, G.W. and Pezdek, K. (1980) 'Cognitive Mapping: Knowledge of Real-World Distance and Location Information'. *Journal of Experimental Psychology: Human Learning and Memory* 6: 13–24.

Faber, L. (1998) *RE:play: Ultimate Games Graphics*. Lawrence King, London.

Fabi, M. (1998) *Wyrm*. Batham Books, New York.

Fabrikant, S.I. (1999) 'Spatial Metaphors for Browsing Large Data Archives'. Unpublished dissertation, Department of Geography, University of Colorado-Boulder, USA.

Fabrikant, S.I. (2000) 'Spatialized Browsing in Large Data Archives'. *Transactions in GIS* 4, 1.

Featherstone, M. and Burrows, R. (1995) *Cyberspace / Cyberbodies / Cyberpunk: Cultures of Technological Embodiment*. Sage Publications, London.

Ferguson, E.L. and Hegarty, M. (1994) 'Properties of Cognitive Maps Constructed from Texts'. *Memory and Cognition* 22: 455–473.

Fernback, J. and Thompson, B. (1995) 'Virtual Communities: Abort, Retry, Failure?' Paper presented at the annual convention of the International Communication Association, Albuquerque, New Mexico. http://www.well.com/user/hlr/texts/VCcivil.html

Fisher, P. and Unwin, D. (forthcoming) *Virtual Reality in Geography*. Taylor and Francis, London.

Fitting, P. (1991) 'The Lessons of Cyberpunk'. In Penley, C. and Ross, A. (eds) *Technoculture*. University of Minnesota Press, Minneapolis, pp. 295–315.

Fleming, D.K. (1984) 'Cartographic Strategies for Airline Advertising'. *Geographical Review* 74: 76–93.

Foote, K.E. and Hubner, D.J. (1995) 'Error, Accuracy, and Precision'. *The Geographer's Craft Project*, Department of Geography, University of Texas, Austin.
http://www.utexas.edu/depts/grg/gcraft/notes/error/error_f.html

Foster, D. (1997) 'Community and Identity in the Electronic Village'. In Porter, D. (ed.) *Internet Culture*. Routledge, London, pp. 23–38.

Foucault, M. (1991) *Discipline and Punishment*. Penguin, London.

Foy, G. (1996) *The Shift*. Batham Books, New York.

Foy, G. (1997) *Contraband*. Batham Books, New York.

Franklin, N., Tversky, B., and Coon, V. (1992) 'Switching Points of View in Spatial Mental Models'. *Memory and Cognition* 20: 507–518.

Frécon, E. and Smith, G. (1998) 'WebPath – A Three-Dimensional Web History'. *Proceedings IEEE Symposium on Information Visualization (InfoVis '98)*. Chapel Hill, NC, USA, pp. 3–10. http://www.comp.lancs.ac.uk/computing/users/gbs/webpath/

Froehling, O. (1997) 'The Cyberspace "War of Ink and Internet" in Chiapas, Mexico'. *The Geographical Review* 87: 291–307.

Froehling, O. (1999) 'Internauts and Guerilieros: The Zapastista Rebellion in Chiapas, Mexico and its Extension into Cyberspace'. In Crang, M., Crang, P., and May, J. (eds) *Virtual Geographies*. Routledge, London, pp. 164–177.

Gaines, B.R., Chen, L.L., and Shaw, M.L. (1997) 'Modeling the Human Factors of Scholarly Communities Support Through the Internet and World Wide Web'. *Journal of the American Society for Information Science* 48, 11: 987–1003.

Garfinkel, S. (2000) *Database Nation: The Death of Privacy in the Twenty-first Century*. O'Reilly and Associates, New York.

Garland, K. (1994) *Mr Beck's Underground Map*. Capital Transport Publishing, Middlesex.

Gärling, T., Book, A., Lindberg, E., and Nilsson, T. (1981) 'Memory for the Spatial Layout of the Everyday Physical Environment: Factors Affecting Rate of Acquisition'. *Journal of Environmental Psychology* 1: 263–277.

Gärling, T., Book, A., and Lindberg, E. (1986) 'Spatial Orientation and Wayfinding in the Designed Environment'. *Journal of Architectural Planning Research* 3: 55–64.

Garton, L., Haythornthwaite, C., and Wellman, B. (1997) 'Studying Online Social Networks'. *Journal of Computer Mediated-Communication* 3, 1. http://www.ascusc.org/jcmc/vol3/issue1/garton.html

Gershon, N. and Eick, S.G. (eds) (1995a) *Proceedings of Information Visualization '95*. IEEE Computer Society Press, Los Alamitos, California.

Gershon, N. and Eick, S.G. (1995b) 'Visualization's New Track: Making Sense of Information'. *IEEE Spectrum* 32, 1: 38–56.

Gershon, N., Eick, S.G., and Card, S. (1998) 'Information Visualization'. *Interactions*: March/April 1998: 9–15.

Gibson, W. (1984) *Neuromancer*. HarperCollins, London.

Gibson, W. (1986) *Count Zero*. HarperCollins, London.

Gibson, W. (1987) *Mona Lisa Overdrive*. HarperCollins, London.

Gibson, W. (1989) 'High Tech High Life. William Gibson and Timothy Leary in Conversation'. *Mondo 2000*, Fall, 7.

Gibson, W. (1992) *Virtual Light*. Penguin, London.

Gibson, W. (1996) *Idoru*. Penguin, London.

Gifford, J.J. (1996) 'Quake Tectonics'. *FEED Magazine*, 28 September 1996.
http://www.feedmag.com/96.09gifford/96.09gifford.html

Gillepsie, A. and Williams, H. (1988) 'Telecommunications and the Reconstruction of Regional Comparative Advantage'. *Environment and Planning A* 20: 1311–1321.

Girardin, L. (1995) 'Mapping the Virtual Geography of the World Wide Web'. Poster presentation at the *Fifth International World Wide Web Conference*, 6–10 May 1999, Paris, France. http://heiwww.unige.ch/girardin/cgv/www5/index.html

Gleeson, B.J. (1996) 'A Geography for Disabled People?' *Transactions of the Institute of British Geographers* 21: 387–396.

Gloor, P. (1997) *Elements of Hypermedia Design: Techniques for Navigation and Visualization in Cyberspace*. Birkhauser, Boston, Mass.

Goldhaber, M.H. (1997) 'Attention Shoppers!' *Wired* 5.12, December 1997.

Golding, P. (1990) 'Political Communication and Citizenship: The Media and Democracy in an Egalitarian Social Order'. In Ferguson, M. (ed.) *Public communication: The new imperatives*. Sage, London.

Golledge, R.G. (1978) 'Representing, Interpreting and Using Cognized Environments'. *Papers and Proceedings of the Regional Science Association* 41: 169–204.

Golledge, R.G. (1991) 'Cognition of Physical and Built Environments'. In Gärling, T. and Evans, G.W. (eds) *Environment, Cognition and Action – An Integrated Approach*. Oxford University Press, New York, pp. 35–62.

Golledge, R.G. (1992) 'Place Recognition and Wayfinding: Making Sense of Space'. *Geoforum* 23: 199–214.

Golledge, R.G., Gale, N., and Richardson, G.D. (1987) 'Cognitive Maps of Cities II: Studies of Selected Populations'. *National Geographical Journal of India* 33: 1–16.

Golledge, R.G., Smith, T.R., Pellegrino, J.W., Doherty, S., and Marshall, S.P. (1985) 'A Conceptual Model and Empirical Analysis of Children's Acquisition of Spatial Knowledge'. *Journal of Environmental Psychology* 5: 125–152.

Golledge, R.G. and Stimson, R.J. (1997) *Spatial Behavior: A Geographic Perspective*. Guildford Press, New York.

Goodchild, M.F. (1989) *The Accuracy of Spatial Databases*. Taylor and Francis, London.

Goodman, S.E., Press, L.I., Ruth, S.R., and Rutkowski, A.M. (1994) 'The Global Diffusion of the Internet: Patterns and Problems'. *Communications of the ACM*, August 1994, 37, 8: 27–31.

Goodwin, M. (1993) 'The City as a Commodity: The Contested Spaces of Urban Development'. In Kearns, R. and Philo, C. (eds) *Selling Places: The City as Cultural Capital, Past and Present*. Pergamon Press, Oxford.

Gorman, S.P. (1998) 'The Death of Distance, But Not the End of Geography: The Internet as a Network'. Paper presented at the *Regional Science Association Conference*, 29 October 1998, Santa Fe, USA.

Graham, S. (1993) 'Changing Communications Landscapes: Threats and Opportunities for UK Cities'. *Cities* 10: 158–166.

Graham, S. and Aurigi, A. (1997) 'Virtual Cities, Social Polarization, and the Crisis in Urban Public Space'. *Journal of Urban Technology* 4, 1: 19–52.

Graham, S., Brooks, J., and Heery, D. (1996) 'Towns on the Television: Closed Circuit TV in British Towns and Cities'. *Local Government Studies* 22, 3: 1–27.

Graham, S. and Marvin, S. (1996) *Telecommunications and the City: Electronic Spaces, Urban Places*. Routledge, London.

Gregory, D. (1994) *Geographical Imaginations*. Blackwell, Oxford.

Griswold, W. (1994) *Cultures and Societies in a Changing World*. Pine Forge Press, Thousand Oaks, CA.

Guha, R.V. (n.d.) *Meta Content Framework: A Whitepaper*. Apple Research. http://www.xspace.net/hotsauce/wp.html

Habermas, J. (1989) *The Structural Transformation of the Public Sphere: An Inquiry into Bourgeois Society* (trans. Burger, T. and Lawrence, F.). MIT Press, Cambridge, MA. (First published 1962).

Haddon, L. (1993) 'Interactive Games'. In Hayward, P. and Wollen, T. (eds) *Future Visions: New Technologies of the Screen*. British Film Institute, London, pp. 123–147.

Hafner, K. (1997) 'The Epic Saga of The Well: The World's Most Influential Online Community (And It's Not AOL)'. *Wired* 5.05, May 1997.
http://www.wired.com/wired/archive/5.05/ff_well_pr.html

Hafner, K. and Lyons, M. (1996) *Where Wizards Stay up Late: The Origins of the Internet*. Simon and Schuster, New York.

Halabi, B. (1997) *Internet Routing Architectures*. New Riders Publishing, Indianapolis.

Hall, K. (1996) 'Cyberfeminism'. In Herring, S. (ed.) *Computer Mediated Communication*. John Benjamins, Amsterdam.

Hall, S.S. (1992) *Mapping the Next Millennium: How Computer-Driven Cartography is Revolutionizing the Face of Science*. Vintage Books, New York.

Handy, S.L. and Mokhtarian, P.L. (1996) 'The Future of Telecommuting'. *Futures* 28: 227–240.

Haraway, D. (1991) *Simians, Cyborgs and Women*. Free Association Press, London.

Hardwick, D.A., McIntyre, C.W., and Pick, H.L. (1976) 'The Content and Manipulation of Cognitive Maps in Children and Adults'. *Monographs of the Society for Research in Child Development* 41: 1–55.

Hardy, H.E. (1993) 'The History of the Net'. Unpublished Masters thesis. Grand Valley State University. http://www.ocean.ic.net/ftp/doc/nethist.html

Hargittai, E. (1998) 'Holes in the Net: The Internet and International Stratification'. *INET '98 Conference: The Internet Summit*, 21–24 July 1998, Geneva, Switzerland.
http://www.isoc.org/inet98/proceedings/5d/5d_1.htm

Harley, J.B. (1989) 'Deconstructing the Map'. *Cartographica* 26: 1–20.

Harpold, T. (1999) 'Dark Continents: Critique of Internet Metageographies'. *Postmodern Culture*, January 1999, 9, 2. http://www.lcc.gatech.edu/~harpold/papers/dark_continents/index.html

Harry, E.L. (1996) *Society of the Mind*. Coronet Books, London.

Hart, J.A., Reed, R.R., and Bar, F. (1992) 'The Building of the Internet: Implications for the Future of Broadband Networks'. *Telecommunications Policy* 16: 666–689.

Hart, R. and Berzok, M.A. (1983) 'A Problem Oriented Perspective on Children's Representation of the Environment'. In Potegal, M. (ed.) *The Neural and Development Bases of Spatial Orientation*, Academic Press, New York.

Harvey, A.S. and Macnab, P.A. (2000) 'Who's Up? Global Interpersonal Temporal Accessibility'. In Janelle, D. and Hodge, D. (eds) *Information, Place and Cyberspace: Issues in Accessibility*. Elsevier, Amsterdam.

Harvey, D. (1989) *The Condition of Postmodernity: An Enquiry Into the Origins of Cultural Change*. Blackwell, Oxford.

Hauben, M. (1995) *The Net and Netizens: The Impact the Net Has on People's Lives*. Preface. Http://www.cs.columbia.edu/~hauben/netbook/

Hawking, S. (1988) *A Brief History of Time*. Bantam Books, London.

Hayles, K. (1993) 'Virtual Bodies and Flickering Signifiers'. *October* 66: 69–91.

Haywood, T. (1998) 'Global Networks and the Myth of Equality: Trickle Down or Trickle Away?' In Loader, B. (ed.) *The Cyberspace Divide*. Routledge, London, pp. 19–34.

He, T. and Eick, S.G. (1998) 'Constructing Interactive Network Visual Interfaces'. *Bell Labs Technical Journal*, April–June 1998, 3, 2: 47–57. http://www.bell-labs.com/user/eick/bibliography/1998/bltj.pdf

Healy, D. (1997) 'Cyberspace and Place: The Internet as Middle Landscape on the Electronic Frontier'. In Porter, D. (ed.) *Internet Culture*. Routledge, London, pp. 55–72.

Hearnshaw, H.M. and Unwin, D.J. (eds) (1995) *Visualisation in Geographical Information Systems*. Wiley, Chichester.

Heart, F., McKenzie, A., McQuillian, J., and Walden, D. (1978) *ARPANET Completion Report*, 4 January 1978, Bolt, Beranek and Newman, Burlington, Mass., USA.

Hendley, R.J., Drew, N.S., Wood, A.M., and Beale, R. (1995) 'Narcissus: Visualising Information'. *Proceedings of InfoVis '95. IEEE Symposium on Information Visualization*, New York, pp. 90–96, 146.

Hendrix, H.V. (1997) *Lightpaths*. Ace Books, New York.

Hepworth, M. (1990a) 'Planning for the Information City: The Challenge and Response'. *Urban Studies* 27: 537–558.

Hepworth, M. (1990b) *Geography of the Information Economy.* Guilford Press, New York.

Herz, J.C. (1997) *Joystick Nation: How Videogames Ate Our Quarters, Won Our Hearts, and Rewired Our Minds.* Little Brown and Company, New York.

Hess, D.J. (1995) *Science and Technology in a Multicultural World: The Cultural Politics of Facts and Artifacts.* Columbia University Press, New York.

Heylighen, F. and Bollen, J. (1996) 'The World Wide Web as a Super-Brain: From Metaphor to Model'. In Trappl, R. (ed.) *Cybernetics and Systems '96,* Austrian Society for Cybernetics, pp. 917–922.

Hillis, K. (1996) 'A Geography of the Eye: The Technologies of Virtual Reality'. In Shields, R. (ed.) *Cultures of the Internet: Virtual Spaces, Real Histories, Living Bodies.* Sage Publications, London, pp. 70–98.

Hillis, K. (1998) 'On the Margins: The Invisibility of Communications in Geography'. *Progress in Human Geography* 22, 4: 543–566.

Hiltz, S.T. and Turoff, M. (1993) *The Network Nation: Human Communication via Computer* (2nd edn). MIT Press, Cambridge, Mass.

Hodgkiss, A.G. (1980) *Understanding Maps: A Systematic History of their Use and Development.* Dawson, Folkestone.

Hoffman, D.L. and Novak, T.P. (1998) 'Bridging the Racial Divide on the Internet'. *Science,* 17 April 1998, 280: 390–391.

Hoffman, E. and Claffy, K. (1997) 'Address Administration in IPv6'. In Kahin, B. and Keller, J. (eds) *Coordinating the Internet.* MIT Press, Cambridge, Mass., pp. 288–308.

Holderness, M. (1998) 'Who are the World's Information-Poor?' In Loader, B.D. (ed.) *The Cyberspace Divide.* Routledge, London, pp. 35–56.

Hollinger, V. (1991) 'Cybernetic Deconstructions: Cyberpunk and Postmodernism'. In McCaffery, L. (ed.) *Storming the Reality Studio: A Casebook of Cyberpunk and Postmodern Fiction.* Duke University Press, London, pp. 203–218.

Holloway, S., Valentine, G., and Bingham, N. (forthcoming) 'Institutionalising Technologies: Masculinities, Femininities and the Heterosexual Economy of the IT Classroom'. *Environment and Planning A.*

Holtzman, D. (1997) 'Domain Names: Will We Run Out?' *InterNIC News,* June 1997.

Holtzman, S.R. (1994) *Digital Mantras: The Languages of Abstract and Virtual Worlds.* MIT Press, Cambridge, Mass.

Holtzman, S.R. (1997) *Digital Mosaics: The Aesthetics of Cyberspace.* Simon and Schuster, New York.

Honkela, T., Kaski, S., Kohonen, T., and Lagus, K. (1998) 'Self-organizing Maps of Very Large Document Collections: Justification for the WEBSOM Method'. In Balderjahn, I., Mathar, R., and Schader, M. (eds) *Classification, Data Analysis, and Data Highways.* Springer-Verlag, Berlin, pp. 245–252.

Huberman, B.A., Pirolli, P.L.T., Pitkow, J.E., and Lukose, R.M. (1998) 'Strong Regularities in World Wide Web Surfing'. *Science* 280: 95–97.

Hubbard, K., Kosters, M., Conrad, D., Karrenberg, D., and Postel, J. (1996) 'Internet Registry IP Allocation Guidelines'. *RFC 2050,* USC/Information Sciences Institute, November 1996. ftp://ftp.ripe.net/rfc/rfc2050.txt

Huffaker, B., Jung, J., Wessels, D., and Claffy, K. (1998) 'Visualization of the Growth and Topology of the NLANR Caching Hierarchy'. Paper presented at *Third International WWW Caching Workshop,* 15–17 June 1998, Manchester, UK. http://www.cache.ja.net/events/workshop/17/kclaffy.html

Huston, G. (1994) 'Observations on the Management of the Internet Address Space'. *RFC 1744,* USC/Information Sciences Institute, December 1994. ftp://ftp.ripe.net/rfc/rfc1744.txt

Huston, G. (1997) 'Do Internet Addresses Have a Value?' *On The Internet,* January/February 1997, 3, 1. http://www.isoc.org/isoc/publications/oti/articles/do.html

Huxor, A. (1997) 'The Role of Virtual World Design in Collaborative Working'. Paper presented at the *Information Visualization '97 Conference,* July 1997, London, UK.

Imielinski, T. and Navas, J.C. (1999) 'Geographic Addressing, Routing, and Resource Discovery with the Global Positioning System'. *Communications of the ACM,* April 1999, 42, 4: 86–92.

Imkem, O. (1999) 'The Convergence of Virtual and Actual in the Global Matrix: Artificial Life, Geo-economics and Psychogeography'. In Crang, M., Crang, P., and May, J. (eds) *Virtual Geographies*. Routledge, London, pp. 92–106.

Imperative! (1999) *US Domain Distribution Summary*. Internet.Org, Imperative! Inc., Pittsburgh. http://www.internet.org/

Interrogate the Internet (1996) 'Contradictions in Cyberspace: Collective Response'. In Shields, R. (ed.) *Cultures of Internet: Virtual Spaces, Real Histories and Living Bodies*. Sage, London, pp. 125–132.

ISC (2000) *Internet Domain Survey*. Internet Software Consortium (ISC), January 2000. http://www.isc.org/ds/WWW-200001/report.html

Ito, M. (1997) 'Virtually Embodied: The Reality of Fantasy in a Multi-user Dungeon'. In Porter, D. (ed.) *Internet Culture*. Routledge, London, pp. 87–110.

Ito, J. and Ito, M. (1996) 'Introduction'. In Hershman Leeson, L. (ed.) *Clicking In*. Bay Press, Seattle, pp. 78–104.

ITU (1997) *Challenges to the Network: Telecoms and the Internet*. International Telecommunications Union (ITU), Geneva.

ITU (1998) *Challenges to the Network: Internet for Development, 1999*. International Telecommunications Union (ITU), Geneva.

Jacobs, J. (1996) *Edge of Empire: Postcolonialism and the City*. Routledge, London.

Jacobson, R.D. and Kitchin, R.M. (1997) 'GIS and People with Visual Impairments or Blindness: Exploring the Potential for Education, Orientation and Navigation'. *Transactions in Geographic Information Systems* 2, 4: 315–332.

Jacobson, R.D., Kitchin, R.M., and Golledge, R.G. (forthcoming) 'Non-visual Virtual Reality for Presenting Geographic Information'. In Fisher, P. and Unwin, D. (eds) *Virtual Reality in Geography*. Taylor and Francis, London.

Jacobson, R.E. (1999) *Information Design*. MIT Press, Cambridge, Mass.

Jameson, F. (1991) *Postmodernism, or, the Logic of Late Capitalism*. Duke University Press, Durham.

Jeffrey, P. and Mark, G. (1998) 'Constructing Social Spaces in Virtual Environments: A Study of Navigation and Interaction'. In Höök, K., Munro, A., and Benyon, D. (eds) *Workshop on Personalised and Social Navigation in Information Space*. Swedish Institute of Computer Science, Stockholm, pp. 24–38.

Jenks, G.F. and Caspall, F.C. (1971) 'Error on Choropleth Maps: Definition, Measurement and Reduction'. *Annals of the Association of American Geographer* 61: 217–244.

Jensen, M. (1998) *Internet Connectivity in Africa*. Report, January 1998. http://demiurge.wn.apc.org/africa/

Jess, P. and Massey, D. (1995) 'The Conceptualization of Place'. In Massey, D. and Jess, P. (eds) *A Place in the World? Places, Cultures and Globalization*. Oxford University Press, Oxford.

Jiang, B. and Ormeling, F.J. (1997) 'Cybermap: The Map for Cyberspace'. *The Cartographic Journal* 34, 2: 111–116.

Johnson, S. (1997) *Interface Culture: How New Technology Transforms the Way We Create and Communicate*. Harper, San Francisco.

Johnson, S. (1999) 'Maps and Legends'. *FEED Magazine*, March 1999. http://www.feedmag.com/column/interface/ci190lofi.html

Johnston, P. (1993) 'Teleworking as an Enabler Factor for Economic Growth and Job Creation in Europe'. Paper presented at *Telematics and Innovation Conference*, 17–19 November 1993, Palma, Majorca.

Jones, S.G. (1995a) *CyberSociety: Computer-Mediated Communication and Community*. Sage Publications, Thousand Oaks, CA.

Jones, S.G. (1995b) 'Understanding Community in an Information Age'. In Jones, S.G. (ed.) *Cybersociety: Computer Mediated Communication and Community*. Sage, London, pp. 10–35.

Kahn, P. (1999) *Mapping websites: Planning Diagrams to Site Maps*. Dynamic Diagrams, 10 January 1999. http://www.dynamicdiagrams.com/seminars/mapping/maptoc.htm

Kaplan, N. and Moulthrop, S. (1994) 'Where No Mind Has Gone Before: Ontological Design for Virtual Spaces'. *ECHT '94*. ACM Press, Edinburgh, pp. 206–216.

Kearns, R. and Philo, C. (1993) 'Culture, History, Capital: A Critical Introduction to the Selling of Places'. In Kearns, R. and Philo, C. (eds) *Selling Places: The City as Cultural Capital, Past and Present*. Pergamon Press, Oxford.

Kelly, K. and Rheingold, H. (1993) 'The Dragon Ate My Homework'. *Wired* 1.03. http://www.wired.com/wired/archive/1.03/muds_pr.html

Khan, K. and Locatis, C. (1998) 'Searching Through Cyberspace: The Effects of Link Density on Information Retrieval from Hypertext on the World Wide Web'. *Journal of the American Society for Information Science* 49, 2: 176–182.

Kiesler, S. (1997) *Culture of the Internet*. Lawrence Erlbaum Associates, New Jersey.

Kim, A.J. (1998) 'Killers Have More Fun: Games Like Ultima Online are Grand Social Experiments in Community Building'. *Wired* 6.05: 94ff.

Kim, H. and Hirtle, S.C. (1995) 'Spatial Metaphors and Disorientation in Hypertext Browsing'. *Behaviour and Information technology* 14: 239–250.

Kirsh, E.M., Phillips, D.W., and McIntyre, D.E. (1996) 'Recommendations for the Evolution of Cyberlaw'. *Journal of Computer Mediated Communications* 2, 2. http://www.ascusc.org/jcmc/vol2/issue2/kirsh.html

Kitchin, R. (1998) *Cyberspace: The World in the Wires*. John Wiley and Sons, Chichester.

Kitchin, R. and Blades, M. (forthcoming) *The Cognition of Geographic Space*. IB Tauris, London.

Kitchin, R. and Freundschuh, S. (2000) *Cognitive Mapping: Past, Present and Future*. Routledge, London.

Kitchin, R. and Tate, N.J. (1999) *Conducting Research in Human Geography*. Longman, Harlow.

Kling, R. and Lamb, R. (1996) 'Bits of Cities: Utopian Visions and Social Power in Placed-based and Electronic Communities'. http://www-slis.lib.indiana.edu/kling/pubs/bitsofcities.html

Kneale, J. (1999) 'The Virtual Realities of Technology and Fiction: Reading William Gibson's Cyberspace'. In Crang, M. and Crang, P. (eds) *Virtual Geographies*. Routledge, London, pp. 205–221.

Kohonen, T. (1995) *Self-Organizing Maps*. Springer-Verlag, Berlin.

Kollock, P. and Smith, M. (1994) 'Managing the Virtual Commons: Cooperation and Conflict in Computer Communities'. http://www.sscnet.ucla.edu/soc/csoc/vcommons.htm

Kollock, P. and Smith, M. (1999) 'Communities in Cyberspace'. In Smith, M. and Kollock, P. (eds) *Communities in Cyberspace*. Routledge, London, pp. 3–28.

Kraak, M-J. and Ormeling, F. (1996) *Cartography: Visualization of Spatial Data*. Longman, Harlow.

Krol, E. and Hoffman, E. (1993) 'FYI on 'What is the Internet?' *Request for Comments: 1462*, May 1993. ftp://ftp.ripe.net/rfc/rfc1462.txt

Kuipers, B. (1978) 'Modelling spatial Knowledge'. *Cognitive Science* 2: 129–153.

Kumar, K. (1995) *From Postindustrial to Postmodern Society: New Theories of the Contemporary World*. Blackwell, Oxford.

Lagus, K., Honkela, T., Kaski, S., and Kohonen, T. (1996) 'Self-organizing Maps of Document Collections: A New Approach to Interactive Exploration'. In Simoudis, E., Han, J., and Fayyad, U. (eds) *Proceedings of the Second International Conference on Knowledge Discovery and Data Mining*. AAAI Press, Menlo Park, CA, pp. 238–243. http://websom.hut.fi/websom/doc/ps/lagus96kdd.ps

Lajoie, M. (1996) 'Psychoanalysis and Cyberspace'. In Shields, R. (ed.) *Cultures of Internet: Virtual Spaces, Real Histories and Living Bodies*. Sage, London, pp. 153–169.

Lamm, S.E., Reed, D.A., and Scullin, W.H. (1996) 'Real-time Geographic Visualization of World Wide Web Traffic'. *Proceedings of Fifth International World Wide Web Conference*, 6–10th May 1996, Paris, France. http://www5conf.inria.fr/fich_html/papers/P49/Overview.html

Lamping, J. and Rao, R. (1995) 'The Hyperpolic Browser: A Focus + Context Technique for Visualizing Large Hierarchies'. *Journal of Visual Languages and Computing* 7, 1: 33–55.

Landers, R. (1997) 'One Million Names and Counting . . .'. *InterNIC News*, March 1997.

Langdale, J.V. (1989) 'The Geography of International Business Telecommunications: The Role of Leased Networks'. *Annals of the Association of American Geographers* 79: 501–522.

Lash, S. and Urry, J. (1994) *Economies of the Sign and Spaces*. Sage, London.

Laurel, B. (1990) *Art of Human-Computer Interface Design*. Addison-Wesley Publishing Company, Reading, Mass.

Lawrence, S. and Giles, C.L. (1998) 'Searching the World Wide Web'. *Science* 280: 98–100.

Lawrence, S. and Giles, C.L. (1999) 'Accessibility of Information on the Web'. *Nature* 400: 107–109.

Laws, G. (1994) 'Aging, Contested Meanings, and the Built Environment'. *Environment and Planning A* 26: 1787–1802.

Lemos, A. (1996) 'The Labyrinth of Minitel'. In Shields, R. (ed.) *Cultures of Internet: Virtual Spaces, Real Histories and Living Bodies*. Sage, London, pp. 33–48.

Lefebvre, H. (1991) *The Production of Space*. Blackwell, Oxford. (Originally published in 1974).

Lessig, L. (1999) *Code and Other Laws of Cyberspace*. Basic Books, New York.

Liben, L. (1981) 'Spatial Representation and Behavior: Multiple Perspectives'. In Liben, L., Patterson, A.M., and Newcombe, N. (eds) *Spatial Representation and Behavior Across the Life Span*. Academic Press, New York, pp. 3–36.

Liben, L.S. (1991) 'Environmental Cognition Through Direct and Representational Experiences: A Life-span Perspective'. In Gärling, T., and Evans, G.W. (eds) *Environment, Cognition and Action – An Integrated Approach*. Plenum Press, New York, pp. 245–276.

Liben, L.S. and Downs, R.M. (1989) 'Understanding Maps as Symbols: The Development of Map Concepts in Children'. In Reese, H.W. (ed.) *Advances in Child Development and Behavior*, 22: 145–201.

Licklider, J.C.R., Taylor, R., and Herbert, E. (1968) 'The Computer as a Communication Device'. *International Science and Technology*, April 1968, 76: 21–31.

Light, J. (1999) 'From City Space to Cyberspace'. In Crang, M., Crang, P., and May, J. (eds) *Virtual Geographies*. Routledge, London, pp. 109–130.

Lin, X. (1992) 'Visualization for the Document Space'. *Proceedings of IEEE Visualization '92*, pp. 274–281. Reproduced in Card, S.K., MacKinlay, J., and Shneiderman, B. (eds) (1999) *Readings in Information Visualization: Using Vision to Think*. Morgan Kaufman, San Francisco, pp. 432–440.

Lin, X. (1997) 'Map Displays for Information Economy Retrieval'. *Journal of the American Society for Information Science* 48, 1: 40–54.

Livingstone, D. (1992) *The Geographical Tradition*. Blackwell, London.

Lloyd, R. (1989a) 'Cognitive Maps: Encoding and Decoding Information'. *Annals of the Association of American Geographers* 79: 101–124.

Lloyd, R. (1989b) 'The Estimation of Distance and Direction from Cognitive Maps'. *The American Cartographer* 16: 109–122.

Lloyd, R. (1993) 'Cognitive Processes and Cartographic Maps'. In Gärling, T. and Golledge, R.G. (eds) *Behavior and Environment: Psychological and Geographical Approaches*. North Holland, London, pp. 141–169.

Lloyd, R. (1997) *Spatial Cognition: Geographic Environments*. Dordrecht, Kluwer.

Loader, B. (1997) 'The Governance of Cyberspace'. In Loader, B. (ed.) *The Governance of Cyberspace*. Routledge, London, pp. 1–19.

Longley, P. and Clarke, G. (1995) *GIS for Business and Service Planning*. GeoInformation International, Cambridge.

Lottor, M. (1992) 'Internet Growth (1981–1991)'. *RFC 1296*, USC/Information Sciences Institute, January 1992. ftp://ftp.ripe.net/rfc/rfc1296.txt

Loy, W. (ed.) (1997) 'US National Report to ICA, 1987'. *The American Cartographer* 14, 3.

Luke, T. (1993) 'Community and Ecology'. In Walker, S. (ed.) *Changing Community: The Graywolf Annual Ten*. Graywolf Press, St Paul, MN, pp. 207–221.

Lupton, D. (1995) 'The Embodied Computer/User'. In Featherstone, M. and Burrows, R. (eds) *Cyberspace, Cyberbodies and Cyberpunk: Cultures of Technological Embodiment*. Sage, London, pp. 97–112.

Lynch, K. (1960) *The Image of the City*. MIT Press, Cambridge, Mass.

Lyon, D. (1988) *The Information Society: Issues and Illusions*. Polity Press, Oxford.

Lyon, D. (1994) *Postmodernity*. Open University Press, Milton Keynes.

Maarek, Y.S., Jacovi, M., Shtalhaim, M., Ur, S., Zernik, D., and Ben, I.Z. (1997) 'WebCutter: A System for Dynamic and Tailorable Site Mapping'. Paper presented at the *Sixth International World Wide Web Conference*, April 1997, Santa Clara, USA.
http://decweb.ethz.ch/WWW6/Technical/Paper040/Paper40.html

McCabe, H. (1999) 'The Net: Enemy of the State?' *Wired News*, 12 August 1999.
http://www.wired.com/news/news/politics/story/21240.html

McCaffery, L. (1991) 'Introduction: In the Desert of the Real'. In McCaffery, L. (ed.) *Storming the Reality Studio: A Casebook of Cyberpunk and Postmodern Fiction*. Duke University Press, London, pp. 1–16.

McClellan, J. (1999) 'Mind Game in the MUD'. *Guardian*, Online section, 28 January 1999: 2–3.

McCormick, B.H., Defanti, T.A., and Brown, M.D. (1987) 'Visualization in Scientific Computing'. *Computer Graphics 21*, 6.

McCreary, S. and Claffy, K.C. (1998) 'How Much of the Internet Address Space is Used?' *Cooperative Association for Internet Data Analysis*, August 1998. http://www.caida.org/IPv4space/

McDowell, L. (1999) *Gender, Identity and Place*. Minnesota University Press, Minneapolis.

MacEachren, A.M. (1992a) 'Learning Spatial Information from Maps: Can Orientation-Specificity be Overcome'. *Professional Geographer* 44: 431–443.

MacEachren, A.M. (1992b) 'Application of Environmental Learning Theory to Spatial Knowledge Acquisition from Maps'. *Annals of the Association of American Geographers* 82: 245–274.

MacEachren, A.M. (1995) *How Maps Work: Representation, Visualization, and Design*. Guildford, New York.

MacEachren, A.M. (1998) 'Cartography, GIS and the World Wide Web'. *Progress in Human Geography* 22, 4: 575–585.

Machover, C. and Tice, S.E. (1994) 'Virtual Reality'. *IEEE Computer Graphics and Applications*, January: 15–16.

Mackenzie, D. and Wajcman, J. (1985) *The Social Shaping of Technology*. Open University Press, Milton Keynes.

McLaren, S. (1999) 'Wall Street Goes 3D'. *BBC News Online*, 15 March 1999.
http://news.bbc.co.uk/hi/english/sci/tech/newsid_297000/297341.stm

McLaughlin, M.L., Osborne, K.K., and Smith, C.B. (1995) 'Standards of Conduct on Usenet'. In Jones, S.G. (ed.) *CyberSociety: Computer-Mediated Communication and Community*. Sage Publications, Thousand Oaks, CA, pp. 90–111.

McLuhan, M. (1964) *Understanding Media: The Extensions of Man*. Macmillan, New York.

Macmillan, B. (1996) 'Fun and Games: Serious Toys for City Modelling in a GIS Environment'. In Batty, M. and Longley, P. (eds) *Spatial Analysis: Modelling in a GIS Environment*. GeoInformation International, Cambridge, pp. 153–165.

McRae, S. (1997) 'Flesh Made Word: Sex, Text and the Virtual Body'. In Porter, D. (ed.) *Internet Culture*. Routledge, London, pp. 73–86.

Maddox, T. (1991) 'Mechanist/Shaper Narratives'. In McCaffery, L. (ed.) *Storming the Reality Studio: A Casebook of Cyberpunk and Postmodern Fiction*. Duke University Press, London, pp. 324–330.

Makower, J. (1986) *The Map Catalogue*. Vintage Books, New York.

Makridakis, S. (1995) 'The Forthcoming Information Revolution: Its Impact on Society and Firms'. *Futures* 27: 799–821.

Malamud, C. (1997) *A World's Fair for the Global Internet*. MIT Press, Cambridge, Mass.

Malmgren, C.D. (1991) *Worlds Apart: Narratology in Science Fiction*. Indiana University Press, Bloomington.

Malone, T.W. and Rockhart, J.F. (1991) 'Computers, Networks and the Corporation'. *Scientific American*, September 1991: 92–99.

Maltz, T. (1996) 'Customary Law and Power in Internet Communities'. *Journal of Computer Mediated Communication 2*, 1. http://www.ascusc.org/jcmc/vol2/issue1/custom.html

Marchionini, G. (1997) *Information Seeking in Electronic Environments*. Cambridge University Press, Cambridge.

Martin, J. (1978) *The Wired Society*. Prentice Hall, Englewood Cliffs, N.J.

Martin, W.J. (1995) *The Global Information Society*. Aslib Gower, London.

Marvin, C. (1998) *When Old Technologies Were New: Thinking About Electric Communication in the Late Nineteenth Century*. Oxford University Press, Oxford.

Marx, G.T. (1988) *Undercover: Police Surveillance in America*. University of California Press, Berkeley.

Massey, D. (1994) *Space, Place and Gender*. University of Minnesota Press, Minneapolis.

Maurer, H. (1996) *Hyperwave: The Next Generation Web Solution*. Addison-Wesley Publishing Company, Reading, Mass.

May, M., Peruch, P., and Savoyant, A. (1995) 'Navigating in a Virtual Environment with Map-acquired Knowledge: Encoding and Alignment Effects'. *Ecological Psychology* 7: 21–36.

Mayer-Kress, G. and Barczys, C. (1995) 'The Global Brain as an Emergent Structure from the Worldwide Computing Network'. *The Information Society* 11, 1: 1–27.

Mehta, M.D. and Plaza, D.E. (1997) 'Pornography in Cyberspace: An Exploration of What's in Usenet'. In Kiesler, S. (ed.) *Culture of the Internet*. Lawrence Erlbaum Associates, New Jersey, pp. 53–67.

Mele, C. (1999) 'Cyberspace and Disadvantaged Communities: The Internet as a Tool for Collective Action'. In Smith, M.A. and Kollock, P. (eds) *Communities in Cyberspace*. Routledge, London, pp. 264–289.

Memarzia, K. (1997) *Towards the Definition and Applications of Digital Architecture*. School of Architectural Studies, University of Sheffield. http://www.shef.ac.uk/students/ar/ara92km/thesis/

Menges, J. (1996) 'Feeling Between the Lines'. *Computer Mediated Communications Magazine* 3, 10.

Miller, S. (1996) *Civilising Cyberspace: Power, Policy and the Information Superhighway*. ACM Press, New York.

Mitchell, W.J. (1995) *City of Bits: Space, Place and the Infobahn*. MIT Press, Cambridge, Mass.

Mitchell, W.J. (1998) 'The New Economy of Presence'. *Environment and Planning B: Planning and Design*, anniversary issue, 1998: 20–21.

Mitterer, J. and O'Neill, K. (1992) 'The End of "Information": Computers Democracy and the University'. *Interchange* 23: 123–139.

Mnookin, J.L. (1996) 'Virtual(ly) Law: The Emergence of Law in LambdaMOO'. *Journal of Computer Mediated Communication* 2, 1. http://www.ascusc.org/jcmc/vol2/issue1/lambda.html

Moar, I. and Bower, G. (1983) 'Inconsistency in Spatial Knowledge'. *Memory and Cognition* 11: 107–113.

Monk, T. and Claffy, K. (1997) 'Internet Data Acquisition and Analysis: Status and Next Steps', *Proceedings of the INET '97 Conference*, 24–27 June 1997, Kuala Lumpur, Malaysia. http://www.isoc.org/isoc/whatis/conferences/inet/97/proceedings/F1/F1_3.HTM

Monmonier, M. (1991) *How to Lie with Maps*. University of Chicago Press, Chicago.

Montgomery, J. (1997) 'Fiber in the Sky: The Orbiting Internet'. *BYTE Magazine*, November: 58–72. http://www.byte.com/art/9711/sec5/art1.htm

Morley, D. and Robins, K. (1995) *Spaces of Identity: Global Media, Electronic Landscapes and Cultural Boundaries*. Routledge, London.

Morningstar, C. and Farmer, R. (1991) 'The Lessons of Lucasfilm's Habitat'. In Benedikt, M. (ed.) *Cyberspace: First Steps*. MIT Press, Cambridge, Mass., pp. 273–301.

Morris, M. and Ogan, C. (1994) 'The Internet as Mass Medium'. *Journal of Computer Mediated Communication* 1, 4. http://www.ascusc.org/jcmc/vol1/issue4/morris.html

Morse, M. (1997) 'Nature Morte: Landscape and Narrative in Virtual Environments'. In Moser, M. and MacLeod, D. (eds) *Immersed in Technology: Art and Virtual Environments*. MIT Press, Cambridge, Mass., pp. 195–232.

Morville, P. (1996) 'Mapping Your Site: A Picture's Worth a Thousand Words'. *Web Review*, September. http://webreview.com/wr/pub/96/09/27/arch/index.html

Mosaic Group (1998) *The Global Diffusion of the Internet Project: An Initial Inductive Study*, March. http://www.agsd.com/gdi97/gdi97.html

Moss, M. (1986) 'Telecommunications, World Cities and Urban Policy'. *Urban Studies* 24: 534–546.

Moss, M.L. and Mitra, S. (1998) 'Net Equity: A Report on Income and Internet Access'. *Journal of Urban Technology* 5, 3: 23–32. http://urban.nyu.edu/research/net-equity/

Moss, M.L. and Townsend, A. (1996) *Leaders and Losers on the Internet*. Taub Urban Research Center, New York University, September 1996. http://urban.nyu.edu/research/l-and-l/

Moss, M.L. and Townsend, A. (1997a) *Manhattan Leads the 'Net Nation*. Taub Urban Research Center, New York University, August 1997. http://urban.nyu.edu/research/domains/

Moss, M.L. and Townsend, A. (1997b) 'Tracking the Net: Using Domain Names to Measure the Growth of the Internet in U.S. Cities'. *Journal of Urban Technology*, December, 4, 3: 47–59.

Moss, M.L. and Townsend, A.M. (1998) 'The Role of the Real City in Cyberspace: Understanding Regional Variations in Internet Accessibility and Utilization'. Paper presented at the *Project Varenius Meeting on Measuring and Representing Accessibility in the Information Age*, November 1998. Pacific Grove, California.

Mukherjea, S. and Foley, J.D. (1995) 'Visualizing the World Wide Web with the Navigational View Builder'. *Proceedings of the Third International World Wide Web Conference*, April 1995, Darmstadt, Germany. http://www.igd.fhg.de/www/www95/proceedings/papers/44/mukh/mukh.html

Munro, A., Hook, K., and Benyon, D. (1999) *Social Navigation: Footprints in the Snow*. Springer-Verlag, Berlin.

Munzner, T. (1998) 'Exploring Large Graphs in 3D Hyperbolic Space'. *IEEE Computer Graphics and Applications* 18, 4: 18–23. http://www-graphics.stanford.edu/papers/h3cga

Munzner, T. and Burchard, P. (1995) 'Visualizing the Structure of the World Wide Web in 3D Hyperbolic Space'. *Proceedings of VRML '95*, December 1995, San Diego, California, pp. 33–38. http://www-graphics.stanford.edu/papers/webviz/

Munzner, T., Hoffman, E., Claffy, K. and Fenner, B. (1996) 'Visualizing the Global Topology of the Mbone'. In Gershon, N. and Eick, S.G. (eds) *Proceedings of the 1996 IEEE Symposium on Information Visualization*, 28–29 October 1996, San Francisco, pp. 85–92. http://www-graphics.stanford.edu/papers/mbone/

Murray, H. (1997) *Hamlet on the Holodeck: The Future of Narrative in Cyberspace*. MIT Press, Cambridge, Mass.

Nadeau, J., Lointier, C., Morin, R., and Descoteaux, M.A. (1998) 'Information Highways and the Francophone World: Current Situation and Strategies for the Future'. *INET '98 Conference: The Internet Summit*, 21–24 July 1998, Geneva, Switzerland. http://www.isoc.org/inet98/proceedings/5f/5f_3.htm

Naisbitt, J. (1984) *Megatrends: Ten New Directions Transforming our Lives*. William Morrow Books, New York.

Negroponte, N. (1995) *Being Digital*. Vintage Books, London.

Neustadt, R.M. (1985) 'Electronic Politics'. In Forester, T. (ed.) *The Information Technology Revolution*. Blackwell, Oxford.

Neystadt, J. (1999) *Israeli Internet – Internet Lines Maps*. IGuide-Israeli Internet Guide. http://www.iguide.co.il/maps.htm

Nielsen, J. (1990) *Hypertext and Hypermedia*. Academic Press, San Diego.

Novak, I. (1999) 'How Mapquest.com Delivered 1.6 Billion Maps over the Web'. *Mapping Awareness*, September, pp. 42–43.

Novak, M. (1991) 'Liquid Architectures in Cyberspace'. In Benedikt, M. (ed.) *Cyberspace: First Steps*. MIT Press, Cambridge, Mass., pp. 225–254.

Novak, M. (1995) 'Transmitting Architecture: TransTerraFirma/TidsvagNoll v2.0'. In Spiller, N. and Pearce, M. (eds) *Architects in Cyberspace*. Academy Editions, London.

NTIA (1995) *Falling Through the Net: A Survey of the 'Have Nots' in Rural and Urban America*. Report by the National Telecommunications and Information Administration (NTIA), July 1995, USA. http://www.ntia.doc.gov/ntiahome/fallingthru.html

NTIA (1998) *Falling Through the Net II: New Data on the Digital Divide*. Report by the National Telecommunications and Information Administration (NTIA), July 1998, USA. http://www.ntia.doc.gov/ntiahome/net2/falling.html

NTIA (1999) *Falling Through the Net: Defining the Digital Divide*. National Telecommunications and Information Administration (NTIA), 8 July 1999. http://www.ntia.doc.gov/ntiahome/fttn99/contents.html

Nua (1998) http://www.nua.ie/index.html

Nua (2000) 'How Many Online?' http://www.nua.ie/surveys/how_many_online/index.html

NWT (1998) *New World Times*. Online archive. http://vrnews.synergycorp.com/nwt/

O'Brien, J. (1999) 'Writing the Body: Gender (Re)production in Online Interaction'. In Smith, M. and Kollock, P. (eds) *Communities in Cyberspace*. Routledge, London, pp. 76–106.

OECD (1998) 'Internet Infrastructure Indicators'. Report by Paltridge S., Directorate for Science, Technology and Industry, Organisation for Economic Co-operation and Development (OECD), October 1998, Paris, France. http://www.oecd.org/dsti/sti/it/cm/prod/tisp98-7e.htm

OECD (1999) *OECD Communications Outlook 1999*. Organisation for Economic Co-Operation and Development, Paris.

O'Lear, S. (1997) 'Electronic Communication and Environmental Policy in Russia and Estonia'. *The Geographical Review* 87, 2: 275–290.

O'Neill, J.E. (1995) 'The Role of ARPA in the Development of the ARPANET, 1961–1972'. *IEEE Annals of the History of Computing* 17, 4: 76–81.

O'Tuathail, G. (1994) 'Shadow Warriors and Electronic Jury: Mexico and Chiapas Revolt in the Geo-financial Panopticon'. *Ecumene* 4: 300–317.

Openshaw, S. (1984) 'The Modifiable Areal Unit Problem'. *Concepts and Techniques in Modern Geography* 38, Geo Books, Norwich.

Papadakakis, N., Markatos, E.P., and Papathanasiou, A.E. (1998) 'Palantir: A Visualization Tool for the World Wide Web'. *Proceedings of the INET '98 Conference*, 21–24 July 1998, Geneva Switzerland. http://www.isoc.org/inet98/proceedings/1e/1e_1.htm

Passini, R. (1992) *Wayfinding in Architecture*. Van Nostrand Reinhold, New York.

Pearson, I. (1998) *The Atlas of the Future*. Routledge, London.

Peet, R. (1986) 'The Destruction of Regional Cultures'. In Johnston, R.J. and Taylor, P.J. (eds) *A World in Crisis? Geographical Perspectives*. Blackwell, Oxford.

Peet, R. (1998) *Modern Geographic Thought*. Blackwell, Oxford.

Penley, C. and Ross, A. (1991) *Technoculture*. University of Minnesota Press, Minneapolis, MN.

Penny, S. (1994) 'Virtual Reality as the Completion of the Enlightenment Project'. In Bender, G. and Druckery, T. (eds) *Culture on the Brink: Ideologies of Technology*. Bay Press, Seattle, pp. 231–248.

Péruch, P. and Lapin, E.A. (1993) 'Route Knowledge in Different Spatial Frames of Reference'. *Acta Acta Psychologica* 84: 253–269.

Péruch, P., Vercher, J.L., and Gauthier, G.M. (1995) 'Acquisition of Spatial Knowledge through Visual Exploration of Simulated Environments'. *Ecological Psychology* 7: 1–20.

Pesce, M. (1995) *VRML: Browsing and Building Cyberspace*. New Riders Publishers, Indianapolis.

Peterson, M.P. (1999) 'Trends in Internet Map Use – A Second Look'. *Proceedings of nineteenth International Cartographic Conference*, 14–21 August 1999, Ottawa, Canada. http://maps.unomaha.edu/MP/Articles/ICA99/Trends99.html

Petrazzini, B. and Kibati, M. (1999) 'The Internet in Developing Countries'. *Communications of the ACM*, June, 42, 6: 31–36.

Piche, D. (1977) 'The Geographical Understanding of Children Aged 5 to 8 Years'. Unpublished Doctoral dissertation, London School of Economics.

Pick, H.L. (1976) 'Transactional-Constructivist approach to Environmental Knowing: A Commentary'. In Moore, G.T. and Golledge, R.G. (eds) *Environmental Knowing*. Dowden, Hutchinson and Ross, Stroudsberg, PA, pp. 185–188.

Pile, S. and Thrift, N. (1995) *Mapping the Subject: Geographies of Cultural Transformation*. Routledge, London.

Pink Village (1998) *Pink Village Info Pages*. http://www.geocities.com/WestHollywood/8382/

Plant, S. (1996) 'On the Matrix: Cyberfeminist Simulations'. In Shields, R. (ed.) *Cultures of Internet: Virtual Spaces, Real Histories and Living Bodies*. Sage, London, pp. 170–184.

Platt, C. (1991) *The Silicon Man*. Wired Books, San Francisco.

Pleitner, S. and Brown, D. (1995) 'Geotraceman: A Visual Traceroute'. *Proceedings of the Asia-Pacific WWW '95 Conference*, 19–21 September 1995, Sydney, Australia.

Plewe, B. (1997) *GIS Online: Information, Retrieval, Mapping then the Internet*. OnWord Press, Santa Fe.

Postel, J. (1994) 'Domain Name System Structure and Delegation'. *RFC 1591*, USC/Information Sciences Institute, March 1994. ftp://ftp.ripe.net/rfc/rfc1591.txt

Poster, M. (1995) *The Second Media Age*. Polity, Oxford.

Poster, M. (1997) 'Cyberdemocracy: Internet and the Public Sphere'. In Porter, D. (ed.) *Internet Culture*. Routledge, London, pp. 201–218.

Press, L. (1997) 'Tracking the Global Diffusion of the Internet'. *Communications of the ACM*, November 1997, 40, 11: 11–17.

Presson, C.C. and Hazelrigg, M.D. (1984) 'Building Spatial Representations through Primary and Secondary Learning'. *Journal of Experimental Psychology: Learning, Memory and Cognition* 10: 716–722.

Pryor, S. and Scott, J. (1993) 'Virtual Reality: Beyond Cartesian Space'. In Hayward, P. and Wollen, T. (eds) *Future Visions: New Technologies of the Screen*. British Film Institute, London, pp. 166–179.

Quarterman, J.S. (1990) *The Matrix: Computer Networks and Conferencing Systems Worldwide*. Digital Press, Cambridge, Mass.

Quarterman, J.S. (1997) 'The Internet Weather Report'. In Staple, G.C. (ed.) *TeleGeography 1997/98: Global Telecommunications Traffic Statistics and Commentary*. TeleGeography, Inc., Washington, DC, pp. 69–72.

Quarterman, J.S. (1998) 'Tracemap'. *Matrix News*, January, 8, 1: 8–10.

Quarterman, J.S., Carl-Mitchell, S., and Phillips, G. (1993) 'Mapping Networks and Services'. *Proceedings of INET '93 Conference*, August 1993, San Francisco, USA.

Quarterman, J.S., Carl-Mitchell, S. and Phillips, G. (1994) 'Internet Interaction Pinged and Mapped'. *Proceedings of INET '94 Conference*, June 1994, Prague Czechoslovakia.

Quarterman, J.S., Carl-Mitchell, S., and Phillips, G. (1995) 'About the MIDS Internet Weather Report (IWR)'. *Matrix News*, August, 5, 8.

Quarterman, J.S. and Hoskins, J.C. (1986) 'Notable Computer Networks'. *Communications of the ACM*, October 29, 10: 932–971.

Rafaeli, S. and Sudweeks, F. (1996) 'Networked Interactivity'. *Journal of Computer Mediated Communications* 2, 4. http://www.ascusc.org/jcmc/vol2/issue4/rafaeli.sudweeks.html

Randall, N. (1997) *The Soul of the Internet: Net Gods, Netizens and the Wiring of the World*. International Thomson Computer Press, London.

Raper, J., Rhind, D., and Shepherd, J. (1992) *Postcodes: The New Geography*. Longman Scientific and Technical, Harlow.

Regian, J.W. and Shebiske, W.L. (1990) 'Virtual Reality: A Instructional Medium for Visual-Spatial Tasks'. *Journal of Communications* 42, 4: 136–149.

Reid, B. (1988) 'Network Maps (DECWRL Netmap)'. *Digital Equipment Corporation (DEC) Western Research Lab*. Palo Alto, California, June 1998. ftp://gatekeeper.dec.com/pub/maps/

Reid, E. (1991) 'Electropolis: Communication and Community on Internet Relay Chat'. Unpublished honours thesis, University of Melbourne. http://people.we.mediaone.net/elizrs/electropolis.html

Reid, E. (1994) 'Cultural Formations in Text-Based Virtual Realities'. Unpublished Masters thesis, University of Melbourne. http://people.we.mediaone.net/elizrs/cult-form.html

Reid, E. (1995) 'Virtual Worlds: Culture and Imagination'. In Jones, S.G. (ed.) *Cybersociety: Computer Mediated Communication and Community*. Sage Publications, London, pp. 164–183.

Reid, E. (1999) 'Hierarchy and Power: Social Control in Cyberspace'. In Smith, M.A. and Kollock, P. (eds) *Communities in Cyberspace*. Routledge, London, pp. 107–133.

Reid, R.H. (1997) *Architects of the Web: 1,000 Days that Built the Future of Business*. John Wiley and Sons, Inc., New York.

Relph, E. (1976) *Place and Placelessness*. Pion, London.

Retsmah (1998) 'It's a Corporation, Not a Country'. Message posted on the *Active Worlds Community Newsgroup*, on 10 May 1998, 13:08:04 GMT. news://news.activeworlds.com/awcommunity/

Rheingold, H. (1991) *Virtual Reality*. Touchstone, New York.

Rheingold, H. (1993) *The Virtual Community: Homesteading on the Electronic Frontier*. Addison-Wesley, New York.

Rheingold, H. (1998) 'Virtual Communities, Phony Communities?' *Second International Harvard Conference on Internet and Society*, 26–29 May 1999, Boston, USA. http://cybercon98.harvard.edu/wcm/rheingold.html

Ribarsky, W., Bolter, J., Op den Bosch, A., and van Teylingen, R. (1994) 'Visualization and Analysis Using Virtual Reality'. *IEEE Computer Graphic and Applications*, January: 10–12.

Richardson, A.E., Montello, D.R., and Hegarty, M. (in press) 'Spatial Knowledge Acquisition from Maps, and from Navigation in Real and Virtual Environments'. *Memory and Cognition*.

Rickard, J. (1996) 'Mapping the Internet with Traceroute'. *Boardwatch Magazine*, December 1996. http://boardwatch.internet.com/mag/96/dec/bwm38.html

Riddell, R. (1997) 'Doom Goes to War: The Marines Are Looking for a Few Good Games'. *Wired*, April 1997, 5.04: 114ff.

Rimmer, P.J. and Morris-Suzuki, T. (1999) 'The Japanese Internet: Visionaries and Virtual Democray'. *Environment and Planning A* 31, 7: 1189–1206.

Roberts, R. (1993) *A New Species: Gender and Science in Science Fiction*. University of Illinois Press, Illinois.

Robertson, N. (1996) 'Stalking the Elusive Usage Data'. *Internet World*, April 1996, 7, 4. http://www.iw.com/1996/04/webwatch.html

Robins, K. (1995) 'Cyberspace and the World we Live in'. In Featherstone, M. and Burrows, R. (eds) *Cyberspace, Cyberbodies and Cyberpunk: Cultures of Technological Embodiment*. Sage, London, pp. 135–156.

Robins, K. and Hepworth, M. (1988) 'Electronic Spaces: New Technologies and the Future of Cities'. *Futures* 20: 155–176.

Robins, K. and Webster, F. (1989) *The Technical Fix*. Macmillan, London.

Robinson, P. (1991) 'Globalization, Telecommunications and Trade'. *Futures* 23: 801–814.

Robinson, A.H., Morrison, J.L., Muehrcke, P.C., Kimerling, A.J., and Guptill, S.C. (1995) *Elements of Cartography* (6th edn.). John Wiley and Sons, Inc., New York.

Roehl, B. (1997) 'Shared Worlds'. *VR News* 6, 9: 10–15.

Roehl, W. (1999) 'Usenet Geography: Modeling Traffic on the alt.sports.baseball.* Hierarchy'. Paper presented at the *Association of American Geographers Conference*, April 1999, Honolulu, Hawaii.

Rose, G. (1993) *Feminism and Geography*. Polity, Cambridge.

Rosenau, P.M. (1992) *Postmodernism and the Social Sciences: Insights, Inroads and Intrusions*. Princeton University Press, Chichester.

Ross, A. (1991) *Strange Weather: Culture, Science and Technology in the Age of Limits*. Verso, London.

Rossney, R. (1996) 'Metaworlds'. *Wired*, June 1996, 4.06: 140ff.

Roszak, T. (1994) *The Cult of Information*. University of California Press, Berkeley.

Rucker, R. (1982) *Software*. Avon Books, New York.

Rucker, R. (1988) *Wetware*. Avon Books, New York.

Ruddle, R.A., Payne, S.J., and Jones, D.M. (1997) 'Navigating Buildings in "Desk-top" Virtual Environments: Experimental Investigations Using Extended Navigational Experience'. *Journal of Experimental Psychology – Applied* 3, 2: 143–159.

Rushkoff, D. (1994) *Cyberia: Life in the Trenches of Hyperspace*. Flamingo, London.

Sack, R. (1980) *Conceptions of Space in Social Thought: A Geographic Perspective*. MacMillan, London.

Salus, P. (1995) *Casting the Net: From Arpanet to Internet and Beyond. . . .* Addison-Wesley, Reading, Mass.

Sardar, Z. (1995) 'alt.civilisations.faq: Cyberspace as the Darker Side of the West'. *Futures* 27: 777–794.

Sarkar, M. and Brown, M.H. (1994) 'Graphical Fish-eye Views'. *Communications of the ACM* 37, 12: 73–84.

Satalich, G.A. (1995) *Navigation and Wayfinding in Virtual Reality: Finding Proper Tools and Cues to Enhance Navigation Awareness*. University of Washington, HIT Lab. http://www.hitl.washington.edu/publications/satalich/home.html

Scanlon, J. (1999) 'Ride the Dow'. *Wired*, June 1999: 176–179.
  http://www.wired.com:80/wired/archive/7.06/nyse_pr.html

Schroeder, R. (1994) 'Cyberculture, Cyborg Post-modernism and the Sociology of Virtual Reality Technologies: Surfing the Soul of the Information Age'. *Futures* 26: 519–528.

Schroeder, R. (1997) 'Networked Worlds: Social Aspects of Multi-user Virtual Reality Technology'. *Sociological Research Online* 2, 4.

Schroeder, R., Heather, N., and Lee, R.M. (1998) 'The Sacred and the Virtual: Religion in Multi-user Virtual Reality'. *Journal of Computer Mediated Communication* 4, 2:

Schuler, D. (1995) 'Public Space in Cyberspace'. *Internet World*, December: 89–95.

Schwartz, A. (1996) 'Comments on MUD Research'. *Journal of MUD Research* 1, 1.
  http://journal.tinymush.org/jomr/v1n1/intro.html

Scullin, W.H., Kwan, T.T., and Reed, D.A. (1995) 'Real-Time Visualization of World Wide Web Traffic'. *Symposium on Visualizing Time-Varying Data*, September 1995.
  http://vibes.cs.uiuc.edu/Publications/Papers/VRWWW.ps.gz

Selby, J. (1995) 'Telecottages in their Context: The Welsh Experience'. Unpublished undergraduate dissertation, University of Wales, Swansea, UK.

Semeria, C. (1997) 'Understanding IP Addressing: Everything You Ever Wanted to Know'. Unpublished 3Com Technical paper. http://www.3com.com/nsc/501302.html

Sempsey, J. (1998) 'When is the MUD too Gooey?' *Journal of MUD Research* 3, 2.
  http://journal.tinymush.org/v3n2/sempsey-comment.html

Sensorium (1997) *Sensorium FAQs*. http://www.sensorium.org/faqs/index.html

Shapiro, A.L. (1995) 'Street Corners in Cyberspace: Keeping On-line Speech Free'. *Nation*, July 3: 10–14.

Shaw, R. (1997) 'Internet Domain Names: Whose Domain is This?' In Kahin, B. and Keller, J. (eds) *Coordinating the Internet*. MIT Press, Cambridge, Mass., pp. 107–134.

Sherman and Judkins, B. (1992) *Glimpses of Heaven, Visions of Hell*. Hodder and Stoughton, London.

Shields, R. (1989) 'Social Spatialisation and the Built Environment: The Case of the West Edmonton Mall'. *Environment and Planning D* 7: 147–164.

Shields, R. (1991) *Places on the Margin: Alternative Geographies of Modernity*. Routledge, London.

Shields, R. (1997) 'Spatial Stress and Resistance: Social Meanings and Spatialisation'. In Benko, G. and Strohmayer, U. (eds) *Space and Social Theory*. Blackwell, Oxford, pp. 186–202.

Shiode, N. and Dodge, M. (1999) 'Visualising the Spatial Pattern of Internet Address Space in the United Kingdom'. In Gittings, B. (ed.) *Innovations in GIS 6: Integrating Information Infrastructure with GI Technology*. Taylor and Francis, London, pp. 105–118.

Shneiderman, B. (1997) *Designing the User Interface: Strategies for Effective Human-Computer Interaction* (3rd edn). Addison-Wesley Publishing Company, Reading, Mass.

Shum, S. (1990) 'Real and Virtual Spaces: Mapping from Spatial Cognition to Hypertext'. *Hypermedia* 2: 133–158.

Siegel, A.W. (1977) 'Finding One's Way Around the Large-scale Environment: The Development of Spatial Representation'. In McGurk, H. (ed.) *Ecological Factors in Human Development*. North Holland, Amsterdam.

Siegel, A.W. and White, S. (1975) 'The Development of Spatial Representation of Large Scale Environments'. In Reese, H. (ed.) *Advances in Child Development and Behavior*. Academic Press, New York, pp. 9–55.

Slouka, M. (1996) *War of the Worlds: The Assault on Reality*. Abacus, London.

Smith, M.A. (1997) 'Measuring and Mapping the Social Structure of Usenet'. Paper presented at the *Seventeenth Annual International Sunbelt Social Network Conference*, 13–16 February 1997, San Diego, California. http://www.sscnet.ucla.edu/soc/csoc/papers/sunbelt97/Sunbelt_Talk.htm

Smith, M.A. (1999) 'Invisible Crowds in Cyberspace: Mapping the Social Structure of the Usenet'. In Smith, M.A. and Kollock, P. (eds) *Communities in Cyberspace*. Routledge, London, pp. 195–219.

Smith, M. and Kollock, P. (1999) *Communities in Cyberspace*. Routledge, London.

Snowdon, D., Fahlén, L., and Stenius, M. (1996) 'WWW 3D: A 3D Multi-User Web Browser'. Paper presented at *WebNet '96*, October 1996, San Francisco, USA.
  http://www.crg.cs.nott.ac.uk/~dns/vr/www3d/webnet96-final.html

Soja, E. (1985) 'The Spatiality of Social Life: Towards a Transformative Retheorisation'. In Gregory, D. and Urry, J. (eds) *Social Relations and Spatial Structures*. MacMillan, London, pp. 90–122.

Soja, E. (1996) *Thirdspace*. Blackwell, Oxford.

Soja, E. (1997) 'Planning in/for Postmodernity'. In Benko, G. and Strohmayer, U. (eds) *Space and Social Theory*. Blackwell, Oxford, pp. 236–249.

Sorkin, M. (1992) *Variations on a Theme Park: The New American City and the End of Public Space*. Hill and Wang, New York.

Southworth, M. and Southworth, S. (1982) *Maps*. Little and Brown, Boston.

Spertus, E. (1997) 'ParaSite: Mining Structural Information on the Web'. *Proceedings of the Sixth International World Wide Web Conference*, 7–11th April 1997, Santa Clara, CA, USA. http://decweb.ethz.ch/WWW6/Technical/Paper206/Paper206.html

Sproull, L.S. and Kiesler, S.B. (1992) *Connections: New Ways of Working in Networked Organization*. MIT Press, Cambridge, Mass.

Squire, S.J. (1996) 'Re-territorializing Knowledge(s): Electronic Spaces and Virtual Geographies'. *Area* 28: 101–103.

Stalder, F. (1998) 'The Logic of Networks: Social Landscapes *vis-à-vis* the Space of Flows'. *Ctheory*, Review 46. http://www.ctheory.com/r46.html

Standage, T. (1998) *The Victorian Internet: The Remarkable Story of the Telegraph and the Nineteenth Century's Online Pioneers*. Weidenfeld and Nicolson, London.

Staple, G.C. (1995) 'Notes on Mapping the Net: From Tribal Space to Corporate Space'. *Telegeography '95*. http://www.telegeography.com/Publications/mapping.html

Staple, G.C. (1997) *TeleGeography 1997/98: Global Telecommunications Traffic Statistics and Commentary*. TeleGeography, Inc., Washington, DC.

Staple, G.C. (1998) *TeleGeography 1999: Global Telecommunications Traffic Statistics and Commentary*. TeleGeography, Inc., Washington, DC.

Staple, G.C. and Mullins, M. (1989) 'Telecom Traffic Statistics – MiTT Matter: Improving Economic Forecasting and Regulatory Policy'. *Telecommunications Policy*, June, 13, 2: 105–128.

Stefik, M. (1996) *Internet Dreams: Archetypes, Myths, and Metaphors*. MIT Press, Cambridge, Mass.

Stein, J. (1999) 'The Telephone: Its Social Shaping and Public Negotiation in Late Nineteenth and Early Twentieth Century London'. In Crang, M., Crang, P., and May, J. (eds) *Virtual Geographies*. Routledge, London, pp. 44–62.

Steinberg, S.G. (1996) 'Multicasting: The Internet's Connected to the Mbone'. *Wired*, June, 2.06: 103.

Stephenson, N. (1992) *Snow Crash*. Bantam Spectra, New York.

Stephenson, N. (1995) *The Diamond Age*. New York: Batham Books.

Stephenson, N. (1996) 'Mother Earth Mother Board'. *Wired*, October, 4.12.

Sterling, B. (1986) 'Introduction'. In Sterling, B. (ed.) *Mirrorshades: The Cyberpunk Anthology*. HarperCollins, London.

Sterling, B. (1988) *Islands in the Net*. Ace Books, New York.

Sterling, B. (1994) *Heavy Weather*. Batham Books, New York.

Sterling, B. (1996) *Holy Fire*. Batham Books, New York.

Stone, A.S. (1991) 'Will the Real Body Please Stand-up?: Boundary Stories About Virtual Cultures'. In Benedikt, M. (ed.) *Cyberspace: First Steps*. MIT Press, Cambridge, Mass., pp. 81–118.

Stonier, T. (1983) *The Wealth of Information: A Profile of the Postindustrial Economy*. Meuthen, London.

Sudweeks, F., McLaughlin, M., and Rafaeli, S. (1998) *Network and Netplay: Virtual Groups on the Internet*. MIT Press, Cambridge, Mass.

Suler, J. (1997) *The Psychology of Avatars and Graphical Space in Multimedia Chat Communities*. Unpublished paper, Rider University, July 1997. http://www.rider.edu/users/suler/psycyber/psyav.html

Sutton, L.A. (1996) 'Cocktails and thumbtacks in the Old West: What would Emily Post Say?' In Cherney, L. and Reise, E.R. (eds) *Wired Women: Gender and New Realities in Cyberspace*. Seal Press, Seattle, pp. 169–187.

Suvin, D. (1979) *Metamophoses of Science Fiction: On the Poetics and History of a Literary Genre*. Yale University Press, New Haven.

Takemura, S., Nishimura, Y., and Ohne, H. (1998) 'Designing a Public Sensory Platform on the Net'. *Proceedings of INET '98 Conference*, 21–24 July 1998, Geneva, Switzerland. http://www.sensorium.org/inet98/index.html

Tauscher, L. and Greenburg, S. (1997) 'How People Revisit webpages: Empirical Findings and Implications for the Design of History Systems'. *International Journal of Human Computer Studies* 47: 97–137.

Taylor, H.A. and Tversky, B. (1992a) 'Description and Depictions of Environments'. *Memory and Cognition* 20: 483–496.

Taylor, H.A. and Tversky, B. (1992b) 'Spatial Mental Models Derived from Survey and Route Descriptions'. *Journal of Memory and Language* 31: 261–292.

Taylor, J. (1997) 'The Emerging Geographies of Virtual Worlds'. *The Geographical Review* 87: 172–192.

Technological Partners (1997) 'Now You See It'. *Business Issues in Technology: Computer Letter* 13, 10, 24 March 1997.

Tepper, M. (1997) 'Usenet Communities and the Cultural Politics of Information'. In Porter, D. (ed.) *Internet Culture*. Routledge, London, pp. 39–54.

Thoen, B. (1997) 'Gain a Sense of Place in Cyberplace'. *GIS World*, December 1997: 32–33.

Thomas, R. (1995) 'Access and Inequality'. In Heap, N., Thomas, R., Einon, G., Mason, R., and MacKay, H. (eds) *Information Technology and Society: A Reader*. Open University Press, Milton Keynes.

Thorndyke, P.W. (1983) 'Spatial Cognition and Reasoning'. In Harvey, J. (ed.) *Cognition, Social Behavior and the Environment*. Lawrence Erlbaum, Hillsdale, NJ, pp. 137–149.

Thorndyke, P.W. and Hayes-Roth, B. (1982) 'Differences in Spatial Knowledge Acquired from Maps and Navigation'. *Cognitive Psychology* 14: 560–589.

Thrift, N. (1996) 'New Urban Eras and Old Technological Fears: Reconfiguring the Goodwill of Electronic Things'. *Urban Studies* 33, 8: 1463–1493.

Thrower, N. (1996) *Maps and Civilization: Cartography Culture and Society*. University of Chicago Press, Chicago.

Thu Nguyen, D. and Alexander, J. (1996) 'The Coming of Cyberspace Time and the End of Polity'. In Shields, R. (ed.) *Cultures of Internet: Virtual Spaces, Real Histories and Living Bodies*. Sage, London, pp. 99–124.

Thurman, R. and Worfolk, P. (1999) 'SaVi: Software for the Visualization of Satellite Constellations'. The Geometry Center, University of Minnesota. http://www.geom.umn.edu/~worfolk/SaVi/

Tlauka, M. and Wilson, P.N. (1996) 'Orientation-free Representations from Navigating Through a Computer Simulated Environment'. *Environment and Behavior* 28: 647–664.

Tobler, W.R. (1970) 'A Computer Movie Simulating Urban Growth in the Detroit Region'. *Economic Geography* (supplement), 46: 234–240.

Toffler, A. (1980) *The Third Wave*. Pan, London.

Tomas, D. (1991) 'Old Rituals for New Space: Rites de passage and William Gibson's Cultural Model of Cyberspace'. In Benedikt, M. (ed.) *Cyberspace: First Steps*. MIT Press, Cambridge, Mass., pp. 31–48.

Treese, W. (1997) *The Internet Index*. http://www.openmarket.com/intindex/

Tufte, E.R. (1990) *Envisioning Information*. Graphics Press, Chesire, Conn.

Turkle, S. (1995) *Life on the Screen: Identity in the Age of the Internet*. Simon and Schuster, New York.

Turner, P. and Turner, S. (1997) 'Distance Estimation in Minimal Virtual Environments'. *Proceedings of UK VRSIG 97*. http://www.brunel.ac.uk/~empgrrb/VRSIG97/proceed/030/30.htm

Tversky, B. (1981) 'Distortions in Memory for Maps'. *Cognitive Psychology* 13: 407–433.

Uncapher, W. (1999) 'Electronic Homesteading on the Rural Frontier: Big Sky Telegraph and its Community'. In Smith, M.A. and Kollock, P. (eds) *Communities in Cyberspace*. Routledge, London, pp. 264–289.

United Nations Human Development Report (1998) United Nations Development Programme. United Nations, New York, http://www.undp.org/hdro/98.htm

Urry, J. (1985) 'Social Relations, Space and Time'. In Gregory, D. and Urry, J. (eds) *Social Relations and Spatial Structures.* St Martin's Press, New York, pp. 20–48.

Vevo (1999) *About the AlphaWorld Map.* Greg Roelofs and Pieter van der Meulen, Advanced Technology Group at the Philips Multimedia Center, Palo Alto, California. http://awmap.vevo.com/about.html

Viégas, F.B. (1999) *Chat Circles Project Page,* Media Lab, MIT. http://www.media.mit.edu/~fviegas/circles/index.html

Viégas, F.B. and Donath, J.S. (1999) 'Chat Circles'. *Proceedings of the Conference on Human Factors in Computing Systems,* 15–20 May 1999, Pittsburgh, USA. http://www.media.mit.edu/~fviegas/chat_circles.pdf

Vilett, R. (1999) *AlphaWorld Maps.* Active Worlds, Inc. http://www.activeworlds.com/satellite.html

Warf, B. (1995) 'Telecommunications and the Changing Geographies of Knowledge Transmission in the Late Twentieth Century'. *Urban Studies* 32: 361–378.

Warf, B. (forthcoming) 'Compromising Positions: The Body in Cyberspace'. In Wheeler, J., Aoyama, Y., and Warf, B. (eds) *Cities in the Telecommunications Age: The Fracturing of Geographies.* Routledge, London.

Warf, B. and Grimes, J. (1997) 'Counterhegemonic Discourses and the Internet'. *The Geographical Review* 87: 259–274.

Warren, R., Warren, S., Nunn, S., and Warren, C. (1998) 'The Future of the Future in Planning: Appropriating Cyberpunk Visions of the City'. *Journal of Planning Education and Research* 18: 49–60.

Waters, M. (1995) *Globalisation.* Routledge, London.

Weijers, T., Meijer, R., and Spoelman, E. (1992) 'Telework Remains "made to measure": The Large Scale Introduction of Telework in the Netherlands'. *Futures* 24: 1048–1055.

Weinberger, D. (1998) 'The Scenic Route'. *Wired.* December: 216–217.

Wellman, B., Salaff, J., Dimitrova, D., Garton, L., Gulia, M., and Haythornwaite, C. (1996) 'Computer Networks as Social Networks'. *Annual Review of Sociology* 22: 211–238.

Wellman, B. and Gulia, M. (1999) 'Virtual Communities as Communities: Net Surfers don't Ride Along'. In Smith, M.A. and Kollock, P. (eds) *Communities in Cyberspace.* Routledge, London, pp. 167–194.

Wexelblat, A. and Maes, P. (1999) 'Footprints: History-Rich Tools for Information Foraging'. Paper presented at *CHI '99 Conference,* 15–20 May 1999, Pittsburgh, USA. http://wex.www.media.mit.edu/people/wex/CHI-99-Footprints.html

Wilbur, S.P. (1997) 'An Archaeology of Cyberspaces: Virtuality, Community, Identity'. In Porter, D. (ed.) *Internet Culture.* Routledge, London, pp. 5–22.

Williams, T. (1996) *Otherland.* Orbit, London.

Wilson, P., Foreman, N., and Tlauka, M. (1997) 'Transfer of Spatial Information from a Virtual to a Real Environment'. *Human Factors* 39: 526–531.

Winston, B. (1998) *Media Technology and Society: A History from the Telegraph to the Internet.* Routledge, London.

Wise, J.A., Thomas, J.J., Pennock, K., Lantrip, D., Pottier, M., Schur, A., and Crow, V. (1995) 'Visualizing the Non-Visual: Spatial Analysis and Interaction with Information from Text Documents'. *Proceedings of IEEE Information Visualization '95,* pp. 51–58, 140. Reproduced in Card, S.K., MacKinlay, J., and Shneiderman, B. (eds) (1999) *Readings in Information Visualization: Using Vision to Think.* Morgan Kaufman Publishers, San Francisco, pp. 442–450.

Witmer, B.G., Bailey, J.H., Knerr, B.W., and Parsons, K.C. (1996) 'Virtual Spaces and Real-world Places: Transfer of Route Knowledge'. *International Journal of Human-Computer Studies* 45: 413–428.

Wood, A.M., Drew, N.S., Beale, R., and Hendley, R.J. (1995) 'HyperSpace: Web Browsing with Visualisation'. *Proceedings of the Third International World Wide Web Conference,* April 1995, Darmstadt, Germany, pp. 21–25. http://www.igd.fhg.de/www/www95/proceedings/posters/35/index.html

Wood, D. (1993) *The Power of Maps.* Routledge, London.

Wood, L. (1999) *Lloyd's Satellite Constellations.* Centre for Communication Systems Research, University of Surrey, UK. http://www.ee.surrey.ac.uk/Personal/L.Wood/constellations/

Woodruff, A., Aoki, P.M., Brewer, E., Gauthier, P., and Rowe, L.A. (1996) 'An Investigation of Documents from the World Wide Web'. Presented at the *Fifth International World Wide Web Conference*, 6–10 May 1996, Paris, France.
  http://epoch.cs.berkeley.edu:8000/~woodruff/inktomi/index.html
Woolley, B. (1992) *Virtual Worlds: A Journey in Hype and Hyperreality*. Penguin Books, London.
Wurman, R.S. (1997) *Information Architects*. Graphis Press Corp., New York.
Zakon, R.H. (1996) *Hobbes' Internet Timeline*.
  http://info.isoc.org/guest/zakon/Internet/History/HIT.html
Zook, M.A. (1998) 'The Web of Consumption: The Spatial Organization of the Internet Industry in the United States'. Paper presented at the *Association of Collegiate Schools of Planning 1998 Conference*, 5–8 November 1998, Pasadena, CA, USA.
  http://socrates.berkeley.edu/~zook/pubs/acsp1998.html
Zook, M.A. (2000) 'Matthew Zook's Page of Domain Name Data'.
  http://socrates.berkeley.edu/~zook/domain_names/index.html
Zook, M.A. (forthcoming) 'The Web of Production: The Economic Geography of Commercial Internet Content Production in the United States'. *Environment and Planning A*.
Zukin, S. (1992) *Landscapes of Power*. University of California Press, Berkeley.

Index